IEE CONTROL ENGINEERING SERIES 42

Series Editors: Professor P. J. Antsaklis
                Professor D. P. Atherton
                Professor K. Warwick

DIGITAL SIGNAL PROCESSING:

principles, devices and applications

Other volumes in this series:

DIGITAL SIGNAL PROCESSING:

principles, devices and applications

**Edited by
N. B. Jones and
J. D. McK. Watson**

Peter Peregrinus Ltd. on behalf of the Institution of Electrical Engineers

Published by: Peter Peregrinus Ltd., London, United Kingdom

© 1990: Peter Peregrinus Ltd.

British Library Cataloguing in Publication Data

ISBN 0 86341 210 6

Printed in England by Short Run Press Ltd., Exeter

Contents

Chapter 1

Introduction—Engineering and Signal Processing

N. B. Jones

1.1 Engineering - Information and Control

Throughout history, but in particular over the last two centuries, engineering has provided outstanding benefits as well as raised threats and challenges. As a direct result of progress in engineering, large numbers of people now lead far richer and more varied lives than could previously have been imagined. The development of engineering skills and of the associated technologies has not been uniform and particular eras have been associated with spectacular developments in particular branches of engineering. Many surges in engineering knowledge have been associated with military and economic rivalry between nations but have often lead to general civilian benefit in the long run. People of the last century, for example, saw revolutionary developments in industrial, railway and marine engineering and more recently we have seen spectacular advances in aeronautical engineering and, of course, in electronics; much of which grew out of military requirements but quickly spread into general civilian use.

The recent extraordinary development of electronics, especially digital micro-electronics, has brought us into a new phase of exciting growth, in particular associated with Information and Control Technology; areas of engineering and mathematics concerned with the acquisition, transmission, processing, storing and display of information and, in particular, with its use in control and decision making at all levels.

Processing of information is as old as the human race and the technology associated with it dates back to before the earliest form of writing to the use of counting beads and to the time of cave drawings. However the invention of printing (1049), and the electronic binary coding of data (1837) and more recently of radio communications (1891) and of the electronic digital computer (1946) has increased the speed of generation and processing of data to such an extent that information and control technology is no longer just concerned with automatic processes replacing manual processes but now provides the opportunity to do entirely new things. It is now, for example, possible to envisage machines which can see, hear, talk and reason as well as machines which just execute required motions, no matter how complex. We have now entered an era where automatic processes are replacing human sensory and cognitive processes as well as manipulative processes.

The primary area of engineering associated with these new opportunities is signal processing since in order to proceed in an intelligent and useful way a machine must assimilate and process an enormous amount of data which arrives in the form of signals from some forms of sensor, either local or remote. It is the successful and efficient processing of these signals which is the key to progress in this new field of endeavour.

For reasons of simplicity and flexibility associated with the binary nature of the electronics, processing of signals is most conveniently done digitally and it is this major area of electronics, information technology and control engineering known as digital signal processing which is the subject of this book.

1.2   The scope of digital signal processing

When data is generated in continuous streams, in the form of signals, it is usually the case that there is too much detail. Thus, before the useful information can be efficiently stored, transmitted, displayed or used for control purposes, it is almost invariably necessary to process it in some way, so as to reduce the total amount of data while maintaining the essential information. For reasons of cost, precision and flexibility it is now generally more appropriate to do this digitally. Only a few years ago digital processing was limited by problems of speed and cost to signals of relatively low information content. However the development of special digital integrated circuits (DSP chips) and associated systems has provided new opportunities for low cost digital processing of mass data in the form of wide band signals.

One good example of the use of such devices is in the precise control of processes and machines by the implementation of sophisticated algorithms, often not easily reproduced using analogue controllers. In communications there is extended scope for noise rejection and error correction if signals are processed digitally. The flexibility and sophistication of DSP is also well illustrated in image processing such as medical axial tomography where complex images are reconstructed using a multitude of signals from an array of sensors and, of course, in the commercially important areas of digital TV and digital sound recording. The control of robots is, in its own right, an appropriate area for real time signal processing for motion control but new possibilities of intelligent robot responses to audio, visual, tactile and other inputs are being realised because of the power of DSP chips allied with new software ideas in Artificial Intelligence for signal analysis and interpretation.

It can be appreciated that digital processing of signals has an extensive range of applications, from the military to the medical, from entertainment to mass production. In many areas of application the advent of these specially designed DSP devices has started a revolution in engineering which will pervade most areas of modern life.

1.3   The scope of this book

This book is a collection of presentations by active workers in the field of DSP. The authors have made their own contributions to the development of this new field and have presented the material at a vacation school run by the Institution of Electrical Engineering at the University of Leicester in March 1990. The chapters of the book are linked by common themes of theory and technology which give unity for continuous reading. However each chapter can be read independently by readers with appropriate backgrounds.

The first part of the book contains the essential fundamentals of the theoretical bases of DSP. However most of the book is devoted to the practicalities of designing hardware and software systems for implementing the desired signal processing tasks and to demostrating the properties of the new families of DSP chips which have enhanced progress so much recently. A book of this nature, which has a significant technological content, will inevitably tend to go out of date as time goes on. However the material has been written so that the fundamentals of the design and processing methods involved can be seen independently of the technology and thereby retain their value indefinitely.

Consideration is given to the use of DSP methods in control, communications, speech processing and biomedical monitoring. Much of the material is illustrated by case studies and it is hoped that, in this way, many of the practical points associated with DSP engineering can be highlighted.

Chapter 2

Devices overview

R. J. Chance

2.1 DIGITAL SIGNAL PROCESSING BEFORE 1980

2.1.1 Genesis

The vast majority of the general purpose computers which we use today are considered to be von Neumann machines. The calculating machine proposed by John von Neumann in 1946 was for a machine with a 40 bit word, an 8 bit instruction field and a 12 bit address [1]. Two instruction/address combinations occupied one word. Its construction used 2300 vacuum tubes and the memory access time was designed to be less than 50 microseconds. It may come as a surprise that its design was the result of the need to build a machine to solve linear differential equations, rather than the wide range of purposes for which we now use computers. The feature of modern general purpose computers that is most associated with von Neumann is the use of a single memory space to hold both instructions and data. In addition to this, most modern machines use a program counter to hold the instruction address and are of the CISC (complex instruction set computer) type where problem orientated instructions are constructed from a sequence of simpler operations. This is the type of computer with which most of us are familiar and refer to as a von Neumann machine.

The earliest digital computers were used to explore the possibilities of processing analogue signals digitally. The basic principle of operation involves the sampling of analogue signals and the representation of the analogue value by a digital code. Obviously, the analysis and simulation of analogue circuits, such as filters, was a rewarding application for such machines. During the 1960s, there was an obvious trend towards faster, cheaper and more complex integrated circuits. One gets the feeling that fundamental researchers during this period were looking forward confidently to the construction of real-time implementations of digital systems, so strongly implied by the term 'signal processing'. In particular, digital filters such as finite impulse response (FIR) and infinite impulse response (IIR) designs were developed and implemented in software in this period. At this time, real-time digital signal processing was not a reality, when judged by modern criteria. Buttle et. al. [2] described a real-time digital low-pass filter, with a cut-off frequency of 0.1 Hz and 20 Hz sampling rate, implemented on a minicomputer. Where a solution exists, there always seems to be a problem to be solved. Even at these low frequencies, geological and electroencephalographic signals were amenable to these digital filters.

2.1.2   Real-time   signal   processing

When one talks of digital signal processing today, the processing of data in real time is usually taken for granted. To the engineer, 'real-time' processing is important when carried out on signals relevant to the market-place. Audio and the many other familiar signals in the same frequency band have appeared as a magnet for DSP development. The 1970s saw the implementation of signal processing algorithms, especially for filters and Fast Fourier transforms (FFT) by means of digital hardware, developed for the purpose. One early system (1968), described by Jackson et. al [3] used shift register delays, bit-serial adders, word-serial multipliers and read only memory for filter coefficients. The Intel 2920 system, which will be described shortly, bears a substantial resemblence to this design. An interesting suggestion, in this paper, was that several of these serial arithmetic units should compute the filters by operating in parallel.

By the late seventies, processing speeds had increased to the point where real-time speech processing could be carried out, at least in research laboratories. The development of the type of stored program computing systems which execute instruction sequences addressed through a program counter, are of particular interest. These bear at least a superficial resemblance to conventional microprocessors but may be used for digital signal processing. It is this type of DSP that dominates the market at the present time.

Certain schemes which could be of very general signal processing value were identified and implemented during this period. For example, the minimum binary representation of realistic analogue values settled down at between 12 and 16 bits. The hardware implications of integer versus floating point representation of magnitudes in calculations lead to many implementations of both systems. The multiplication function was seen to be a critical limiting step. Use of pipelining at both the bit and the word level was used in multiplier design and the first 'single cycle' multipliers were implemented in the early seventies [4].

2.1.3   Early   sequential   program   digital   signal   processors

In the mid 1970s, several research institutions produced stored program processors, specifically designed for signal processing work. The technology used in signal processors used in advanced research was bipolar, and with emitter coupled logic (ECL), the raw performance sometimes exceeded that of current DSPs. A description of some of these signal processors is given by Allen [5] and is summarised in fig 2.1.

Processor	SSP	DVT	FDP	SPS-41
no. of i/cs	500	470	10000	1400
multiply(ns)	150	212	450	1000
cycle time	100	53	150	200
memory(data)	2 K	.5 K	2 K	*
memory(prog)	1 K	1 K	1 K	*

* see text

Some digital signal processors circa 1975

FIG 2.1

Although the architectures of these processors are varied, they owe very little to the CISC/ von Neumann approach. Even the provision of a program counter is not considered sacrosanct. A common feature is the provision of multiple memory spaces of very small size accessed through multiple data busses of bit lengths which vary even within one machine. Multiple arithmetic-logic units are also notable features. As one might expect, all give an outstanding multiply-accumulate performance.

2.1.3.1 The SSP, (Small Signal Processor) used 24 bit wide data memory and a 52 bit program word, containing a jump to the address of the next instruction. There was no program counter; this obviously assumes that only small programs will be run on the machine. A multiply accumulate module, able to calculate (X Y + Z), was part of the design.

2.1.3.2 The DVT (Digital Voice Terminal) used an instruction pipeline and a separate data memory space. This is usually known as the Harvard architecture and is used in many modern DSP devices. It allows the overlapping of the fetching and decoding of instructions and the reading and writing of data from separate program and data memory spaces. The use of the same 16 bit word length for both program and memory spaces and the ability to 'stand alone' without a host computer make this machine look rather similar to many modern monolithic DSPs.

2.1.3.3 The FDP (Fast Digital Processor) was dependent on communication with a Univac host computer. Parallelism was exploited by using four arithmetic units (AU), with 18 x 18 multiply and add functions. These units supported AU to AU communication and were also connected to two data memory spaces. The separate program memory space allowed the simultaneous fetching of two instructions. It is hardly surprising that this machine was said to be rather demanding to program, if the best use was to be made of the potential parallelism.

2.1.3.4 The SPS-41 (Signal-Processing Systems 41) is the only TTL machine in this group (the others being ECL), having a 200 ns cycle time. The design is complex, with three totally different designs of processor specialising in input/output, arithmetic and addressing functions. All three had multiple memory spaces with word lengths varying from 16 to 38 bits. The speed was achieved by the use of four multipliers and six adders. It is of interest that a high level language was used to co-ordinate complex prewritten machine code routines, such as the FFT, which might be expected to be commonly used in signal processing.

These digital signal processors are quite different from the 8 bit monolithic microprocessors of the day, in architecture and arithmetic performance; they really look extremely similar to current general purpose monolithic DSPs.

2.1.4 Bit slice technology

The development of large scale integration of bipolar devices over the 1970s reached the stage where significant parts of a digital signal processing system, for example an 8 x 8 bit multiplier could be put on one chip. However, at the time, it was difficult to expand the word length of a multiplier or adder to a more practical size, such as 32 bits, because of the large number of external connections necessary for the output and two input operands. At the time, a package pin count of much more than 50 was not practicable. Multiplexing input/output signals produces an undesirable speed degradation. Although improved, the package pin count limitation is still affecting the design of digital processors.

A solution, developed in the late seventies, was to construct components which are used as 'slices' through a word. Thus the 4 bit 2901 ALU/register slice, as developed by Advanced Micro Devices Inc. (AMD), provides 4 bits of data input, 4 bits of data output and several control signals. The 2901 can perform several of the fundamental arithmetic and logical operations, selected by control inputs. It can be expanded to virtually any word length basically by connecting the control inputs in parallel. Connections between the slices allow the propogation of the carry bits. In fact, a substantial set of compatible LSI parts, built on the bit slice concept has been produced. The result is that a microprogrammed computer can be designed around standard parts, but possess a specialised instruction set and architecture. Bit slice technology is still relevant today as a means of realising a state of the art computer tailored to a particular function.

2.2 Progress since 1980

By about 1980, the technology of the microprocessor was well advanced in the shape of well-known general purpose 8 bit processors such as the Intel 8080, Motorola 6800 and

Zilog Z80. Such devices use a von Neumann architecture, where instructions and data are stored in the same memory space and delivered to the processor on a single bus. The instructions are complex sequences taking multiple clock cycles. This applies not only to basic operations not supported in the instruction set or in hardware, such as multiply and divide, but also to apparantly simple addressing modes. They are reasonably suitable as targets for high level language compilers or general purpose tasks written in assembler language. In spite of the sophistication of these general purpose microprocessors, and their much improved performance since 1980, they are unable to perform digital signal processing operations anything like as efficiently as the specialised hardware of bit slice and other architectures designed for the purpose. The 8 bit data busses and internal registers are not of adequate word length to handle DSP arithmetic without multiple operations. A 16 bit multiply/ accumulate operation may take several dozen machine cycles on such processors. The low efficiency of multiply and accumulate operations is accompanied by a memory space, unecessarily large for DSP applications and complex memory addressing modes and other instructions aimed at high level language implementation. At this time, we are comparing execution times for a multiply/accumulate operation of around 200 nanoseconds for specialised bipolar technology, aimed at DSP, with around a millisecond for a general purpose microprocessor.

2.2.1   The first monolithic DSPs

In the late seventies and early eighties, several silicon suppliers marketed solutions to the problem of creating a digital signal processing environment on a single chip. The functions envisaged were digital filtering operations, Fast Fourier Transforms (FFT) and waveform generation at audio frequencies. Applications within the telecommunications industry were seen as the providers of the high volume market necessary to support such developments. Three of these early dedicated DSPs will now be described: the Intel 2920, the NEC uPD7720 and the Texas Instruments TMS32010. These are listed chronologically and it is not difficult to imagine that one can see a trend, when examining the differences between them.

2.2.2   The Intel 2920

The Intel 2920 [6] is a digital processor but in fact supports analogue to digital conversion on the chip, so that inputs and outputs are analogue. This is a nice solution to the problem of limited external connections. Fig 2.2 shows the architecture of the 2920, which bears some similarities to the 1968 system of Jackson [3]. A 192 word program memory of 24 bits holds instructions and filter coefficients. Data is transferred between inputs, a 40 word random access scratch-pad, ALU and outputs. The data word length was 25 bits. The layout of an instruction word is shown in Fig 2.3. It can be seen that this DSP bears virtually no

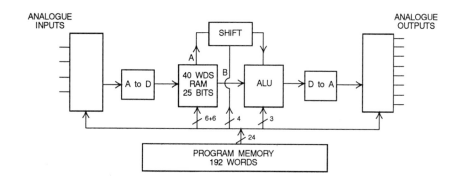

The Intel 2920 digital signal processor (circa 1980)

FIG 2.2

resemblance to the conventional general purpose micro-processor. Notice the shift field, so important for scaling in an integer arithmetic DSP environment.

ALU instruction	A address	B address	shift code	analogue instruction
3	6	6	4	5

The 2920 24 bit instruction word

FIG 2.3

Closer inspection only increases this impression. The very small instruction set (around 16) uses only one addressing mode, direct, and has no branch instructions. The whole program is simply repeated when the end of program memory is reached. At a cycle time of 400 nanoseconds, this gives a 13 KHz. sampling rate. Although there is no explicit multiply instruction, the function may be performed by using a conditional instruction execution facility.

The provision of on chip analogue to digital conversion has neatly solved the problem of packaging a 24 bit ALU without the pin count becoming impossibly large. Although the provision of A to D on a digital signal processor chip is now the exception, most DSP manufacturers are known to possess the capability. The lack of a perceived high volume market is undoubtedly the reason.

The 2920 obviously owes its design more to the dedicated signal processing systems designed in the sixties and sev-enties from raw hardware than to the general purpose mi-croprocessor. An understanding of the ancestry of the DSP helps to explain some of the strange features and unex-pected omissions from the instructions and architectures of so many more modern monolithic DSPs, compared to normal microprocessors.

2.2.3 The NEC 7720

A second DSP produced around 1980 is the NEC 7720 [7]. The first surprising feature is that this device is listed in the 1982 catalogue as a peripheral device rather than a processor. An overview of the internal architecture is shown in fig 2.4. This is, at first sight, much more similar to a conventional microprocessor than the 2920. There is program memory with a program counter to address the current instruction and a 16 bit ALU with two 16 bit accumulators.

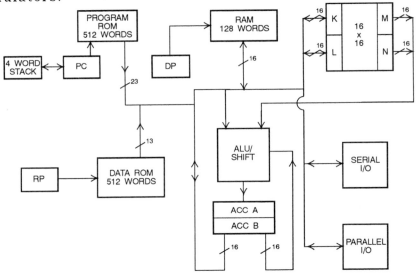

The NEC 7720 digital signal processor (circa 1981)

FIG 2.4

The instruction set supports subroutine calls and branch instructions. However, this similarity is only superficial. It is indeed a microprocessor peripheral, supporting no external memory at all either for program or data. There is a 16 x16 bit multiplier which executes in only one 250 ns machine cycle, a very unusual feature at the time. Secondly, the architecture is not von Neumann and quite unlike contemporary general purpose microprocessors. There are no less than four separate address spaces: Programs reside in 512 words of 23 bits, a small address space but a fairly large instruction word. 128 words of RAM of 16 bits, are quite separate from program memory space. We have 512 words of read only memory using 13 bit words. This address space is clearly aimed at the storage of filter coefficients. The last address space is the stack. The conventional computer programmer would be surprised to find that a separate address space is used for the stack and amazed to find that it was only 4 words deep. Again we can see the influence of the dedicated signal processing architectures of the seventies, such as those in fig 2.1, rather than an extension of the conventional microprocessor.

As in the case of the 2920, a shifter is provided to scale values. Serial input and output ports are provided, with the expectation that A to D and D to A converters, especially the codecs, used in the world of telecommunications will be used. In addition, an 8 bit data bus is designed to be interfaced to a general purpose microcomputer. In this

OPERATION   BITS

subroutine return   22-21
select ALU operand (RAM, register etc.)   20-19
ALU operation   18-15
select accumulator 1 or 2   14
modify RAM address pointer   13- 9
modify ROM address pointer   8
move data (source)   7- 4
move data (destination)   3- 0

(A) instruction word layout

OP MOV @,KLM,SI   transfer two operands
ADD   ACCA,M   perform addition
DPINC   modify RAM pointer
M 1   modify RAM pointer
RET   return from subroutine

(B) example of one instruction

The NEC 7720 parallelism, as seen by the programmer

FIG 2.5

context, the 7720 can be used as a 'hardware subroutine', able to perform functions such as the FFT much faster than the host computer.

A major feature of the 7720 can be seen by examining the instruction word layout of fig 2.5A, explained by the example in fig 2.5B. Several simultaneous operations are expected to be carried out by the DSP in one cycle and specified as such by the programmer. This high degree of parallelism, visible to the programmer, is a feature of previous and subsequent digital signal processors. It is not a feature of general purpose von Neumann machines. The difficulty of allowing the programmer to specify the most efficient operations in any conventional type of high or low level programming language is a problem at the present time.

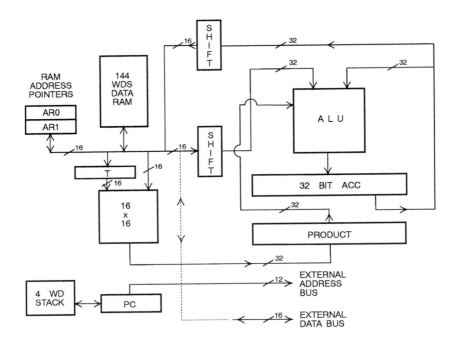

The   Texas   Instruments   TMS32010   (circa   1983)

FIG   2.6

2.2.4   The   Texas   Instruments   TMS32010

The   last   example   of   a   single   chip   DSP   from   the   early
eighties   is   the   TMS32010   [8].   This   device   really   does   look
quite   like   a   conventional   microcomputer   (fig   2.6).   Although
separate   program   and   memory   spaces   are   used   so   that   instruction   fetch   and   execution   can   be   overlapped,   both   are
16   bits   wide.   The   significance   of   this   is   increased   by   the
fact   that   the   architecture   is   not   truly   Harvard;   it   is   possible
to   transfer   data   between   program   memory   and   data
memory.   The   external   16   bit   data   bus   is   used   for   input   /
output   data   transfer   or   for   off   chip   program   memory   (but
not   data   memory   expansion).   This   has   some   practical   advantages: some   or   all   of   program   memory   may   be   read/   write
and   be   used   for   either   program   or   data.   Programs   may   thus
be   loaded   into   RAM,   either   in   a   development   system   or   by
a   host   computer.   Nevertheless,   the   TMS32010   is   marked
out   as   a   device   with   signal   processing   ancestry   by   the   16
x   16   multiplier,   executing   in   one   cycle   and   the   barrel
shifter,   able   to   scale   values   during   an   accumulator   load
operation,   with   no   time   penalty.

A		transfer data between RAM and accumulator
A		perform shift on the transferred data
A B C	G	modify RAM address pointer
A B C	G	select new address pointer
	F	load product to accumulator
B		accumulate product
	E	subtract product
C		multiply register x RAM
B		load multiplicand and register
B	G	move data to RAM address + 1
D		branch

Irregular control over TMS32010 parallelism
with 6 instructions (A to G)

FIG 2.7

The same sort of parallelism is provided as in the NEC 7720, but presented to the programmer in a completely different way. In the TMS32010, particular combinations of parallel operations aimed at a particular task are specified by one mnemonic, with optional address pointer control fields, as shown in fig 2.8. This makes the TMS32010 more like a CISC machine to program than the uPD7720, although the multiple operations are carried out by concurrent hardware rather than a series of clocked operations.

INSTRUCTION   ACTIONS

LTA   *+,1   load multiplier input register
                    add product to accumulated value
                    move data in memory
                    increment RAM pointer
                    select RAM pointer 1

TMS32010 instruction executing in 1 machine cycle

FIG 2.8

Contrast this with fig 2.5. The TMS32010 instructions, which incorporate parallelism, are orientated to specific signal processing algorithms such as FIR filter and other multiply/accumulate operations. Therefore the programmer does not have so much control over the parallel operations, although the assembler language has a more conventional appearance. For example, the loading of the T (multiplicand) register and the addition of the product from the accumulator can be combined in one machine cycle, whereas a product subtraction cannot, although the subtraction is supported separately.

2.3 RECENT DEVELOPMENTS

In the past few years, DSPs have attracted far more commercial interest than previously. This interest has been spread very widely and has been the spur, not only for increased development activity, but also for some new looks at the functions that could be performed by DSPs.

2.3.1   Increased   specifications

Since 1983, there has been a huge increase in the number of monolithic digital signal processors, with manufacturers such as AMD, AT&T, Fujitsu, Hitachi, Inmos, NEC, Philips, Texas Instruments etc. often producing a whole range of devices. Most of these only support integer arithmetic. There have been some increases in processor clock speeds from the 5 million instructions per second (MIPS) of the TMS32010 to the 13.5 MIPS of the Motorola DSP96002. However, the changes in architecture have produced far greater increases in performance. Whereas the TMS32010 performs 2.5 million integer multiplications/s, the increased parallelism of the 96002 gives over 40 million per second in floating point. Address spaces have also increased enormously, with the latest designs addressing several megabytes.

The pinout problem has largely been overcome and several recent devices are able to address simultaneously several external memory or input/output busses. Obviously there is an increasing practical problem in doing this as processor cycle times reduce still further. Because of this, there is still great pressure to increase amounts of on chip memory.

As one might have anticipated, costs, especially of the more mundane performers, have dropped to the point where they can be used in consumer products such as children's toys.

2.3.2   Changes   in   emphasis   and   direction

In addition to these easily measurable changes in DSP performance, it is possible to imagine that one can discern some more subtle ones. DSPs do seem to be looking progressively more and more like conventional processors. Although non-von Neumann architectures are still with us, since the TMS32010, program and data memory on DSPs have used the same word length and a communication path between them has been established. Memory spaces are also getting as large as those of the latest conventional processors. Even the simpler DSP devices have very high clock speeds, when compared to conventional microprocessors. The result of this has been that designers of systems with no signal processing content have often found the DSP to be the most cost-effective available. Thus many programs have been written for DSPs, which never perform a multiplication but do attempt to perform general purpose computing functions with a very specialised and unsuitable set of instructions. In addition, many DSP applications are associated with general purpose interfacing and control functions, where the highest speed is not essential and for which a higher level language than assembler would be attractive. These are all good reasons for the improving performance of DSPs in general programming.

General purpose instructions have certainly been added to the instruction sets of modern DSPs, not only for their use in themselves, but also in their use in enabling efficient compilers to be constructed. C is the language which is most widely supported, although others, particularly Pascal and Forth have a special position.

The AT&T DSP32 is a DSP, first produced as early as 1984, which has set several precedents in DSP architecture and surely has influenced the thinking behind many of the latest designs. Firstly, it supports floating point arithmetic. Not only does this reduce the need for the programmer to keep track of scaling manually, but it is a prerequisite for the effective use of a high level language for writing arithmetic algorithms. Secondly, it uses a von Neumann architecture, with a common memory space for program and data (although there are problems associated with this). The support of register indirect addressing with a post increment and decrement, similar to the C language construct '*p++', is not particularly unusual in DSPs. However the fact that this can operate, not only on a 32 bit word, but also on 16 bit 'integers' and 8 bit 'characters' certainly is. If DSP chips are becoming much more similar to general purpose computers, then the DSP32 is one of the DSPs that pointed the way.

2.3.3 The volume market

Another trend which might be discerned in the current DSP scene is the targetting of specific high volume production areas. The Texas Instruments TMS320C14 DSP is based on the early TMS32010 processor, but uses specialised on chip peripheral support for pulse width modulated waveform genaration, aimed at induction motor control. This type of product has not yet found its feet and it might be that the on chip analogue to digital converter, as seen on the Intel 2920, will re-appear here.

2.4 EXTRA MAINSTREAM ACTIVITIES

Although the specialised monolithic digital signal processor is the most commercially relevant, other approaches to signal processing exist.

2.4.1 Bit slice

The development of bit slice technology has continued in the nineteen eighties, with the development of 8 bit slices. Bit slice costs are greater than the use of a monolithic DSP. However, when exploring unusual architectures or for the attainment of the ultimate in processing speed, a bit slice design may still be the best choice.

2.4.2 Vector processors

The parallel processing of different data with the same operation has been used in mainframe computing and is now being applied commercially to DSP environments.

2.4.3 systolic arrays

One way of using the space on a VLSI device is to construct a large number of interconnected simple processing elements. Many signal processing problems can be solved in this way. A few systolic arrays are commercially avaiable.

2.4.4 Multiprocessors

Parallel processing has been a feature of almost all the signal processing systems that have been mentioned. Most incorporate some support for colaboration with other processors. We have also mentioned the difficulty of controlling parallel processes to attain maximum efficiency. Many attempts are of course being made to solve this problem in general purpose computing. The Transputer is a sophisticated example, being supported by the language Occam specifically designed for parallel processing. The fact that a parallel solution is obvious in many signal processing algorithms makes the operation of concurrent processors particularly attractive.

2.5 CONCLUSION

This survey of the development of the modern digital signal processor has concentrated on the mainstream DSP developments which have given us incredibly powerful and low-cost arithmetic processors. The view presented here is that the modern DSP is not historically derived from the conventional general purpose von Neumann processor, but from a separate 'species' of digital signal processors. Any similarities to the general purpose processor are of quite recent origin, driven by market forces and only skin deep. The DSP should in no way be seen as an enhanced performance general purpose microprocessor. The concentration on filtering, FFT and similar applications has meant that electronic engineers, rather than computer scientists have controlled DSP development so that software developments have often lagged behind the von Neumann machines. Nevertheless, mainstream DSP development seems to be in the process of implementing the lessons learned from other computing activities.

Meanwhile, we can have access to an arithmetic capability of collosal price/ performance ratio suitable for many new applications. In 1936, F.H. Wales said "I cannot conceive that anybody will require multiplications at the rate of 40,000, or even 4000 per hour". There are expected to be customers for the Motorola DSP96002, with a 40 Mflops capability. Watch this space!

REFERENCES

[1] Estrin G., 'A description of the electronic computer at the institute for advanced studies', Association for Computing Machinery, 1952, vol 9, p 95.

[2] Buttle, A., Constantinides A.G. & Brignell, J.E. 'Online digital filtering', 1968, June 14th, pp252-253.

[3] Jackson, L.B. Kaiser, J.F., McDonald H.S., 'Implementation of digital filters', 1968, IEEE Int. Conf. digest, March 1968, p213.

[4] Allen J. & Jensen E.R. 'A new high speed multiplier design', Res. Lab Electron., M.I.T., Quart. progress report, 1972, Apr 15 105.

[5] Allen, J 'Computer architecture for signal processing', IEEE Proc., 1975, 63,4,624-633

[6] Component Data Catalog, Intel Corp. Santa Clara CA., 1981, pp 4-45 to 4-55.

[7] NEC Electronics catalog, peripherals: uPD7720 digital signal processor, 1982, pp 551-565.

[8] TMS32010 user's guide, Texas Instruments, 1983

Chapter 3

Discrete signals and systems

N. B. Jones

3.1 Classification of Signals & Systems

In order to proceed with a discussion of system performance and design and to extend it towards the objectives of this book it is useful to start by classifying the major types of signals and systems of interest. Particularly useful groupings are into the three classes, continuous, discrete and digital.

Continuous signals are those which are defined at all points in time and only contain isolated discontinuities such as steps. These signals are, for analytical purposes regarded as being resolved in amplitude to an infinite degree of precision, only limited in practice by the precision of the instruments used to measure them and by the intrinsic noise level.

A discrete signal is one which is defined only at isolated discrete points in time, its value at other times being either unknown or assumed to be zero. A discrete signal thus contains only discontinuities and is best described analytically as a set of impulses. Most discrete signals are defined at equally spaced points in time but examples exist of unequal spacing which can successfully be processed (1). This book considers only equally spaced discrete signals. The amplitude of a discrete signal is normally considered to be an infinitely resolved function as with a continuous signal but the digital signal is a discrete signal which is an important exception to this and is classified separately.

Another special case of a discrete signal is a point process which is described only by its times of occurrence and does not have an associated amplitude variable. Only unequally spaced point processes are of interest. Never-the-less conventional digital signal processing can be used successfully as an approximate method of analysing such processes (2). No further consideration will be given to point processes in this book.

The most important sub-class of discrete signals is the set of digital signals. These are discrete signals, equally spaced in time, which are also quantised in amplitude. These signals then are completely definable numerically since both the dependent and independent variables are quantised. For this reason they are pre-eminent in signal processing since they can have the correct form for direct manipulation in a digital computer, and it is special purpose digital computing elements for manipulating such signals which are the main subject of this book.

It is worth noting in passing that an important field of engineering is concerned with the processing of non-digital discrete signals. Hardware associated with this has some of the advantages of flexibility of digital systems and yet can be used to process essentially analogue signals directly (3). These devices are not considered in this book except as ancillary devices to purely digital processors.

As with signals, processing systems can be classified into continuous, discrete and digital; again with digital systems being special cases of discrete systems. The classification of systems is entirely in line with that of the signals being processed and where more than one type is involved the system is regarded as hybrid.

There is also an important classification of systems for present purposes into linear and non-linear and into time invariant and time varying. Linear systems are those which obey the principle of superposition (4) and time invariant (shift invariant) systems are those whose responses are not effected by the absolute value of the time origin of the signals being processed (5). This book is almost exclusively devoted to digital, linear, time invariant signal processing.

Much of the design of digital signal processing algorithms and devices can be conducted by treating digital signals as discrete signals, that is without reference to amplitude quantisation. This property is then introduced as a secondary factor effecting the accuracy and resolution of the computations.

3.2 Signal Conversion

Some signals, such as the record of the number of aircraft movements per hour in Heathrow airport, are intrinsically discrete (and in this case digital as well) but most signals of interest are intrinsically continuous and need to be sampled to make them discrete and also quantised to make them digital. This conversion from continuous to digital signal is normally achieved by the use of analogue to digital converters. It is not the purpose of this chapter to describe the engineering of these converters but it is important to understand that they can be modelled by an ideal sampler followed by an ideal quantiser. Details of the design of such devices and their departure from this ideal model can be found in reference 6.

3.2.1 Sampling

The ideal sampling process applied to the continuous signal $x(t)$ is conveniently modelled by the process of multiplying $x(t)$ by $d(t)$ to give the sampled signal $x_s(t)$

$$x_s(t) = x(t).d(t) \qquad\qquad [3.1]$$

where
$$d(t) = \sum_{k=-\infty}^{\infty} \delta(t\text{-}kT) \qquad\qquad [3.2]$$

and $d(t)$ is known as a Dirac comb, since $\delta(t\text{-}kT)$ is a Dirac delta function at the point kT where k is an integer and T is the sampling inteval.

This yields

$$x_s(t) = \sum_{k=-\infty}^{\infty} x_k \delta(t-kT) \qquad\qquad [3.3]$$

where $x(kT)$ has been written as x_k for brevity.

It seems reasonable to conclude that since $x_s(t)$ is only known at the sample points $t=kT$ then some of the information in the original signal $x(t)$ is lost. Inparticular it would not be expected that the value of $x(t)$ between the sample points could be deduced from the set x_k alone. This is generally true. However for one important class of signals it is possible to reconstruct the signal $x(t)$ from the samples x_k, at least in theory. If $x(t)$ has no components with frequencies greater than some limiting value F Hz then $x(t)$ is given at all values of t from the samples x_k by Whittaker's interpolation formula (7).

$$x(t) = \sum_{k=-\infty}^{\infty} x_k \frac{\text{Sin } \pi f_s(t-kT)}{\pi f_s(t-kT)} \qquad\qquad [3.4]$$

where f_s (= $1/T$) is the sampling frequency in Hz. This formula only applies provided that $f_s > 2F$. The limiting sampling frequency $f_s=2F$ is known as the Nyquist rate.

It is interesting to note that to reconstruct $x(t)$ accurately in practice requires an infinite number of samples and an ideal low-pass continuous filter, which are both physically unrealisable. However, although the original signal cannot be reconstructed from the samples in reality even if it is bandwidth limited it is comforting to know that essentially no information has been lost in the sampling process in these circumstances.

On the other hand, if the signal is not bandwidth limited, or if it is and the sampling rate is set below the Nyquist frequency, information is lost in the process. This is manifested as an ambiguity in the frequency of components of the sampled signal derived from those components of $x(t)$ with frequencies greater than $f_s/2$. The errors introduced this way are known as aliasing errors. A component at frequency f, where for example $f_s/2 < f < f_s$, would add to any components at frequency $f_s- f$ and be indistinguishable from them. In general the frequency of the false (aliased) components arising from a true component at frequency f ($>f_s/2$) is Nf_s-f if $(N-1/2)f_s< f < Nf_s$ and is $f-Nf_s$ if $Nf_s< f < (N+1/2)f_s$, where N is an integer.

Precautions against aliasing involve increasing the sampling rate and attempting to reduce the spectral spread of $x(t)$ by analogue low-pass filtering before the sampler.

3.2.2 Quantisation

Once the sampling process is complete the next step is quantisation. This process allocates binary numbers to samples. In practice sampling and quantising usually take place together but the processes are essentially separate. In order to quantise the signal the gain of the converter is adjusted so that the signal occupies most of the range R but never exceeds it. In this situation an n-bit converter will allocate one of $N=2^n$ binary numbers to each sample by comparing the size of the sample to the set of voltage levels mR/N (m=0,N) where the difference between successive voltage levels is:

$$d = \frac{R}{N-1} \text{ volts}$$

It can be shown (8) that the quantisation process results in the introduction of extra noise into the signal with RMS value

$$\frac{d}{2\sqrt{3}} \text{ volts}$$

3.2.3 Other Conversion Errors

Aliasing errors and quantisation noise are intrinsic to the conversion process and exist in the conditions described above even if the converter is perfectly engineered.

Other errors due to imperfections in the converter may also need to be accounted for. These include aperture uncertainty errors, arising from jitter in the sampling process, and non-linearity errors due to uneven quantisation of the amplitude range R. No attempt is made to discuss these here and the reader is referred to reference (9) for details.

3.3 Linear Processing of Discrete Signals

As already mentioned the theoretical basis of DSP has developed by treating digital signals as discrete signals initially. The quantisation of amplitude while being of fundamental importance is considered as a secondary problem effecting the practicalities of implementing the designs.

In this section the fundamental terminology and techniques of linear processing of discrete signals are presented. The use of these for designing DSP systems and the effect of quantisation on the designs are covered in later chapters.

3.3.1 Difference Equations and z-transforms

Linear processing of discrete signals clearly involves manipulations of input samples x_k to produce output samples y_k using operations which do not violate the principle of superposition. Thus only additions and subtractions of samples are allowed and the multiplication of samples by constants. No other processes are valid.

The generalised equation for linear processing of samples can thus be written

$$y_k + a_1 y_{k-1} + a_2 y_{k-2} + \ldots = b_0 x_k + b_1 x_{k-1} + \ldots \qquad [3.5]$$

Such an equation is known as a difference equation and is analogous to the differential equation governing linear, continuous signal processing. In fact difference equations can be constructed from differential equations by using some finite difference approximants to derivatives such as

$$\frac{dy}{dt} = \frac{y_{k-1} - y_k}{T}$$

where T is the interval between samples.

Differential equations can be converted into algebraic equations by the use of integral transforms such as the Laplace Transform. Similar mathematical convenience is afforded in the analysis and design of discrete systems by the use of a transform called a z-transform defined as

$$X(z) = \sum_{k=0}^{\infty} x_k \, z^{-k} \qquad [3.6]$$

where $z = \exp(sT)$ and s is the Laplace complex variable. Equation [3.6] is the Laplace Transform of equation [3.3] assuming $x(t)=0$ for $t<0$.

This transform has the property that the z-transform of $x_{k-1} = z^{-1}X(z)$ (provided $x_k=0$ for $k<0$). This property, which is easily proved by shifting the summation count by one in the definition of the transform of x_{k-1}, is, in some ways, equivalent to the statement that the Laplace transform of the derivative of a continuous signal is the Laplace transform of the signal itself multiplied by s.

The major consequence of the use of this transform is that [3.5] can be transformed into

$$(1 + a_1 z^{-1} + a_2 z^{-2} + ...)Y(z) = (b_0 + b_1 z^{-1} + ...)X(z) \qquad [3.7]$$

an algebraic equation allowing $Y(z)$ to be computed from $X(z)$ and a knowledge of the parameters a_r and b_r of the processor.

Details of how to evaluate z-transforms and their inverses are given in the next chapter.

3.3.2 Convolution sums and the z-transform

A signal processor (filter) can be known either in terms of its structure and parameters (e.g. as a difference equation) or in terms of how it responds to a standard input signal. When using this second description it is traditional to define a system in terms of its response to a unit impulse or its response to a sinusoid over a range of frequencies.

If the impulse response of a discrete system is the sequence h_k ($k=0$, ∞) then a unit impulse at time $t=nT$ elicits as a response the sequence $h_{(k-n)}$ starting at $t=nT$. Similarly a single sample x_n at time nT elicits a response sequence $x_n h_{(k-n)}$ starting at $t=nT$. Given that any input is a sequence of impulses of weight x_n occuring at points $t=nT$ ($n=0$, ∞) then, after consideration of equation [3.3], it can be seen that the response to the sequence x_k ($x_k = 0$ for $k< 0$) is:

$$y_k = \sum_{n=0}^{\infty} x_n h_{(k-n)} \qquad [3.8]$$

Equation [3.8] is an example of a fundamental operation known as a convolution sum.

As Laplace transformation converts a convolution integral into a product we find that z-transformation converts a convolution sum into a product, as follows:

$$\sum_{k=0}^{\infty} y_k z^{-k} = \sum_{k=0}^{\infty} \sum_{n=0}^{\infty} x_n h_{(k-n)} z^{-k}$$

$$= \sum_{m=-n}^{\infty} \sum_{n=0}^{\infty} x_n h_m z^{-n} z^{-m}$$

$$= \sum_{m=0}^{\infty} h_m z^{-m} \sum_{n=0}^{\infty} x_n z^{-n}$$

in which the lower limit is set to zero since in general the impulse response h_m cannot exist before the impulse arrives at the input.

Thus, from the definition of the z-transform [3.6], the above can be written

$$Y(z) = H(z) X(z) \qquad [3.9]$$

Similar results can be derived for sequences, including impulse response sequences, which are not zero for k<0. While an analagous situation normally implies that the system is not physically realisable in continuous processing, for discrete processing it merely implies a time delay and the ideas are still useful in practical design situations. Readers are referred to Oppenheim & Schafer (10) for discussion of double-sided z-transforms which are required in this situation.

Equation [3.7] provides a means of computing the z-transform, $Y(z)$, of the output if the z-transform, $X(z)$, of the input and the coefficients of the difference equation are known. This equation can be rewritten as:

$$\frac{Y(z)}{X(z)} = \frac{\sum_{n=0}^{N} b_n z^{-n}}{1 + \sum_{m=1}^{M} a_m z^{-m}} \qquad [3.10]$$

Equation [3.8] provides a means of relating the input and output z-transforms via the z-transform, $H(z)$, of the impulse response. This equation can be rewritten as

$$\frac{Y(z)}{X(z)} = H(z) \qquad [3.11]$$

By comparing [3.10] and [3.11] it can be seen

$$H(z) = \frac{\displaystyle\sum_{n=0}^{N} b_n z^{-n}}{1 + \displaystyle\sum_{m=1}^{M} a_m z^{-m}} \qquad [3.12]$$

and this is known as the transfer function of the processing system.

Transfer functions of systems combined with z-transforms of signals can be used to compute the systems' responses to those signals. Methods of formulating and manipulating z-transforms and then of evaluating the discrete signals from them are described in the next chapter.

3.3.3 Fourier transforms and Frequency Response Functions

It is often important to describe signals in terms of their spectral content and to describe systems in terms of their ability to process sine waves. This is because sinusoidal waves are the only non-transient waves which are unaltered in shape by linear processing and because the changes in the size and relative phase of the sinusoidal components of signals are traditionally used to describe the dynamics of many processors, particularly wave filters.

In continuous processing the spectral content of a signal x(t) is described by its Fourier Transform:

$$X(j\omega) = \int_{-\infty}^{\infty} x(t) \exp (j\omega t)\, dt \qquad [3.13]$$

and the frequency response of a system with transfer function G(s) is easily shown to be G(jω) (11). Problems exist in that [3.13] often does not converge and other descriptions of spectral content, such as the power-density spectrum (12) need to be invoked.

If x(t) is sampled as described by equation [3.3] the spectrum is seen to be:

$$X_s(j\omega) = \int_{-\infty}^{\infty} \sum_{k=-\infty}^{\infty} x_k \delta(t-kT) \exp (-j\omega t)\, dt$$

$$= \sum_{k=-\infty}^{\infty} x_k \exp (-j\omega kT) \qquad [3.14]$$

If x(t) = 0 for t < 0 then [3.13] is the Laplace transform of x(t) with s replaced by jω. Similarly for x_k =0 when k<0 the lower limit of [3.14] is zero and the equation is the z-transform of the sequence x_k with z replaced by exp(jωT). Equation [3.14] is a Fourier series and so the spectrum of the sampled signal $X_s(j\omega)$ is periodic; with period $2\pi/T$ rads/s.

In order to represent $X_s(j\omega)$ in a computer this function must also be turned into a discrete set of samples, say X_n, and by considering the inverse Fourier transform in terms of the samples X_n it can be seen, by an analagous argument to that above, that the time signal must also now be periodic as well. Thus the process of quantising time and frequency to allow the signal and its Fourier transform to both be represented in discrete form defines that both the time sequence and its spectrum will be periodic.

This interesting constraint on the nature of the discrete signal x_k and its Fourier transform X_n implies that there is only a finite number of different terms in the Fourier series defining x_k and also in that defining X_n. Thus the upper limit of summation in equation [3.14] and of its inverse must be finite or the sums will not converge.

The distance between the spectral lines defined by the sequence X_n is the inverse of the repetition period of x_k; that is:

$$\delta f = \frac{1}{NT} \text{ Hz}$$

[3.15]

where N is the number of samples x_k to be included in the calculation of X_n.

Similarly the time interval between samples x_k is the inverse of the period of the sequence X_n; that is

$$T = \frac{1}{M \delta f}$$

[3.16]

where M is the number of samples to be included in one period of the spectral sequence.

Comparing [3.15] and [3.16] shows that N=M; that is the number of samples in the time sequence x_k equals the number of samples in the spectral sequence X_n.

Hence the N spectral lines X_n are spaced out by $2\pi/NT$ rad/s and equation [3.14] and its inverse can be written:

$$X_n = \sum_{k=0}^{N-1} x_k \exp\left(-2\pi j \frac{nk}{N}\right)$$

[3.17]

$$x_k = \frac{1}{N} \sum_{n=0}^{N-1} X_n \exp\left(2\pi j \frac{nk}{N}\right)$$

[3.18]

where [3.18] can be established from [3.17] by using the orthogonality condition

$$\sum_{m=0}^{N-1} \exp\left(2\pi j m \frac{(n-k)}{N}\right) = 0 \qquad \text{for } n \neq k$$

[3.19]

Equations [3.17] and [3.18] are a discrete Fourier transform (DFT) pair. More will be said about these equations and their implementation in Chapter 6.

Consideration of equations [3.12] and [3.14] show that the frequency response function of a discrete filter is periodic since the frequency response is derived by putting $z=\exp(j\omega T)$ which is itself periodic in ω of period $2\pi/T$. In this way the frequency response function of a discrete system is radically different from that of a continuous system which leads to design complications which are considered in Chapters 7 to 10.

3.4 Summary

In order to make progress in the design of DSP systems it is important to understand the basic features of the analogue to digital conversion process and the fundamental principles of linear processing of discrete signals.

The key issues in signal conversion are the sampling rate and the number of bits of resolution to be used in the digital version of the signal. Some fundamental criteria for choice of these parameters have been presented here, but it must be remembered that other issues, of cost and the characteristics of the later processing system may also be important.

The essential basis for understanding of design in DSP systems is that of linear analysis, the central feature of which is the z-transform. This concept provides a means of reducing difference equations to algebraic equations and of "unscrambling" convolution sums. A powerful concept arising from this is the z-transfer function which defines the relationship between the z-transforms of the input and the output sequences in a DSP system.

A useful method for describing linear systems is by means of the frequency response function. The frequency response function of a discrete system can be derived from the transfer function by the substitution $z=\exp(j\omega T)$ and is periodic. The associated transform, the discrete Fourier transform (DFT), is a convenient way of describing the spectral content of discrete signals and is a very important algorithm in digital signal processing.

3.5 References

1). Roberts, J.B. "Spectral Analysis of Signals sampled at Random Times" In:- Digital Signal Processing Ed. N.B. Jones, Peter Peregrinus 1982, 194-212.

2). Lago, P.J.A. & Jones, N.B. "Turning points spectral analysis of the interference myoelectric activity". Med. & Biol. Eng. & Comput. 1983, 21, 333-342.

3). Regan, T."Introducing the MF10 - a versatile Monolithic active filter" National Semi Conductor Application Nat. 307, 1982, 217-227.

4). Oppenheim, A.V. & Willsky, A.S. "Signals & Systems", Prentice-Hall, 1983, p.74

5). Oppenheim, A.V. & Willsky, A.S. "Signals & Systems", Prentice-Hall, 1983, p.75

6). Sheingold, D.H. "Analog-Digital Conversion Handbook", (Ed.)Prentice-Hall, 1986, 169-367.

7). Oppenheim, A.V. & Schafer, R.W. "Digital Signal Processing", Prentice-Hall, 1975, p.29.

8). Jones, N.B. "Signal Conversion". In Digital Signal Processing. Ed. N.B. Jones, Peter Peregrinus, 1982, 71-72.

9). Sheingold, D.H. "Analog-Digital Conversion Handbook" (Ed.) Prentice-Hall, 1986, 344-367.

10). Oppenheim, A.V. & Schafer, R.W. "Digital Signal Processing", Prentice-Hall, 1975, p.46.

11). Oppenheim, A.V. & Willsky, A.S. "Signals & Systems", Prentice-Hall, 1983, p.213.

12). Oppenheim, A.V. & Schafer, R.W. "Digital Signal Processing", Prentice-Hall, 1975, 390-391.

Chapter 4

Z-transforms

A. S. Sehmi

4.1   INTRODUCTION

Many signals are represented by a function of a continuous-time variable. Typical examples of these signals are radar, sonar, thermal noise, and various bioelectric potentials. In many cases these signals can be manipulated by analogue processing techniques incorporating amplifiers, filters, envelope and peak detectors, clippers and so on. The advent of digital computers and their inherent flexibility makes it more convenient to use a signal representation scheme that is a function of a discrete-time variable. A set of numbers at equally spaced sampling intervals T and with amplitudes corresponding to those of the continuous signal, is the discrete signal representation used.

The chapter begins with a discussion of types of signals encountered in digital systems. It will be shown that in the discrete-time domain most digital systems can be represented by a linear difference equation and that the system response is facilitated by transforming the difference equation into the z-domain. This leads to the definition of the system transfer function and finally the frequency behaviour of the system.

There is a correspondence between the properties of signals and the analytical methods used in the continuous analogue world (convolution integral, differential equations, complex s-domain and Laplace transforms, transfer functions and frequency response) and those in the discrete digital world. Those familiar with the fundamentals of continuous-time systems and their analysis will find this chapter straight-forward.

4.2   SOME FUNDAMENTAL Definitions

4.2.1   Discrete Systems

A functional representation of a discrete system is shown in Fig.[4.1]. The inputs and outputs of this system are discrete sequences $\{x(n)\}$ and $\{y(n)\}$ respectively. In such a system, time is quantised and the input and output signal amplitudes are defined only at discrete instants in time. So $t = nT$, where T is the interval between sampled values and $-\infty \leq n \leq +\infty$.

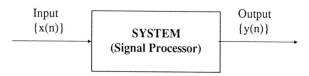

Fig.[4.1] A functional block diagram of a discrete system

4.2.2   Sequences

The input and output signals are sequences of values defined at integer values of T. In a discrete system the magnitudes of the members in the sequence are continuous. The usual mathematical representation of sequences is:

$\{x(n)\}$ or $\{x(nT)\}$ where $N_1 \le n \le N_2$.

4.2.3   Digital Systems

In a digital system the input and output sequence members assume values quantised in both time and magnitude. The magnitude is encoded digitally in some binary manner using standard methods of analogue to digital conversion.

4.2.4   Linear Shift Invariant Systems

To simplify our discussions we will restrict ourselves to discrete linear shift invariant systems. The definition of a linear system can be expressed with the following equations:

System Input	System Output
$Ax_1(n)$	$Ay_1(n)$
$Bx_2(n)$	$By_2(n)$
$Ax_1(n) + Bx_2(n)$	$Ay_1(n) + By_2(n)$

One way of visualising linearity is that a graph of the input vs. output for a system must always be a straight line through the origin.

Shift invariance means that the response of a system to a given input does not vary with time. Mathematically we represent shift invariance in the following manner:

System Input	System Output
x(n)	y(n)
x(n-n$_0$)	y(n-n$_0$)

A consequence of linearity and shift invariance is that the output of a digital (or discrete) system can be expressed as the convolution sum of the input and the unit impulse response of the system. Mathematically the output is given by:

$$y(n) = \sum_{k=-\infty}^{\infty} x(k)\, h(n-k) \qquad (4.1)$$

where {h(n)} is the unit impulse response - i.e. the response of the system to a sequence that only consists of a unit sample at n = 0 and is zero for all other values of n. The convolution sum in equation (4.1) is often written in the form (also see chapter three, equation (3.8)):

$$y(n) = x(n) * h(n) \qquad (4.2)$$

4.3   LINEAR CONSTANT COEFFICIENT DIFFERENCE EQUATIONS

We have seen that digital signal processing is achieved by applying an input sequence to a digital system and determining the values of an output sequence. Let us now consider what processing can take within the digital system. In reality we can only perform the following rather limited operations if we are to restrict ourselves to linear shift invariant systems:

• [a] Store the present and past values of a sequence.
• [b] Multiply sequences by a constant.
• [c] Add or subtract sequences.

It is therefore seen that the output from any discrete or digital system can be represented by the following difference equation (also see chapter three, section 3.3.1)

$$y(n) = \sum_{r=0}^{\infty} a_r x(n-r) - \sum_{k=0}^{\infty} b_k x(n-k) \qquad (4.3)$$

where x(n-r) represents the present and past values of the input sequence, y(n) is the present value of the output sequence and y(n-k) signifies past values of the output sequence. In digital filter design we will always assume that a_r and b_k are constants that do not vary with time. In adaptive signal processing systems both a_r and b_k are time varying.

It should be noted that the difference equation shown in equation (4.3) both represents the system in mathematical terms and can be used for system implementation. It shows what data is to be stored and how that data must be manipulated to compute the present value of the output.

A system which can be represented by equation (4.3) is referred to as a *recursive* or *infinite impulse response* (IIR) system. The word recursive is used because the present output is obtained recursively from past values of the output, and if a unit impulse is applied to the system then the feedback term

$$\sum_{k=0}^{\infty} b_k x(n-k)$$

leads to a unit impulse response sequence that is infinitely long in theory.

If b_k in equation (4.3) is equal to zero for all values of k then the following difference equation results:

$$y(n) = \sum_{r=0}^{\infty} a_r x(n-r) \tag{4.4}$$

A system defined by this type of difference equation is referred to as a *non recursive* or *finite impulse response* (FIR) system. Since there are no feedback terms in equation (4.4) we see that FIR systems are inherently stable.

Digital filter design involves determining the values of a_r and b_k in equation (4.3) to produce a digital filter which realises a given magnitude and phase characteristic. The design of such filters is facilitated if we use the z-transformation.

4.4   Z-TRANSFORMATION

The z-transformation of a sequence is a way of converting a time series into a function of a complex variable z. It has the property of enabling the solutions to linear constant coefficient difference equations to be obtained via algebraic manipulation. it also enables designers to get a visual appreciation of the time and frequency behaviour of a discrete system from pole-zero diagrams which can thereby circumvent a lot of mathematical analysis.

The z-transform of a sequence x(n) is defined as

$$X(z) = \sum_{n=-\infty}^{\infty} x(n)z^{-n} \tag{4.5}$$

For most practical sequences x(n)=0 for n≤0 and the two sided z-transform becomes the one sided z-transform defined as

$$X(z) = \sum_{n=0}^{\infty} x(n)z^{-n} \qquad\qquad (4.6)$$

We will assume from now on that we are only dealing with the one sided z-transform.

Example:

What is the z-transform of the sequence $\{1, 0.5, 0, 3, 0.2\}$?

$$X(z) = 1 + 0.5z^{-1} + 3z^{-3} + 0.2z^{-4}$$

This very simple example illustrates a very useful property of the z-transform. A physical interpretation of z^{-1} is a time delay of one time period in a sequence.

Example:

Obtain the z-transform of the geometric sequence $x(n) = g^n$.

$$X(z) = \sum_{n=0}^{\infty} g^n z^{-n}$$

By using the binomial theorem it is observed that

$$X(z) = (1 - gz^{-1})^{-1} = \frac{z}{(z-g)}$$

Note that $X(z)$ only converges for $|z| > |g|$. The region in the complex z-plane over which the z-transform of a sequence converges is known as the region of convergence and the region over which the transform diverges is known as the region of divergence.

From the z-transform of the geometric sequence it is possible to obtain the z-transform of some other useful sequences. It is straight forward to see that the z-transform of a sequence $x(n)=e^{jn\omega}$ is given by:

$$X(z) = \frac{z}{z - e^{j\omega}}$$

Example:

Obtain the z-transform of the sequence $x(n) = \cos n\omega$

$$x(n) = \frac{1}{2}[\,e^{jn\omega} + e^{-jn\omega}\,]$$

and from the z-transform of an exponential sequence it is seen that

$$X(z) = \frac{1}{2}\left[\frac{z}{z - e^{j\omega}} + \frac{z}{z - e^{j\omega}}\right]$$

$$X(z) = \frac{z^2 - z\cos\omega}{z^2 - 2z\cos\omega + 1}$$

It is left as an exercise to show that when

$$X(z) = \frac{z\sin\omega}{z^2 - 2z\cos\omega + 1}$$

A table of some of the common sequences and their z transform can be found in [1, 2].

4.5   PROPERTIES OF THE Z-TRANSFORM

4.5.1   Linearity

If we assume the sequence

$$x(n) = Ax_1(n) + Bx_2(n)$$

then the z-transform of the sequence is given by

$$\sum_{n=0}^{\infty}\left[Ax_1(n) + Bx_2(n)\right]z^{-n}$$

$$= A\sum_{n=0}^{\infty}x_1(n)z^{-n} + B\sum_{n=0}^{\infty}x_2(n)z^{-n}$$

$$X(z) = A\,X_1(z) + B\,X_2(z) \tag{4.7}$$

Hence the z-transform is a linear operator.

4.5.2   Shifting Property

Assuming

$$y(n) = x(n - k)$$

i.e. the sequence y(n) is the sequence x(n) delayed by k time periods.

$$Y(z) = \sum_{n=0}^{\infty} x(n-k)z^{-n} \quad = x(-k) + x(1-k)z^{-1} + x(2-k)z^{-2} + \ldots \qquad x(-1)z^{-(k-1)} + x(0)z^{-k} + x(1)^{-(k+1)} + \ldots$$

but the sequence $y(n) = 0$ $n < 0$

$$Y(z) = \left[x(0)z + x(1)z^{-1} + x(2)z^{-2} \right] z^{-k}$$

$$Y(z) = X(z)z^{-k} \tag{4.8}$$

4.5.3 Convolution in the Z-Domain

We know that the output sequence of a linear shift invariant system is given by the convolution sum:

$$y(n) = \sum_{k=0}^{\infty} x(k)\, h(n-k)$$

The z-transform of this output sequence is

$$Y(z) = \sum_{n=0}^{\infty} \left[\sum_{k=0}^{\infty} x(k)\, h(n-k)z^{-n} \right]$$

Using the shifting theorem it is seen that

$$Y(z) = H(z) \sum_{k=0}^{\infty} x(k)z^{-k}$$

and from the definition of the z-transform

$$Y(z) = H(z)\, X(z) \tag{4.9}$$

This result is very important in discrete system theory. It is observed that convolution in the discrete time domain is equivalent to multiplication in the complex z-domain. The converse is also true, that convolution in the z-domain is equivalent to multiplication in the discrete time domain.

$H(z)$ is the z-transform of the unit impulse response $h(n)$ and can also be written as

$$H(z) = \frac{Y(z)}{X(z)} \tag{4.10}$$

$H(z)$ is the output of a system divided by the input of the system and is usually called the transfer function of the system.

These are the basic properties of the z-transform that we require to understand digital theory at this stage. A more detailed discussion on the properties of the z-transform can be found in [1, 2].

4.6 THE INVERSE Z-TRANSFORM

Obviously we require to be able to take a z-transform $X(z)$ and manipulate this to obtain the discrete time sequence $x(n)$. There are three ways of doing this:

- Long division
- Partial fractions
- Contour integration

We will concentrate on the first two methods and briefly mention the third.

4.6.1 Long Division Method

The z-transform of a sequence is usually a polynomial in z and by dividing the numerator by the denominator a series expansion is obtained which yields the individual terms in the time sequence. This method is easy but does not produce close form expressions for the sequence.

Example:

$$X(z) = \frac{2z^2 - 0.752}{z^2 - 0.752 + 0.125}$$

Hence the first three terms in $X(z)$ are

$$X(z) = 2 + 0.75z^{-1} + 0.3125z^{-2} + \dots$$

and the first three terms in the discrete time series are

$$x(n) = \{2, 0.75, 0.3125, \dots\}$$

4.6.2 Partial Fraction Method

If $X(z)$ can be expanded by partial fractions then it is possible to use a table of z-transforms to obtain a closed form solution for $x(n)$.

Example:

We will use the equation from the previous example.

$$X(z) = \frac{2z^2 - 0.75z}{z^2 - 0.75z + 0.125}$$

$$= \frac{2z^2 - 0.75z}{(z - 0.5)\,(z - 0.25)}$$

Expanding $X(z)$ by partial fractions

$$X(z) = \frac{Az}{z - 0.5} + \frac{Bz}{z - 0.25}$$

$$X(z) = \frac{z^2(A + B) - z(0.25A + 0.75B)}{(z - 0.5)(z - 0.25)}$$

By equating coefficients it is readily seen that $A = B = 1$ and therefore

$$X(z) = \frac{z}{z - 0.5} + \frac{z}{z - 0.25}$$

furthermore, we have seen that this will give

$$x(n) = 0.5^n + 0.25^n$$

Notice that this expression for x(n) contains the first three terms evaluated by the long division method.

4.6.3   Contour Integration Method

Using the Cauchy integral theorem from complex variable theory it can be shown that [2]

$$x(n) = \frac{1}{2\pi j} \int_c X(z) z^{n-1}\, dz$$

where c is a closed conour within the region of convergence of $X(z)$ and encircling the origin in the z-plane. The integral canbe evaluated using the residue theorem and hence:

$$x(n) = \sum \left[\text{residues of } X(z) z^{n-1} \text{ at the poles inside } C \right]$$

In most cases the first two methods will be adequate to evaluate the inverse z-transform.

4.6 DIFFERENCE EQUATIONS AND Z-TRANSFORMS

Digital systems can be represented in the discrete time domain by linear constant coefficient difference equations (see chapter three). The application of z-transform techniques to a difference equation enables a relationship to be found in the z-domain between the input and output of a system. This relationship is called a transfer function.

Consider a digital filter described by the following difference equation:

$$y(n) = \sum_{r=0}^{M} a_r x(n-r) - \sum_{k=1}^{N} b_k y(n-k) \qquad (4.11)$$

Using the z-transform shifting theorem we see that

$$Y(z) = \sum_{r=0}^{M} a_r X(z)z^{-r} - \sum_{k=1}^{N} b_k Y(z)z^{-k}$$

$$Y(z)\left[1 + \sum_{k=1}^{N} b_k z^{-k} \right] = X(z) \sum_{r=0}^{M} a_r z^{-r}$$

The transfer function of a system is by definition the ratio of the output of the system over the input of the system, i.e.

$$H(z) = \frac{Y(z)}{X(z)} = \left[\frac{\displaystyle\sum_{r=0}^{M} a_r z^{-r}}{1 + \displaystyle\sum_{k=1}^{N} b_k z^{-k}} \right] \qquad (4.12)$$

The transfer function $H(z)$ in equation (4.12) is the ratio of two polynomials in z. The values of the complex number z which satisfy the equation

$$\sum_{r=0}^{M} a_r z^{-r} = 0$$

are called the zeros of $H(z)$ and the values of z which satisfy the equation

$$1 + \sum_{k=1}^{N} b_k z^{-k} = 0$$

are termed the poles of $H(z)$.

Poles and zeros of a transfer function is an important concept in both analogue and digital system design since the position of these singularities in either the s-domain (analogue) or the z-domain (digital) provides the system designer with much useful information.

Equation (4.12) is the transfer function of an IIR filter. The transfer function of an FIR filter is simply

$$H(z) = \frac{Y(z)}{X(z)} = \sum_{r=0}^{M} a_r z^{-r} \qquad (4.13)$$

Equation (4.13) shows that an FIR filter only has zeros.

The magnitude and phase response of filters and systems in general is of interest. These responses can be obtained from the transfer function H(z). The information needed can be obtained by considering the relationship that exists between the s-domain the z-domain.

4.7   RELATIONSHIP BETWEEN S-PLANE AND Z-PLANE

When analysing continuous systems we find that they are represented in the time domain by linear differential equations with constant coefficients, and that the analysis of such systems is greatly facilitated if we use the Laplace transformation and consider such systems in the complex s-domain.

Discrete systems are represented by linear difference equations which have been transformed into the complex z-domain. We will now consider the relationship between these two complex planes.

From the shifting theorem we know that z^{-1} has a physical meaning - it represents a time delay of one sampling interval T in a sequence. From Laplace transform theory a time delay of T in any signal is represented in the s-domain by e^{sT}. Hence

$$z^{-1} = e^{-sT} \qquad (4.14)$$

But s is a complex number denoted by $s = \sigma + j\omega$ and so

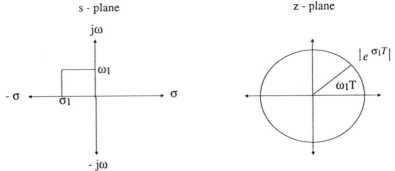

Fig.[4.2] Transformation of s-plane to z-plane

$$z = e^{(\sigma + j\omega)T}$$

$$= e^{\sigma T} e^{j\omega T}$$

So a point $\sigma_1 + j\omega_1$ in the s-plane transforms to a point $| e^{\sigma_1 T} | \angle\omega_1 T$ in the complex z-domain. This is illustrated in Fig.[4.2].

We can now consider the mapping between the s-domain and the z-domain for several important regions in the s-domain. When a transfer function in the s-domain is evaluated along the $j\omega$ axis then the frequency response of the system can be determined. Therefore it is important to relate the $j\omega$ axis in s-plane to the z-plane. When $\sigma = 0$ equation (4.14) becomes

$$z = e^{j\omega T}$$

The first important point is that the $j\omega$ axis in the s-domain maps onto a unit circle in the z-domain. However $e^{j\omega T}$ is a periodic function and the frequencies along the $j\omega$ axis between $0 \le \omega \le {}^{2\pi}\!/_T$ map onto the complete circle. Similarly the $j\omega$ axis frequencies between ${}^{2\pi}\!/_T \le \omega \le {}^{4\pi}\!/_T$ also map onto the complete unit circle in the z-domain. We therefore have the following important results:

- The frequency behaviour of a digital system can be determined by evaluating the transfer function H(z) along the unit circle in the z-domain.
- The frequency behaviour of a digital system is periodic. The value of the frequency response of the system at ω_1 is identical to that at $\omega_1 + {}^{2\pi}\!/_T$. It is therefore imperative to ensure that input signals fed into a digital filter are band limited to frequencies in the range $0 \le \omega \le {}^{\omega_s}\!/_2$ where ω_s is the sampling frequency.

In a continuous system the poles of any transfer function must be restricted to the left half of the s-plane if the system is to be stable, i.e. $\sigma < 0$ for all poles. What is the implication of this in the z-domain ?

$$z = |e^{\sigma T}| \angle j\omega T$$

Therefore, (a strip ω_s wide in ..) the left half of the s-plane ($\sigma < 0$, $|z| < 1$) maps inside the unit circle in the z-domain and a digital system will be stable if the poles of the transfer function lie inside the unit circle. The right half of the s-plane ($\sigma > 0$, $|z| > 1$) maps outside the unit circle in the z-plane. The imaginary axis in the s-plane ($\sigma = 0$, $|z| = 1$) transforms to the circumference of the unit circle in the z-plane. The complete mapping between the s-plane and the z-plane is shown in Fig.[4.3].

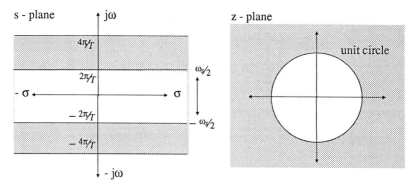

Fig.[4.3] Correspondence between regions in s-plane and z-plane

4.8   THE FREQUENCY RESPONSE FROM Z-PLANE POLES and Zeros

The transfer function of any digital system is given by

$$H(z) = \left[\frac{\displaystyle\sum_{r=0}^{M} a_r z^{-r}}{1 + \displaystyle\sum_{k=1}^{N} b_k z^{-k}} \right] \tag{4.15}$$

The frequency response is determined by evaluating the transfer function along the unit circle and is therefore given by

$$H(e^{j\omega T}) = \left[\frac{\displaystyle\sum_{r=0}^{M} a_r e^{-j\omega rT}}{1 + \displaystyle\sum_{k=1}^{N} b_k e^{-j\omega kT}} \right] \tag{4.16}$$

However it is also possible to express the transfer function in terms of its poles and zeros.

$$H(z) = \left[\frac{G \displaystyle\prod_{r=1}^{M} (z - \alpha_r)}{\displaystyle\prod_{k=1}^{N} (z - \beta_k)} \right] \tag{4.17}$$

where:

G is a gain constant
α_r is the rth zero of H(z)
β_k is the kth pole of H(z)

Once again to obtain the frequency response we must set $z = e^{j\omega T}$ in equation (4.17).

$$H(e^{j\omega T}) = \left[\frac{G \prod\limits_{r=1}^{M} (e^{j\omega T} - \alpha_r)}{\prod\limits_{k=1}^{N} (e^{j\omega T} - \beta_k)} \right] \qquad (4.18)$$

At any particular frequency ω_i, the terms ($e^{j\omega_i T} - \alpha_r$) in the numerator of equation (4.18) represent vectors from the rth zero to a point $\omega_i T$ on the unit circle and similarly the terms ($e^{j\omega_i T} - \beta_k$) in the denominator are vectors from k^{th} pole to the same point $\omega_i T$ on the unit circle. This is illustrated in Fig.[4.4] for M = 2, N = 2.

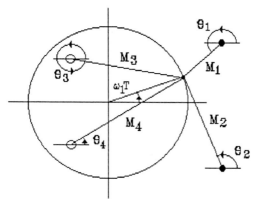

Fig.[4.4] Illustration of pole and zero vectors in the z - plane

The magnitude response at a frequency ω_1 is given by

$$| H(e^{j\omega_1 T}) | = \frac{G M_1 M_2}{M_3 M_4} \qquad (4.19)$$

and the phase response is given by

$$\angle H(e^{j\omega_1 T}) = (\theta_1 + \theta_2) - (\theta_3 + \theta_4) \qquad (4.20)$$

4.9   CONCLUSION

In this chapter the z-transform representation of discrete signals and systems has been developed. The basic concepts have been reviewed, and the use of the transform in evaluating the frequency response has been shown. It is seen that the z-transform method of solving difference equations is analogous to the Laplace transform method; i.e. both techniques first transform the equations to polynomial representations in the corresponding complex variable and then invert them to obtain a closed-form solution.

In the next chapter the relationship between the z-transform and the discrete Fourier transform will be established.

4.10   REFERENCES

1. J. Candy, *Signal Processing: The Modern Approach*, McGraw-Hill, 1988.
2. L. Rabiner and B. Gold, *Theory and Application of Digital Signal Processing*, Prentice-Hall, New Jersey, 1975.

Chapter 5

Fourier series and the discrete Fourier transform

A. S. Sehmi

5.1 INTRODUCTION

This chapter introduces the real and complex Fourier series. The Fourier integral is developed as a method for extracting spectral information from signals in the continuous time domain. The Fourier transform is developed and leads to the discrete Fourier transform. Important properties of the discrete Fourier transform and issues in its practical usage are explained. The discrete Fourier transform is related to the z-transform (chapter four) and other uses for it are highlighted.

5.2 FOURIER ANALYSIS

Fourier's theorem states that any single valued periodic function, with fundamental period of repetition T, can be represented by an infinite series of sines and cosines which are harmonics of the fundamental repetition interval, i.e. :

$$x(t) = \frac{a_0}{T} + \frac{2}{T} \sum_{n=1}^{\infty} \Big[a_n \cos(2\pi nft) + b_n \sin(2\pi nft) \Big] \qquad (5.1)$$

where $f = {}^1/T$ is the fundamental frequency. The response of a system having x(t) as input is evaluated by superposition over the individual responses due to each of the Fourier components of x(t).

The following two mathematical rules will help in an understanding of Fourier (harmonic) analysis

- During the fundamental period T the *average* of the product of two *different* harmonics is zero.

For example consider the two harmonics $[\, A_1 \cos(2\pi mft)\,]\, [\, A_2 \cos(2\pi nft + \phi)\,]$. Where $n \neq m$ the product gives

$$\frac{1}{2}A_1 A_2 [\, \cos((m-n)\, 2\pi ft - \phi) + \cos((m+n)\, 2\pi ft + \phi)\,]$$

the average of the first term over period $[\,(m-n)f\,]^{-1}$ is zero and the average of the second term over period $[\,(m+n)f\,]^{-1}$ is also zero. The average of the expression must therefore be zero over $T = f^{-1}$; the period of repetition.

$$\frac{1}{T}\int_{-T/2}^{T/2} [\,A_1\cos(2\pi mft)\,]\,[\,A_2\cos(2\pi nft + \phi)\,]\,dt = 0 \quad (5.2)$$

- During the fundamental period T the *average* of the product of two harmonics of the *same* frequency is $= (^1/2$ the product of their peak amplitudes multiplied by the cosine of their phase difference).

Given $[\,A_1\cos(2\pi nft)\,]\,[\,A_2\cos(2\pi nft + \phi)\,] = \frac{1}{2}A_1A_2[\,\cos\phi + \cos(4\pi nft + \phi)\,]$, and that the average of the term $\cos(4\pi nft + \phi)$ over the period $T = f^{-1}$ is zero which gives the result

$$\frac{1}{T}\int_{-T/2}^{T/2} [\,A_1\cos(2\pi nft)\,]\,[\,A_2\cos(2\pi nft + \phi)\,]\,dt = \frac{1}{2}A_1A_2\cos\phi \quad (5.3)$$

From these two basic rules of harmonic analysis, it can be seen that if $x(t)$ in equation (5.3) is averaged over the interval T, all sine and cosine terms will equate to zero, i.e.:

$$\frac{1}{T}\int_{-T/2}^{T/2} x(t)\,dt = \frac{a_0}{T}, \text{ therefore, } a_0 = \int_{-T/2}^{T/2} x(t)\,dt$$

which gives the value of the zero frequency (dc) term in the Fourier series.

In order to find the value of a_1, $x(t)$ is multiplied by $\cos 2\pi ft$ and the average of the resulting product is taken. Using equation (5.3) the average

$$\frac{1}{T}\int_{-T/2}^{T/2} \left(\frac{2}{T}a_1\cos 2\pi ft\right)(\cos 2\pi ft)\,dt = \frac{2a_1}{2T} = \frac{a_1}{T}$$

By equation (5.3) the average of all other terms in the product ($x(t)\cos 2\pi ft$) are zero, therefore we can say

$$\frac{1}{T}\int_{-T/2}^{T/2} x(t)\cos 2\pi ft\,dt = \frac{a_1}{T}, \text{ and it follows that}$$

$$a_n = \int_{-T/2}^{T/2} x(t)\cos 2\pi nft\,dt \quad (5.4)$$

and $b_n = \int_{-T/2}^{T/2} x(t) \sin 2\pi n f t \; dt$ (5.5)

Note the Fourier series of x(t) is sometimes written

$$x(t) = A_0 + \sum_{n=1}^{\infty} \left[A_n \cos 2\pi n f t + B_n \sin 2\pi n f t \right]$$

with $A_0 = \dfrac{a_0}{T}$, $A_n = \dfrac{2}{T} a_n$, $B_n = \dfrac{2}{T} b_n$.

The Fourier series of x(t) contains an infinite number of sines and cosines. This can be reduced to a more compact form by use of trigonometric identities.

$$\cos(2\pi n f t + \phi_n) = \cos 2\pi n f t \cos \phi_n - \sin 2\pi n f t \sin \phi_n$$

Therefore $(a_n^2 + b_n^2)^{1/2} \cos(2\pi n f t + \phi_n) =$

$(a_n^2 + b_n^2)^{1/2} \left[\cos 2\pi n f t \cos \phi_n - \sin 2\pi n f t \sin \phi_n \right]$ and by letting

$$\cos \phi_n = \frac{a_n}{(a_n^2 + b_n^2)^{1/2}} \;, \quad \sin \phi_n = \frac{-b_n}{(a_n^2 + b_n^2)^{1/2}} \;, \quad \tan \phi_n = \frac{-b_n}{a_n}$$

$$a_n \cos 2\pi n f t + b_n \sin 2\pi n f t = (a_n^2 + b_n^2)^{1/2} \cos(2\pi n f t + \phi_n)$$

The Fourier series of x(t) can thus be written

$$x(t) = \frac{a_0}{T} + \frac{2}{T} \sum_{n=1}^{\infty} C_n \cos(2\pi n f t + \phi_n)$$ (5.6)

where $C_n = (a_n^2 + b_n^2)^{1/2}$ (5.7)

$$\phi_n = \tan^{-1} \left[- \frac{b_n}{a_n} \right]$$ (5.8)

The graph of C_n versus frequency is known as the *amplitude spectrum* of x(t). A graph of $\tan^{-1}[-b_n / a_n]$ versus frequency is known as the *phase spectrum* of x(t).

Example :

The voltage across a 1 ohm resistor is given by $V(t) = A_n \cos 2\pi nft + B_n \sin 2\pi nft$, the average power developed in the resistor is $\dfrac{A_n^2 + B_n^2}{2}$ watts. If $A_n = \dfrac{2a_n}{T}$ and $B_n = \dfrac{2b_n}{T}$,

the average power is given by $\dfrac{4}{T^2} \dfrac{a_n^2 + b_n^2}{2} = \dfrac{2}{T^2} C_n^2$

The square of the amplitude spectrum is thus a measure of the power dissipated in a resistor of 1 ohm at the different frequencies n=0, 1 ,2 ,3, etc.

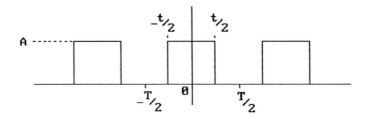

Fig.[5.1] Periodic rectangular pulse train

Example :

Find the Fourier Series of a rectangular pulse train of period T and pulse width t_1. (Fig.[5.1]). Using equation (5.4)

$$a_0 = \int_{-T/2}^{T/2} x(t)\, dt = \int_{-t_1/2}^{t_1/2} A\, dt = A\, t_1$$

The (average) dc value is $A t_1 / T$.

$$a_n = \int_{-T/2}^{T/2} x(t) \cos 2\pi nft\, dt = \int_{-t_1/2}^{t_1/2} x(t) \cos 2\pi nft\, dt$$

$$a_n = \frac{A}{2\pi nf}\Big[\sin (\pi nft_1) - \sin (- \pi nft_1) \Big]$$

$$a_n = \frac{A}{\pi nf} \sin \pi nft_1 \text{, and similarly}$$

$$b_n = \frac{A}{2\pi nf}\Big[\cos (\pi nft_1) - \cos (- \pi nft_1) \Big] = 0$$

Here $C_n = (a_n^2 + b_n^2)^{1/2} = a_n$ and $\phi_n = \tan^{-1} 0 = 0$. The amplitude spectrum is:

$$\left| C_n \right| = \left| a_n \right| = \left| At_1 \frac{\sin (\pi nft_1)}{\pi nft_1} \right| \qquad\qquad (5.9)$$

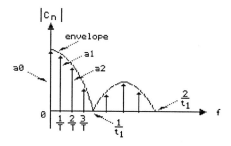

Fig.[5.2] Amplitude spectrum of a periodic rectangular pulse train

$| C_n |$ is shown in Fig.[5.2]. The envelope of the amplitude spectrum is a $\left| \dfrac{\sin x}{x} \right|$ function. This function is unity at $x = 0$ and has zeros at $(x - n\pi)$. Equation (5.9) has zeros where $f = \dfrac{n}{t_1}$. The amplitude of the dc component is $\left| \dfrac{A t_1}{T} \right| = \left| \dfrac{a_0}{T} \right|$ and the amplitudes of the harmonics are $\left| \dfrac{2a_n}{T} \right|$.

If $t_1 = {}^T/_2$ the pulse becomes a square wave and its amplitude spectrum is given by

$$C_n = \frac{A T}{2} \frac{\sin (\pi nf^T/_2)}{\pi nf^T/_2} \text{, but } T = {}^1/f \text{, so}$$

$$\left| C_n \right| = \left| \frac{A}{2f} \frac{\sin (\pi n/_2)}{\pi n/_2} \right| \qquad\qquad (5.10)$$

Here the even harmonics vanish, as shown in Fig.[5.3].

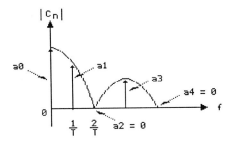

Fig.[5.3] **Amplitude spectrum of a periodic square wave**

5.3   THE COMPLEX FOURIER SERIES

In order to find the amplitude spectrum of a periodic signal both a_n and b_n must first be determined. The amplitude of a particular component C_n can be directly calculated from $x(t)$ by representing $\cos n\omega t$ and $\sin n\omega t$ in their complex exponential forms ($\omega = 2\pi f$ radians).

$$\cos n\omega t = \frac{e^{jn\omega t} + e^{-jn\omega t}}{2} \quad , \quad \sin n\omega t = \frac{e^{jn\omega t} - e^{-jn\omega t}}{2}$$

using equation (5.6) it can be shown that

$$x(t) = \frac{a_0}{T} + \frac{1}{T}\sum_{n=1}^{\infty}\left[(a_n - jb_n)\, e^{jn\omega t} + (a_n + jb_n)\, e^{-jn\omega t}\right] \qquad (5.11)$$

if $C_n = (a_n - jb_n)$ its complex conjugate $C_n^* = (a_n + jb_n)$, and we know that

$$a_n = \int_{-T/2}^{T/2} x(t)\cos n\omega t\, dt \quad , \quad b_n = \int_{-T/2}^{T/2} x(t)\sin n\omega t\, dt$$

therefore, $(a_n - jb_n) = x(t)\displaystyle\int_{-T/2}^{T/2} (\cos n\omega t - j\sin n\omega t)\, dt = \int_{-T/2}^{T/2} x(t)\, e^{-jn\omega t}\, dt$

and similarly $C_n^* = \int_{-T/2}^{T/2} x(t)\, e^{jn\omega t}\, dt$, hence $C_n^* = C_{-n}$, and

$$x(t) = \frac{a_0}{T} + \frac{1}{T} \sum_{n=1}^{\infty} \left[C_n\, e^{jn\omega t} + C_{-n}\, e^{-jn\omega t} \right]$$

since, $\displaystyle\sum_{n=1}^{\infty} C_{-n}\, e^{-jn\omega t} = \sum_{n=-1}^{-\infty} C_n\, e^{jn\omega t}$, and $C_0 = \int_{-T/2}^{T/2} x(t)\, e^{j0}\, dt = a_0$, it follows that

$$x(t) = \frac{1}{T} \sum_{n=-\infty}^{\infty} C_n\, e^{jn\omega t} \qquad (5.12)$$

This is the complex form of the Fourier Series. The negative frequencies are fictitious and they arise from the use of the complex form for sines and cosines. The real form of the Fourier series can be derived by letting $C_n = \left| C_n \right| e^{j\phi_n}$ in equation (5.12)

$$x(t) = \frac{1}{T} \sum_{n=-\infty}^{\infty} \left| C_n \right| e^{j(n\omega t + \phi_n)}$$

$$x(t) = \frac{C_0}{T} + \frac{1}{T} \sum_{n=1}^{\infty} \left[\left| C_n \right| e^{j(n\omega t + \phi_n)} + \left| C_n \right| e^{-j(n\omega t + \phi_n)} \right]$$

$$x(t) = \frac{a_0}{T} + \frac{2}{T} \sum_{n=1}^{\infty} \left| C_n \right| \cos(n\omega t + \phi_n) \qquad (5.13)$$

Equations (5.12) and (5.13) are equivalent. In the complex case the amplitude of a harmonic at both positive and negative frequencies is given by C_n / T.

In the real form the amplitude of each harmonic is $^2/T\, |C_n|$. The corresponding harmonic amplitudes in the complex case are half those in the real case. This is to be expected as only the positive frequency harmonics exist in reality. The positive and negative components summate to give the true value.

5.4   EVEN AND ODD FUNCTIONS

x(t) is an *even* function if x(t) = x(-t), an example is shown in Fig.[5.4(a)]. This function is symmetric about the t=0 axis. Any function of time which is symmetrical about t=0 is *even*.

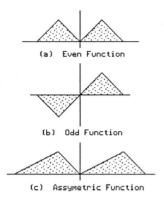

(a) Even Function

(b) Odd Function

(c) Assymetric Function

Fig.[5.4] Illustration of even, odd and assymettric functions

$x(t)$ is an *odd* function if $x(t) = -x(-t)$, an example is shown in Fig.[5.4(b)]. This function is skew-symmetric about the $t=0$ axis. Any function of time which is skew-symmetric about $t=0$ is *odd*.

A function which has no symmetry about the $t=0$ axis is neither odd nor even. An example of this function is the sawtooth waveform in Fig.[5.4(c)].

Odd and even functions have properties which may be used to simplify Fourier analysis. The three forms of the Fourier Series we have have encountered so far are

$$x(t) = \frac{a_0}{T} + \frac{2}{T} \sum_{n=1}^{\infty} \left[a_n \cos(n\omega t) + b_n \sin(n\omega t) \right] \qquad (5.14)$$

$$x(t) = \frac{a_0}{T} + \frac{2}{T} \sum_{n=1}^{\infty} C_n \cos(n\omega t + \phi_n) \qquad (5.15)$$

$$x(t) = \frac{1}{T} \sum_{n=-\infty}^{\infty} C_n e^{jn\omega t} \qquad (5.16)$$

5.4.1 Even Functions

An even function will have a Fourier Series which contains cosines terms only, the cosine being an even function itself. We know that by definition $x(t) = x(-t)$ and this can only be true if all $b_n = 0$, so:

- in equation (5.14) all values of $b_n = 0$,
- in equation (5.15) $|C_n| = a_n$ and all $\phi_n = 0$, and
- in equation (5.16) C_n is a real quantity.

5.4.2 Odd Functions

An odd function will have a Fourier Series which contains sine terms only, the sine being an odd function itself. We know that by definition $x(t) = -x(-t)$ and this can only be true if all $a_n = 0$, so:

- in equation (5.14) all values of $a_n = 0$,
- in equation (5.15) $|C_n| = b_n$ and all $\phi_n = \pi / 2$, and
- in equation (5.16) C_n is an imaginary quantity.

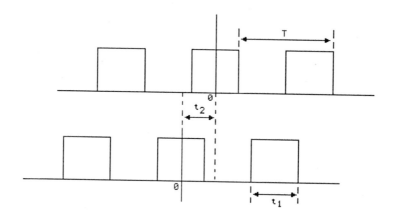

Fig.[5.5] Time axis shifting in pulse train

Many functions which are non-symmetric about $t=0$ can be made either odd or even by addition or subtraction of a suitable delay. The Fourier Series can thus be greatly simplified by addition of a time delay. The changing of the time axis will not affect the relative amplitudes of the frequency components but it will affect the phase angles. Shifting of the time axis is illustrated in Fig.[5.5].

In the first case the waveform is symmetric about $t=0$ and the Fourier Series is given by

$$x(t) = \frac{a_0}{T} + \frac{2}{T}\left[C_1 \cos(2\pi ft) + C_2 \cos(4\pi ft) + C_3 \cos(6\pi ft) + \dots \right]$$

In the second case the waveform is skew-symmetric about $t=0$ and the values of C_n will be complex and each component will have a phase shift ϕ_n, i.e.

$$x(t) = \frac{a_0}{T} + \frac{2}{T}\left[C_1 \cos(2\pi ft + \phi_1) + C_2 \cos(4\pi ft + \phi_2) + \dots \right]$$

If a delay of $-t_2$ seconds is added to the second pulse train it will be identical with the first. This delay is equivalent to a phase shift in each component.

5.5   THE FOURIER INTEGRAL

The amplitude spectrum of a periodic signal is given by the Fourier Series expansion. There are a large number of signals which are non-periodic, (e.g. a single pulse) and a knowledge of their amplitude spectra is often necessary e.g. to transmit such signals over a communications network. Spectral information of non-periodic signals can be obtained from the Fourier Integral.

Consider the periodic pulse trains of Fig.[5.5]. If the period T is allowed to approach infinity the pulse train approximates a single (i.e. non periodic) pulse. With $\omega = 2\pi f$ in equation (5.16)

$$x(t) = \frac{1}{T} \sum_{n=-\infty}^{\infty} C_n\, e^{j2\pi nft} \text{ , where } f = {}^1\!/\mathrm{T}$$

$$C_n = \int_{-T/2}^{T/2} x(t)\, e^{-j2\pi nft}\, dt$$

If we let $\Delta f = (n + 1)f - nf = {}^1\!/\mathrm{T}$ which is the spacing between harmonics, then the Fourier series can be written

$$x(t) = \sum_{n=-\infty}^{\infty} C_n\, e^{j2\pi nft}\, \Delta f$$

As $T \to \infty$, $\Delta f \to 0$, the discrete harmonics in the Fourier Series merge and a continuous spectrum results, i.e. the amplitude spectrum will contain all possible frequencies. C_n is now defined for all frequencies and becomes a continuous function of f.

$$\lim_{T \to \infty} C_n = X(f)$$

Since there are now an infinite number of *harmonics* the summation sign can be replaced by an integral.

$$x(t) = \int_{-\infty}^{\infty} X(f)\, e^{j2\pi nft}\, df$$

n now has all possible values between $\pm\infty$ and the integral is simply written as

$$x(t) = \int_{-\infty}^{\infty} X(f)\, e^{j2\pi ft}\, df \qquad\qquad (5.17)$$

The value of X(f) is that of C_n as $T \to \infty$, i.e.

$$X(f) = \int_{-\infty}^{\infty} x(t) \, e^{-j2\pi ft} \, dt \qquad\qquad (5.18)$$

The integrals in equations (5.17) and (5.18) are known as Fourier Transform pairs.

Example :

Find the Fourier transform of a pulse with amplitude A and duration t_1 seconds.

From equation (5.18), $X(f) = \int_{-t_1/2}^{t_1/2} A \, e^{-j2\pi ft} \, dt = \dfrac{A}{-j2\pi f}\left[e^{-j2\pi ft_1} - e^{j2\pi ft_1} \right]$

$$X(f) = \frac{A}{\pi f} \sin \pi ft_1 = At_1 \frac{\sin \pi ft_1}{\pi ft_1}$$

This is the sinc function (cf. equation (5.9)) and its spectrum is plotted in Fig.[5.6]. It should be noted that the spectral envelope of a single pulse of width t_1 is the same as that of a periodic pulse train of the same pulse width.

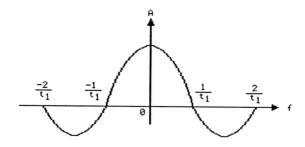

Fig.[5.6] Continuous spectrum of a single pulse of width t_1

The Fourier transform is also applicable to periodic signals, i.e.

$$x(t) = \frac{1}{T} \sum_{n=-\infty}^{\infty} C_n \, e^{j2\pi n(f - nf_0)t} \quad , \quad f_0 = \frac{1}{T}$$

So, $X(f) = \dfrac{1}{T} \displaystyle\int_{-\infty}^{\infty} \left[\sum_{n=-\infty}^{\infty} C_n \, e^{j2\pi n(f - nf_0)t} \right] dt$

$$X(f) = \frac{1}{T} \sum_{n=-\infty}^{\infty} C_n \left[\int_{-\infty}^{\infty} e^{j2\pi n(f-nf_0)t} \, dt \right]$$

$$X(f) = \frac{1}{T} \sum_{n=-\infty}^{\infty} C_n \, (f - nf_0)$$

The Fourier transform of a periodic signal therefore consists of a set of impulses located at harmonic frequencies of the signal. The weight (area) of each impulse is equal to the corresponding coefficient amplitude of the complex Fourier series.

5.6   THE SAMPLING THEOREM

The discrete Fourier transform (DFT) is a numerical approximation to the continuous Fourier transform. When using the DFT to estimate the spectrum of a signal it is important to be aware of the error involved. To demonstrate this error the DFT will be derived from the continuous Fourier transform. We will see later that the DFT is a special case of the z-transform (see chapter four).

It is not possible for digital processors to handle continuous signals. The DFT is specifically designed to operate on samples of a continuous signal. The sampling theorem states that any signal band limited to f Hz is completely defined by samples taken at a rate of at least 2f per second. A formal proof of the sampling theorem is not required here but a knowledge of the effect of sampling on the spectrum is.

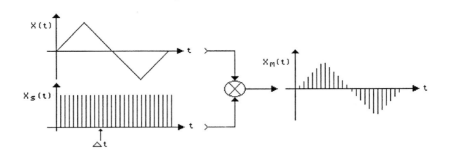

Fig.[5.7] Illustration of the sampling process

Referring to Fig.[5.7] we may regard the sampling process as a multiplication of the continuous signal x(t) by a periodic pulse train $x_S(t)$. The pulse train, being periodic, may be expanded in a Fourier series thus

$$x_s(t) = a_0 + 2a_1 \cos(\omega_s t) + 2a_2 \cos(2\omega_s t) + \ldots, \text{ where } \omega_s = 2\pi f_s$$

Letting $x(t) = \cos \omega t$ the sampled version $x_m(t) = x(t) x_s(t)$ is

$$x_m(t) = a_0 \cos \omega t + a_1 \cos(\omega_s - \omega) t + a_1 \cos(\omega_s + \omega) t + \ldots$$

$$\ldots + a_n \cos(n\omega_s - \omega) t + a_n \cos(n\omega_s + \omega) t$$

The spectrum of the sampled signal contains sum and difference components between the spectrum of $x(t)$ and harmonics of the sampling frequency, as shown in Fig.[5.8].

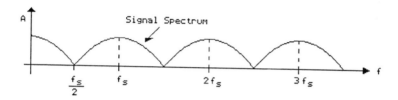

Fig.[5.8] The spectrum of a sampled signal

The diagram shows that in order to avoid distortion (aliasing) the sampling frequency f_s must be at least twice the expected frequency f in the signal. In practice $x(t)$ will be represented at the sampling instants by single numerical values. This is equivalent to sampling $x(t)$ by impulses spaced uniformly at intervals of Δt. The spectrum of such a signal has components of equal amplitude spaced at frequency intervals of $n/\Delta t$.

5.7 THE DISCRETE FOURIER TRANSFORM

Consider once again the Fourier series representation of the periodic signal $x(t)$. If we assume that $x(t)$ is band-limited above $f_x = x/T$ it implies that there are no discontinuities in the waveform and $C_n = 0$ for $n > x$. An example of such a waveform and its spectrum is shown in Fig.[5.9].

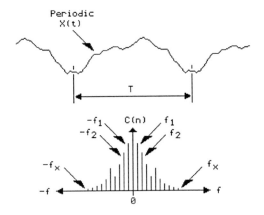

Fig.[5.9] Bandlimited periodic waveform and corresponding spectrum

Since x(t) has no frequency components above f_x, the sampling theorem states that x(t) can be completely defined by samples taken at intervals of $t = 1/(2f_x)$ (strictly speaking the signal should be sampled at $t = 1/(2f_{x-1})$). This would however produce aliasing between f_x and $(f_s - f_x)$). In fact f_s can have any value $> 2f_{x-1}$, though it is convenient to let $f_s = 2f_x$.

The effect of sampling is to make the spectrum of x(t) periodic with period $2f_x$. This is shown in Fig.[5.10] where the sampled version of x(t) is denoted x(kΔt) and the corresponding amplitude spectrum is denoted by C(nΔf).

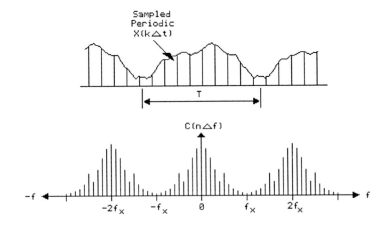

Fig.[5.10] Spectral periodicity arising from sampled data

The Fourier series for x(kΔt) is

$$x(k\Delta t) = \sum_{n=-(x-1)}^{(x-1)} C(n)\, e^{\,j\frac{2\pi nk\Delta t}{T}}, \quad \text{(the constant } {}^{1}/_{T} \text{ has been omitted)}$$

Over the interval $-(x-1) < n < (x-1)$, $C(n)$ is identical to $C(n\Delta f)$, i.e.

$$x(k\Delta t) = \sum_{n=-(x-1)}^{(x-1)} C(n\Delta f)\, e^{\,j\frac{2\pi nk\Delta t}{T}} \tag{5.19}$$

$$k = 0, \pm 1, \pm 2, \ldots, \pm {}^{N}\!/_{2} - 1$$

If N samples are taken during the interval T then $T = N\Delta t$. But $t = {}^{1}\!/_{2f_x} = {}^{T}\!/_{2x}$, so $N = 2x$. Equation (5.19) can therefore be rewritten

$$x(k\Delta t) = \sum_{n=-({}^{N}\!/_{2}+1)}^{({}^{N}\!/_{2}+1)} C(n\Delta f)\, e^{\,j\frac{2\pi nk}{N}}$$

Both $C(n\Delta f)$ and $e^{\,j\frac{2\pi nk}{N}}$ are periodic, hence the range of the summation may be changed and the last equation usually written

$$x(k) = \frac{1}{N} \sum_{n=0}^{N-1} C(n)\, e^{\,j\frac{2\pi nk}{N}} \tag{5.20}$$

$$k = 0, 1, 2, \ldots, N-1$$

which is illustrated in Fig.[5.11].

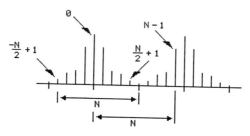

Fig.[5.11] Changing the range of summation in the two-sided DFT

Equation (5.20) contains no explicit frequency or time scale - the variables k, n and N simply take on numerical values. One difficulty arises in that should the number of samples N double, so will the number of frequency components in the spectrum. The average power in the signal x(t) is independent of N so in order to obey Parseval's theorem equation (5.20) is shown, intentionally, preceeded by the factor $^1/_N$. In practice this factor can be omitted as the ratios of the various values of C(n) are generally of greater significance than their actual values.

The complementary transform is

$$C(n) = \sum_{k=0}^{N-1} x(k)\, e^{-j\,\frac{2\pi nk}{N}} \tag{5.21}$$

$$n = 0, 1, 2, \dots, N-1$$

This relationship can be verified by substituting equation (5.20) in equation (5.21)

$$C(n) = \sum_{k=0}^{N-1} \frac{1}{N}\left[\sum_{r=0}^{N-1} C(r)\, e^{j\,\frac{2\pi rk}{N}}\right] e^{-j\,\frac{2\pi nk}{N}}$$

$$C(n) = \left[\frac{1}{N}\sum_{r=0}^{N-1} C(r)\right]\left[\sum_{k=0}^{N-1} e^{j\,\frac{2\pi k}{N}(r-n)}\right] = C(n)$$

since, by orthogonality the rightmost term is N when r = n, and zero otherwise.

Equations (5.21) and (5.20) are called the discrete Fourier transform (DFT) and the inverse discrete Fourier transform (IDFT) respectively.

5.8   PROPERTIES OF THE DFT

The properties of the DFT depend very much on the type of waveform which is analysed and care is required in interpretation of results.

5.8.1   Band Limited Periodic Waveforms

The results obtained for this type of waveform depend strongly on the duration of the sampled waveform and on the position of the samples. Assume initially that the N samples cover a whole number of periods of x(t). This is the particular case for which the DFT was developed since no discontinuities are produced when the N samples are made periodic. Under these circumstances the DFT produces the exact replica of the continuous transform. A waveform of this type, and its spectrum is shown in Fig.[5.12].

In the Fig.[5.12] there were four complete cycles of a cosine wave in the truncation interval T. This gives rise to two components in the DFT corresponding to the fourth harmonic (when written in the form of equation (5.17) the DFT produces a spectrum mirrored about $f_s/2$). The mirror image about $f_s/2$ must be taken into account when the IDFT is to be calculated from a sampled frequency function.

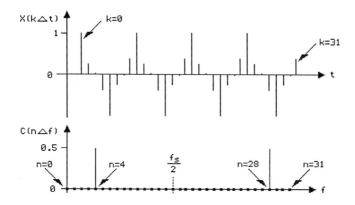

Fig.[5.12] The DFT of a sinusoid when the truncation interval contains four complete cycles

If the *truncation interval* does not contain an integer multiple of periods, a discontinuity appears in the signal when it is made periodic. (The DFT produces the spectrum of a signal which is periodic over the interval $T = N\Delta t$). The truncation of the cosine wave is equivalent to multiplication by a rectangular pulse as shown in

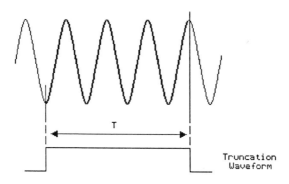

Fig.[5.13] Non-periodic truncation of a cosine wave with a rectangular window function

Fig.[5.13]. The truncation waveform has a spectral density of the form $sinc\, f = \dfrac{\sin f}{f}$. The amplitude spectrum of the truncated signal is given by the convolution of the original spectrum (in this case a pair of delta functions at $\pm f_0$) with the sinc function. This causes a spreading of the original spectrum due to the infinite duration of the sinc function and is called *leakage*.

Truncation removes any band-limitation that may have been performed (because of discontinuities introduced at either end of the truncated interval), and hence it violates the sampling theorem. A distortion called *aliasing* is introduced into the spectrum because the the repeated spectra of the sampled waveform overlap.

Aliasing can be reduced either by decreasing leakage (i.e. by increasing the truncation interval) or by increasing the sampling frequency ([1] has a good discussion of these DFT pitfalls).

5.8.2   General Periodic Waveforms

This type of signal is not band-limited, a typical example being a square wave. When sampled such waveforms produce aliasing in the frequency domain. There are no general rules for choice of sampling frequency and some experimentation is necessary. As a guide it is usual to make the sampling frequency at least ten times the 3dB bandwith of the signal to be processed.

Special care is required if a discontinuity exists within the truncation interval. If the discontinuity does not coincide with a sampling point the exact position of the discontinuity (and thus the pulse width in the case of a square wave) will be undefined. This will result in spectral errors.

If possible, when a discontinuity coincides with a sampling instant the sample should be given the mean value at the discontinuity. The use of the mean value is justified in terms of the inverse Fourier transform.

5.8.3   Random Waveforms

Such waveforms are not periodic and are usually not band-limited either. It is necessary to first truncate the waveform and to make the truncated waveform periodic. This can produce considerable leakage and the spectrum obtained via the DFT requires careful interpretation.

5.8.4   Spectral Resolution

This is a function of the truncation interval T and need not necessarily be related to the spectral components of the sampled waveform. The spectral resolution can be increased at the expense of processing-time by augmenting the data points with zeros. This does not of course affect leakage which is dependent on truncation of the true signal sequence. It should be noted that adding zeros to the signal $x(t)$ will not resolve hidden spectral components despite the increase in resolution. This is because zero padding only enhances previously visible components by an interpolation of spectral values using discrete sinusoidal functions [1]. True resolution increases are only achieved by a reducing the sampling interval.

5.8.5   Leakage Reduction

Leakage in the frequency domain results from the convolution of the signal spectrum and the spectrum of the truncating waveform. The spectrum of a rectangular truncating waveform has a sinc function envelope with zeros at intervals of $1/T$.

Clearly if T (which is not necessarily = NΔt) is increased the width of the sinc function is reduced thereby reducing leakage. It is not always feasible to increase T as this requires a pro rata increase in the number of samples N.

One alternative is the use of a window function [2] which has an amplitude spectrum which decays faster than the sinc function. A widely used truncating waveform is the raised cosine pulse also known as the *Hanning Window function*. The pulse waveform is given by

$$x(t) = \frac{1}{2} \left(1 - \cos 2\pi t / T \right)$$

and this has a spectral density

$$X(f) = \frac{T}{2} \frac{\sin \pi f T}{\pi f T \left(1 - f^2 T^2 \right)}$$

The effect of the window is to remove the discontinuities at the edges of the truncation interval. Other window functions also exist, e.g. the triangular window function, Kaiser and the 50% raised cosine function.

The significant feature of the these window functions is that the side lobes decrease faster than the rectangular case. The penalty paid for this reduction, however, is the increase in the width of the main lobe. In general the smaller the amplitude of the side lobes the greater the width of the main lobe.

The effect of increasing the width of the main lobe is to reduce he effective resolution of the DFT, i.e. the spectral peak becomes broader and less well defined. There is a trade-off therefore between leakage reduction and spectral resolution.

5.9   THE DFT AS A BANK OF FILTERS

The distortion produced by the DFT has been described in terms of leakage and aliasing. The basic limitation of the DFT in this respect is convincingly demonstrated by regarding the DFT operator as a bank of linear digital filters to which the sampled input is applied.

Consider the evaluation of the nth Fourier coefficient

$$C(n) = \sum_{k=0}^{n-1} x(k) e^{-j \frac{2\pi n k}{N}}$$

This is equivalent to the output of a non-recursive digital filter (see chapter eight) in which the tap weights are given by

$$A(k) = e^{-j \frac{2\pi n k}{N}}$$

The z transform of the filter impulse response is (chapter four)

$$A(z) = \sum_{k=0}^{n-1} e^{-j\frac{2\pi nk}{N}} z^{-k}, \text{ and noting that}$$

$$\sum_{k=0}^{n-1} a^n z^{-k} = \frac{1 - a^N z^{-N}}{1 - a z^{-1}}, \text{ gives}$$

$$A(z) = \left[\frac{1 - z^{-N}}{1 - z^{-1} e^{-j\frac{2\pi n}{N}}} \right]$$

Taking the inverse transform gives the frequency response ([2])

$$A(\omega) = e^{-j\frac{\omega(N-1)}{2}} e^{-j\frac{\pi n}{N}} \frac{\sin N\omega/2}{\sin(\omega/2 + Nn\pi)}$$

This function essentially describes the frequency response of N digital filters. The input to each filter is the wideband original signal. The response of each of these filters is similar in shape to a sinc function. The amplitude at any one filter output will be the sum of all components falling within the filter passband. This clearly demonstrates the effect of performing a DFT on a non band-limited signal.

5.10   RELATIONSHIP BETWEEN THE Z-TRANSFORM AND DFT

From equation (4.6), we have the z-transform of x(n)

$$X(z) = \sum_{n=0}^{N-1} x(n) z^{-n}$$

Evaluation of this equation at the point $z = e^{j(2\pi/N)k}$, i.e. at a point on the unit circle with angle $2\pi k/N$, gives

$$X(e^{j(2\pi/N)k}) = \sum_{n=0}^{N-1} x(n) e^{-j(2\pi/N)nk} \qquad (5.22)$$

Comparison of the DFT equation (5.21) with equation (5.22) shows equivalence. Thus the DFT coefficients of a finite duration sequence are the values of the z-transform of that same sequence of N evenly spaced points around the unit circle. More importantly, the DFT coefficients of a finite duration sequence are a unique representation of that sequence because the IDFT (equation (5.20)) can be used to reconstruct the desired sequence exactly from the DFT coefficients. So the DFT transform pair can be used to represent finite duration sequences.

5.11   OTHER USES OF THE DFT

Before we leave the subject of the DFT, it must be stated that the DFT is calculated in the computer by a very efficient algorithm called the fast Fourier transform (FFT, see chapter six). The FFT takes advantage of the symmetry of calculation on the unit circle. It reduces the N^2 calculations required by the DFT to $N\log_2 N$ calculations. Since the FFT provides a very efficient method of computing the DFT and IDFT it becomes attractive to use the DFT in computing other functions.

Power Spectrum

The power spectrum of $x(t)$ is the square of its amplitude spectrum

$$S(n) = |\, C(n)\, |^2$$

Autocorrelation

The autocorrelation function of a waveform $x(t)$ is the inverse Fourier transform of its power spectrum. The power spectrum can be obtained from the DFT and the IDFT can then be used to compute the autocorrelation function $R(k)$.

$$R(k) = \frac{1}{N} \sum_{m=0}^{N-1} x(m)\, x(k+m) \;,\; k = 0, 1, \dots, N-1$$

$$R(k) = \sum_{n=0}^{N-1} S(n) e^{j\frac{2\pi nk}{N}}$$

Convolution

The discrete convolution of two sequences x_1 and x_2 is

$$F(m) = \sum_{i=0}^{M} x_1(i)\, x_2(m-i) \;,\; m = 0, 1, 2, \dots, N-1$$

The maximum lag being $N=1$. The FFT algorithm may be employed to advantage by computing the DFT of x_1 and x_2 multiplying them together and taking the inverse transform of the product to give $F(m)$. i.e. convolution in time domain = multiplication in frequency domain.

5.12   CONCLUSION

It has been shown that under certain conditions the DFT produces an acceptable approximation to the continuous Fourier transform of a signal x(t). The properties of the DFT and several of its pitfalls in the practical use have been discussed and methods of coping with these problems have been put forward. The FFT algorithm enables the DFT to be become a very powerful tool for the digital signal processor.

5.13   REFERENCES

1. J. V. Candy, *Signal Processing: The Modern Approach*, McGraw-Hill, 1988.
2. L. Rabiner and B. Gold, *Theory and Application of Digital Signal Procesing*, Prentice-Hall, New Jersey, 1975.
3. E. Brigham, *The Fast Fourier Transform*, Prentice-Hall, New Jersey, 1974.
4. J. V. Candy, *Signal Processing: The Model-Based Approach*, McGraw-Hill, 1986.

Chapter 6
The FFT and other methods of discrete Fourier analysis
S. Kabay

6.1 INTRODUCTION

The Fourier transform, or spectrum, of a continuous signal x(t) is defined as:

$$X(j\omega) = \int_{-\infty}^{\infty} x(t)\, e^{-j\omega t} dt \qquad (6.1)$$

both x(t) and $X(j\omega)$ may be complex functions of a real variable. The basic property of the Fourier transform is its ability to distinguish waves of different frequencies that have been additively combined. For instance, a sum of sine waves overlapping in time, transforms into a weighted sum of impulses which, by definition, are non-overlapping. In signal processing terminology, the Fourier transform is said to represent a signal in the frequency domain, and ω, the argument of the Fourier transform, is referred to as the angular frequency.

The discrete Fourier transform (DFT) is used to approximate the Fourier transform of a discrete signal. If we take the continuous function x(t) and represent it as a sequence of N samples x(nT), where $0 \leq n \leq N-1$ and T is the inter-sample time interval. The time function given by:

$$x_s(t) = \sum_{n=0}^{N-1} x(nT)\, \delta(t-nT) \qquad (6.2)$$

where $\delta(t)$ is the Kronecker delta function, represents an impulse train $x_s(t)$ whose amplitude is modulated by x(t) at intervals of T (Oppenheim and Schafer, p82, 1989). Similarly, let the spectrum $X(j\omega)$ be represented by $X(k\omega)$, $0 \leq k \leq N-1$, where ω is the chosen increment between samples in the frequency domain. The DFT becomes:

$$X_s(k\omega) = \sum_{n=0}^{N-1} x(nT)\, e^{-j\omega Tnk} \qquad (6.3)$$

where $\omega = 2\pi/NT$. The latter formula yields a periodic sequence of numbers with period N. Alternatively, the discrete Fourier transform may be thought of as an evaluation of the z-transform of the finite sequence x(nT) at N points in the z-plane, all equally spaced along the unit circle at angles of $k\omega$ radians (Rabiner and Gold, p390, 1975). In digital signal processing we consider only discrete samples of both the time function and the spectrum and only a finite number of samples of each.

Direct calculation of the DFT requires N^2 complex multiplications. In the past, such intensive calculations was limited to large computer systems. However, major developments in a class of efficient algorithms known as the fast Fourier transform (FFT) for computing the DFT has dramatically reduced the computational load imposed by the method. First reported by Cooley and Tukey (1965), the FFT algorithm now has many implementations with various levels of performance and efficiency.

The FFT is based on the concept of sub-dividing a large computational problem into a large number of sub-problems which can be solved more easily. This approach has allowed for computation of the DFT on relatively inexpensive computers. The implementation of a particular FFT algorithm will depend on both the available technology and the intended application. In general, the speed advantage of the FFT is such that, in many cases, it is more efficient to implement a time-domain calculation by transforming the analysis into the frequency-domain, and inverse transforming the result back into the time-domain.

This chapter will consider the radix-2 FFT algorithm where the total number of data samples is an integer factor of two and discusses its application to spectral analysis of signals.

6.2 RADIX-2 FFT ALGORITHM

The direct evaluation of the DFT equation (6.3) requires N^2 complex multiplications and additions, and for large N (>1024 points) this direct evaluation can be too time-consuming. The FFT has dramatically reduced the number of computations to $N\log_2(N)$, where N is an integer power of two. For example, with a sample length of N=1024, a computational saving of 99 % is achieved. Efficient application of the algorithms requires that N be highly composite, but for most problems it is possible to impose this restriction on the data to be transformed so that FFT algorithms may be used.

Using the notation $W = e^{-j\omega T}$, then equation (6.3) becomes,

$$X_k = \sum_{n=0}^{N-1} x_n W^{nk} \tag{6.4}$$

and,
$$W = e^{-j2\pi/N} \tag{6.5}$$

Fast algorithms can be developed for any value of N that can be expressed as a product of prime factors; however, the algorithms suffer from the limitation of being specific to that particular value of N. One variation on the FFT is that due to Winograd (Burrus and Parks, 1985), which uses more additions but fewer multiplications than the FFT. However, this method offers no real advantage, since DSP devices are now as efficient at multiplication as they are at addition. For cases where N is not a power of two, it is more efficient to pad the sequence with zeros up to the nearest power of two.

6.3 IN-PLACE FFT ALGORITHMS

In-place FFT algorithms make optimal usage of memory since they only require enough storage to hold the DFT input sequence. As the calculation proceeds between stages, the results are overwritten to the original input buffer. Two classes of in-place algorithm exist. In the *decimation in time* (DIT) method, the transforms of shorter sequences, each composed of every *rth* sample, are computed and then combined into one big transform. The *decimation in frequency* (DIF) method where short pieces of the sequence are combined in *r*-ways to form *r* short sequences, whose separate transforms taken together constitute the complete transform.

6.3.1 Decimation in Time

The clearest derivation of the FFT algorithm is that of Danielson and Lanczos (1942), who showed that the DFT of a sequence of length N, can be considered as the DFTs of two shorter sequences each of length N/2. The two sequences g_l and h_l are composed of the even and odd numbered points of the original sequence respectively. Then,

$$g_l = x_{2l}$$

$$\text{for } l = 0, 1, ..., (N/2)-1. \tag{6.6}$$

$$h_l = x_{2l+1}$$

The DFTs of these sequences may be regarded as periodic sequences with period N/2. These DFTs may be written:

$$G_k = \sum_{l=0}^{N/2-1} g_l W^{2lk} \tag{6.7}$$

$$H_k = \sum_{l=0}^{N/2-1} h_l W^{2lk} \tag{6.8}$$

Since we are interested in the DFT of the whole sequence, which can be written in terms of g_l and h_l as:

$$X_k = \sum_{l=0}^{N/2-1} (g_l W^{2lk} + h_l W^{(2l+1)k}) \tag{6.9}$$

$$\Rightarrow \qquad X_k = \sum_{l=0}^{N/2-1} (g_l W^{2lk} + W^k h_l W^{2lk})$$

$$\Rightarrow \qquad X_k = G_k + W^k H_k \tag{6.10}$$

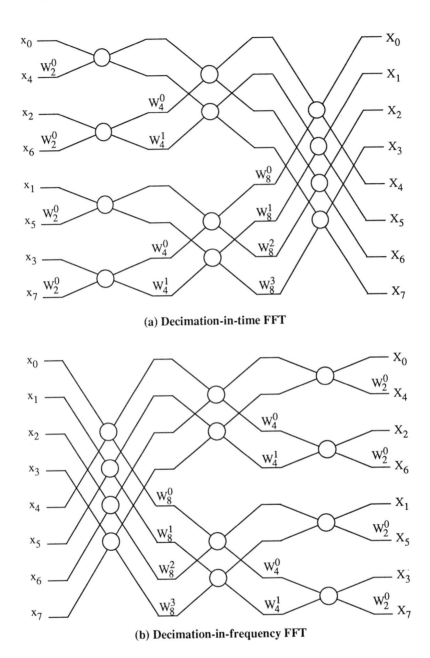

(a) Decimation-in-time FFT

(b) Decimation-in-frequency FFT

Fig. 6.1 The two classes of in-place FFT algorithms.
The signal flow graphs show the decomposition of stages for an 8-point
radix-2 FFT algorithm.

The last relationship has important computational implications. The direct method of computing G_k and H_k would require $(N/2)^2$ each, and an extra N operations are required to form X_k, for a total of $N+(N/2)^2$ operations. Direct computation of X_k would require N^2, and so, for large N, the last relationship reduces the computation by a factor of 2.

The index k runs from 0 to N-1. However, G_k and H_k have a period N/2 and need to be computed only for the range 0 to N/2-1. Hence, a full description of X_k is given by:

$$X_k= G_k+W_k.H_k \qquad (0{\le}k{\le}N/2\text{-}1) \qquad (6.11)$$

$$X_k= G_{k-N/2}+W_k.H_{k-N/2} \qquad (N/2{\le}k{\le}N\text{-}1) \qquad (6.12)$$

The real power of the above relationship is realised when it is used recursively. Where N is an integer power of 2, it follows G_k and H_k can in turn be reduced in the same way as the computation of X_k. Thus, G_k and H_k is computed from the N/4-point transform of the odd and even numbered sequences. In this way, the computation of X_k is successively reduced until we reach the stage where N/2 DFT computations are required of a two sample sequence. Fig. 6.1(a) shows the data flow diagram for an 8-point DIT FFT calculation.

6.3.2 Decimation in Frequency

The DFT of a sequence x_l of length N points is considered as the DFTs of two sequences of length N/2 points, g_l and h_l, where:

$$g_l= x_l$$
$$\text{for } l=0,1,...,(N/2)\text{-}1. \qquad (6.13)$$
$$h_l= x_{l+N/2}$$

The N-point DFT X_k may now be written in terms of g_l and h_l:

$$X_k= \sum_{l=0}^{N/2-1} (g_l W^{lk}+h_l W^{(l+N/2)k})$$

$$\Rightarrow \qquad X_k= \sum_{l=0}^{N/2-1} (g_l+h_l e^{-j\pi k})W^{lk} \qquad (6.14)$$

Taking the even and odd frequency components of X_k separately and replacing k by 2k, then:

$$X_{2k}= \sum_{l=0}^{N/2-1} (g_l+h_l)(W^2)^{lk} \qquad (6.15)$$

and replacing k by 2k+1:

$$X_{2k+1}= \sum_{l=0}^{N/2-1} [(g_l-h_l)W^l](W^2)^{lk} \qquad (6.16)$$

Equations (6.15) and (6.16) are the results of N/2-point DFTs performed on the sum and difference of the first and second halves of the original input sequence. Fig. 6.1(b) shows the data flow diagram for an 8-point DIF FFT calculation.

6.3.3 Summarising the DIT and DIF Algorithms

In the DIT algorithm, the first stage of the DFT was expressed as the sum of two half-length DFTs composed of even and odd samples of the original sequence. This method produces output (frequency) samples that are in correct sequence. The DIF algorithm calculates two half-length DFTs composed of the first and second halves of the original sequence. In this case, the output is composed of the even and odd samples respectively. Thus, DIT refers to grouping the input sequence into even and odd samples, whereas DIF refers to grouping the output sequence into even and odd samples.

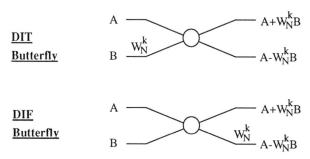

Fig. 6.2 Radix-2 butterfly operations.

There are two features worth noting from Fig. 6.1. First, the DFT calculation makes repeated use of the *butterfly* operation shown in Fig. 6.2. The butterfly is fundamental to the FFT and its execution speed can be used to benchmark the performance of the FFT algorithm. Secondly, the data sequence is scrambled during the course of the calculation. In the DIT algorithm the input is bit-reversed and the output is in correct order. Also, the DIT butterfly is *twiddled* before the DFT. The reverse is true for the DIF algorithm. Fig. 6.2 shows the DIT and DIF butterfly representations.

6.4 BIT-REVERSAL

For in-place computation of the radix-2 FFT, the data sequence must be scrambled to correct for the repeated movement of odd-numbered members of the input data to the end of the sequence during the development of the algorithm. The data scrambling is known as *bit-reversal* because the sample whose time index is given by the binary number $a_2 a_1 a_0$ must be exchanged with the sample at location $a_0 a_1 a_2$. Many digital signal processor devices   incorporate bit-reversed addressing to overcome the compromise between speed, memory and versatility associated with bit-reversal in software.

Other variations to the radix-2 FFT algorithm, which use a more complex data flow arrangement, eliminate bit-reversal altogether (Stockham, 1966). Although both the input data and the resulting spectrum are in natural order, extra memory is required to duplicate the input data sequence.

6.5 FFT HARDWARE PERFORMANCE

Although general-purpose microprocessors have increased in speed, the multiply-and-accumulate instruction - central to the FFT butterfly calculation - still remains relatively slow. A wide variety of specialised DSPs now exist for handling computation intensive signal processing tasks such as the FFT. The Texas Instruments' TMS320 family of fixed-point processors have an instruction set that is particularly suited to signal processing. For example, a pipelined MAC instruction adds the contents of the previous product to the accumulator and then simultaneously reads two values and multiplies them in a single cycle (100 ns). These devices have made it possible to implement real-time spectral analysis and process control.

There are implementation constraints associated with fixed-point arithmetic processors. Since the internal data representation on such systems is of finite precision, care must be taken to reduce the roundoff noise caused by quantisation errors (Rabiner and Gold, p295, 1975). Such errors can be minimised by careful scaling of intermediate results. Block floating-point (BFP) scaling can be used to maximise the dynamic range of a fixed-point register. Instead of normalising each number individually (true floating point), the BFP representation has a fixed exponent associated with an entire block of data. This scheme is memory efficient and well-suited for FFT implementations. The new generation of DSPs, such as the TMS320C30 and the AT&T DSP32C, have a floating-point arithmetic unit. This eliminates any worry about overflows and the accuracy of results - at a price!

Hardware	Execution Time (ms)
TMS320C30 (33MHz)	3.750
TMS320C25 (40MHz)	15.552
Sun 3/260 (25MHz)	104
MicroVAX II (VMS)	302
Compaq 386 (16MHz)	798
IBMPC-AT(12.5 MHz)	1250
PDP-11	1400

Table 6.1

Table 6.1 shows the execution speed for a 1024-point complex FFT for various computer systems (All machines have a floating point accelerator where appropriate). This is intended to provide an informal comparison of the order of performance across architectures.

6.6 APPLICATION OF THE FFT

The increased speed with which the DFT can be calculated has meant an increase in the number of applications. This section will look into one application of the FFT and mention the considerations regarding the implementation of the FFT method.

The enhanced performance of the FFT is partly due to the redundancy produced by the periodicities in the twiddle factor term W_N^{kn}. Further speed improvements are possible by reducing butterfly operations with twiddle factors W^0 and $W^{N/2}$ to simple additions. Special case butterflies can also be used with twiddle terms $W^{N/4}$, $W^{3N/4}$ and $W^{N/8}$.

A look-up table approach to fetching twiddle factors is far more efficient than calculating individual twiddle terms during the main FFT. Also, as the magnitudes of the angles of the twiddle factors θ are always less than π radians, a look-up table containing $\sin |\theta|$ will be sufficient for computing the FFT. The imaginary part of the twiddle factor is the negative of this sine value.

Another consideration is the implementation of special radix-4 butterflies to perform the first two stages of a radix-2 FFT in a single-pass (Papamichalis and So, 1986). Also, the use of in-line coded algorithms will perform faster than looped algorithms since the extra processing associated with testing and branching is eliminated. However, the trade-off between performance speed and increased code length will depend on the hardware architecture and resources available to the programmer.

In general, the data to be processed is a real function of time. Since the FFT algorithm expects complex input data, the imaginary component of the FFT input must be set to zero. This approach is inefficient in that the algorithm will perform many multiplications with zero. Two methods exist which make efficient use of the extra storage and execution time provided by the complex FFT algorithm when dealing with real data. First, the imaginary part of the complex input sequence can be used to more efficiently compute the FFT of two real functions simultaneously. In the second method, the real input sequence can be cleverly packed into a complex sequence of half its length and a complex FFT performed on this shorter length. A more detailed discussion is given by Brigham (1988).

The most obvious application of the FFT is in estimating the power spectrum of a signal. This is achieved by simplifying the basic operations of autocorrelation and Fourier transformation (Cooley et. al., 1967). However, consideration must be given to the errors introduced by this method. The spectral resolution δf for a signal of maximum frequency fmax is determined by the FFT length N, where

$$\delta f = \left(\frac{f_{max}}{N}\right) \qquad (6.17)$$

Since the use of the FFT requires finite-length signals, windowing must be applied in advance of the analysis. The spectral resolution of this method increases with the window length. Truncation or windowing of a signal will inherently produce sharp discontinuities in the time-domain, or equivalently results in a $\sin(x)/x$ function with characteristic side-lobes in the frequency domain (Oppenheim and Schafer, p702, 1989). The side-lobes are responsible for the distortion which tends to leak energy across spectral components. Spectral leakage can be significantly reduced by selecting a window function with minimal side-lobe characteristics. However, the more one reduces leakage, the lower the spectral resolution due to smearing of the spectrum.

The periodogram uses the FFT to estimate the power spectrum of a stationary random signal (Oppenheim and Schafer, p730, 1989). This is calculated from the squared magnitude of the FFT of a segment of the signal. For large data sequences, the accuracy of the periodogram estimate can be improved by dividing the signal into shorter segments and averaging the associated periodograms. Alternatively, the FFT can be used to estimate the autocorrelation function. Applying a window to the autocorrelation estimates followed by the FFT will result in the smoothed periodogram. This is a good spectral estimate.

A further discussion of the FFT and its application to identifying respiratory system dynamics of patients receiving anaesthesia can be found in Chapter 22 of this book.

6.7 HIGH-RESOLUTION SPECTRAL ANALYSIS

The major limitation of FFT-based spectral analysis is the associated spectral sampling. Since the spectrum is computed for a discrete set of sample frequencies, important features in the spectrum may not be evident in the sampled spectrum. This error can be easily reduced by improving the spectral resolution δf, which is inversely proportional to the FFT sample length. A more persistent source of error is the distortion due to leakage caused by windowing. Although the smearing effect of spectral leakage can be reduced by selecting appropriately tapering window functions, the frequency resolution will always be reduced.

Whilst the FFT approach can be successfully applied to stationary data, where an increased record-length and frequency resolution are beneficial for spectral analysis, this would not be applicable for time-varying data where the record-length must be kept sufficiently short such that the signal characteristics are approximately stationary. Spectral analysis of non-stationary signals, such as doppler flow and speech, involves a trade-off between time and frequency resolution. A shorter record-length increases our ability to track spectral characteristics in time, however, the result is a decreased frequency resolution.

These limitations can be alleviated by using parametric spectrum estimation methods. The general approach is to represent the discrete signal x_n as the output of a linear filter excited at the input by white noise. Since the output from the model is only an approximation to the actual data x_n, we must strive to minimise some error criterion between the model and x_n. A spectral estimate of x_n can then be determined from the parameters that characterise the model.

The parametric method of spectral analysis has several advantages over the periodogram:

- provides smoother estimates than periodograms
- improved spectral resolution for short data records
- provides an analytical expression for the spectrum unlike the periodogram which yields tabulated values.

Formulating the mathematical model of a system is a process of system identification, a discipline in itself. The accuracy of the parametric method depends on the order and number of parameters of the model. Many methodologies exist for selecting a model for a specific situation, however, at present no single accepted method seems to be available. Some models appear to be more appropriate for specific situations than others. For example:

- the second-order statistics of a discrete autoregressive (AR) model seems well-suited to cases involving random stationary oscillations (Gersch and Liu, 1976).
- on the other hand, it would be difficult to use an AR model for a signal containing zeros in its spectrum since a high order model would be required. An autoregressive moving average (ARMA) model would be more efficient for this task.
- for transient and deterministic data with time-varying spectral components, the Prony method of decomposing a time series into a sum of exponentials would be suitable (Hildebrand, 1956).

The following sections will review some of the most common parametric spectrum estimation methods currently in use. The approach is based on a fundamental theorem by Wold (1938).

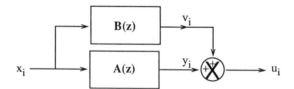

Fig. 6.3 A dynamic system with inputs
$x_{i\,i}$, **output** y_i, **disturbance** v_i **and overall output** u_i.

From the model of the dynamic system shown in Fig. 6.3, we assume that y is a stationary stochastic process given by

$$u_i = y_i + v_i \qquad (6.18)$$

where y_i is a deterministic process (predictable from its past value s) and v_i is a non-deterministic moving average (MA) process given by:

$$v_i = \sum_{k=0}^{q} b_k x_{i-k} \qquad (6.19)$$

Both y_i and v_i are stationary and mutually uncorrelated signals. Equation (6.19) describes a time series model with a finite set of q parameters. Applying the Z-transform to we get an expression for the transfer function B(z), where:

$$V(z) = \sum_{k=0}^{q} b_k z^{-k} X(z) = B(z)X(z) \qquad (6.20)$$

6.7.1 Autoregressive (AR) Model

Autoregressive spectral analysis is the most widespread alternative to the periodogram. The derivation follows on from equation (6.20), by defining $A(z) = 1/B(z)$ then:

$$Y(z)A(z) = X(z) \tag{6.21}$$

where $A(z)$ is a polynomial in z and $B(z)$ is assumed to be invertible.

Then:

$$X(z) = (1+a_1 z^{-1}+a_2 z^{-2}+...)Y(z)$$

$$\Rightarrow \qquad y_i = -\sum_{k=1}^{p} a_k y_{i-k} + x_i \tag{6.22}$$

From equation (6.22), we see that the present y_i is a linear combination of the past outputs y_{i-k} and the present input x_i. This can be considered as a linear prediction equation for y_i, with x_i as a prediction error. Thus, $A(z)$ is a pre-whitening filter, transforming a signal y into white noise x. The inverse power spectral density of y is then given by the squared frequency response of $A(z)$.

6.7.2 Autoregressive-Moving Average (ARMA) Model

According to Wold's theorem, many *real* processes will be composed of a deterministic and stochastic part. This can be modelled as a combination of moving average and autoregressive processes (ARMA). From equations (6.18), (6.19) and (6.22):

$$y_i = -\sum_{k=1}^{p} a_k y_{i-k} + \sum_{k=0}^{q} b_k x_{i-k} \tag{6.23}$$

$$\Rightarrow \qquad Y(z) = \frac{\sum b_k z^{-k}}{\sum a_k z^{-k}} X(z) = \frac{B(z)}{A(z)} X(z) \tag{6.24}$$

where $A(z)$ defines the AR terms and $B(z)$ the MA terms. The transfer function of equation (6.24) can be described as all poles, all zeros and pole-zero models for appropriate values of A and B. The power spectrum is obtained by evaluating equation (6.24) for $z^{-1} = e^{-j\omega T}$ around the unit circle.

6.7.3 Maximum Entropy (ME) Method

The maximum entropy method aims to overcome the window truncation of data outside of the observation interval. The AR method avoids truncation by extrapolating the autocorrelation function beyond the known lags. Burg's (1975) method searches for a signal with maximum entropy (thus making the minimum assumptions outside of the observation interval) whose power density spectrum is whitest and an autocorrelation function approaching that of the observed signal. The spectral estimate derived using the ME method is equivalent to that of the AR estimate (Van den Bos, 1971).

The ME method results in a higher-frequency resolution spectral estimate, for a given length of data, as compared to classical spectrum analysis. This is due to the avoidance of data truncation - a feature of all parametric methods.

6.7.4 Maximum Likelihood Estimation (MLE)

Maximum likelihood estimators of model parameters are based on the idea of using observed data, together with an assumed probability distribution function with unknown parameters, to estimate the parameters such that they maximise the probability of having obtained the specific observed data. Hence, we identify the probability of the data given the parameters, as the probability of the parameters given the data. This approach is purely heuristic and has no formal mathematical basis.

6.7.5 Summary of Parametric Methods

Parametric spectral analysis results in a mathematical model which can be used for classification, identification or other purposes. An analytical description of the process can lead to some insight about the structure of the system under analysis.

The parametric method achieves a relatively high data reduction. The result is often a smoothed, high resolution spectral estimate derived from a limited data set. However, the recursive nature of parametric analysis results in a computational effort that is comparable with Fourier methods.

6.8 BIBLIOGRAPHY

Brigham, E.O. (1988): *The fast Fourier transform and its applications*, Prentice Hall, Englewood Cliffs, NJ.

Burg, J.P. (1975):" Maximum entropy spectral analysis". *PhD Thesis*, Stanford University.

Burrus, C.S. and Parks, T.W. (1985): *DFT/FFT and computation algorithms theory and implementation*, John Wiley and Sons, New York.

Cooley, J.W., Lewis, P., and Welch, P. (1967): "The fast Fourier transform and its applications". *IBM Res. Paper*, RC-1743.

Cooley, J.W. and Tukey, J.W. (1965): "An algorithm for the machine computation of complex Fourier series". *Math. Comput.*, **19**: 297-301.

Danielson, G.C. and Lanczos, C. (1942): "Some improvements in practical Fourier analysis, and their application to x-ray scattering from liquids". *J. Franklin Inst.*, **233**: 365-380.

DSP Committee (1976): *Selected papers in digital signal processing*, IEEE ASSP, IEEE Press, New York.

DSP Committee (1979): *Programs for digital signal processing*, IEEE ASSP, IEEE Press, New York.

Gersch, W. and Liu, R.S.-Z. (1976): "Time series methods for the synthesis of random vibration systems". *Trans. ASME, J. Appl. Mech.*, **43**(1): 159-165.

Hildebrand, F.B. (1956): *Introduction to numerical analysis*, McGraw-Hill, New York.

Jones, N.B. (Ed.) (1982): *Digital signal processing*, IEE Control Engineering Series **22**, Peter Peregrinus, England.

Linkens, D.A. (1982): "Non-Fourier methods of spectral analysis", In: *Digital signal processing*, (Ed.) N.B. Jones, IEE Control Engineering Series **22**, 213-254.

Oppenheim, A.V. and Schafer, R.W. (1989): *Discrete-time signal processing*, Prentice Hall, Englewood Cliffs, NJ.

Papamichalis, P. and So, J. (1986): "Implementation of fast Fourier transform algorithms with the TMS32020". *DSP Applications with the TMS320 Family*, Texas Instruments, p32.

Rabiner, L.R. and Gold, B. (1975): *Theory and application of digital signal processing*, Prentice Hall, Englewood Cliffs, NJ.

Stockham, T.G. (1966): "High speed convolution and correlation". *AFIPS Proc.*, **28**: 229-233.

Van den Bos, A. (1971): "Alternative representation of maximum entropy spectral analysis". *IEEE Trans.*, **IT-17(4)**: 493-494.

Wold, H. (1938): *A study in the analysis of stationary time series*, Almqvist and Wiksell, Upsalla.

Chapter 7

Introduction to digital filters

T. J. Terrell and E. T. Powner

7.1 INTRODUCTION

Digital signal processing (DSP) is an established method of filtering electrical waveforms and digital images, and it is an important topic in a number of diverse fields of science and technology [1]. For example, some practical applications of digital filters are:

* Biomedical signal processing (eeg and ecg signals).

* Audio signal processing (speech signals).

* Sonar signal processing (submarine detection).

* Radar signal processing (aircraft detection).

* Seismic signal processing (oil exploration).

* Image processing (HDTV sampled-data processing).

* Digital control systems (autopilot).

The realisation of the many practical applications has been made possible by the increased availabilty and falling costs of sophisticated very-large-scale integrated (VLSI) circuits. In particular, the ubiquitous microprocessor and associated peripheral chips provide the means of implementing relatively simple and cost-effective digital filters [2]. However, since 1979 the development of self-contained programmable DSP chips [3-6] has provided a practical means of implementing many real-time signal processing operations.

The inherent advantages of digital filters are:

* They do not drift.

* They can handle very low frequency signals.

* Frequency response characteristics can be made to approximate closely to the ideal.

* They can be made to have no insertion loss.

* Linear phase characteristics are possible.

* Filter coefficients are easily changed to enable adaptive performance.

* Performance accuracy can be controlled by the designer.

A typical digital filter system is shown schematically in Figure 7.1.

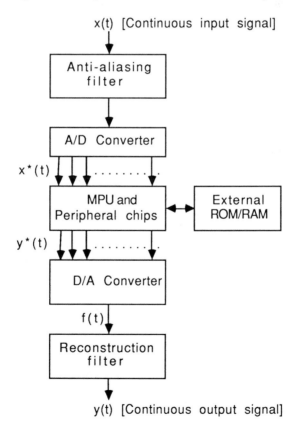

Figure 7.1   Typical Digital Filter System

The input signal, $x(t)$, is sampled regularly at instants T seconds apart (sampling period = T seconds). Each sample is converted to a corresponding binary-word representation, thereby forming the digital input sequence, $x^*(t)$. The digital processor (MPU) operates on the sequence $x^*(t)$ according to a predefined transfer function or algorithm to form the output sequence $y^*(T)$. The sequence $y^*(T)$ is converted to a train of end-on pulses, $f(t)$, whereby the area of each pulse is equal to T times the respective sequence value. To recover the continuous output, $y(t)$, the train of pulses, $f(t)$, is passed through a reconstruction filter (e.g. an R-C analogue lowpass filter). The anti-aliasing filter bandlimits the input signal so that the sampling theorem is observed, i.e. it is required that $\omega_s \geq 2\omega_H$, where $\omega_s = \frac{2\pi}{T}$ is the radian sampling frequency, and ω_H is the highest radian frequency component of the input signal. The frequency response function of the sampled input signal, $x^*(t)$, is given by

$$X^*(j\omega)\frac{1}{T} = \sum_{n=-\infty}^{\infty} X\left[j\left(\omega - n\omega_s\right)\right] \tag{7.1}$$

and the practical significance of this mathematical statement is illustrated in Figure 7.2. It may be observed that:

* The magnitude/frequency response function, $\left|X^*(j\omega)\right|$, is repeated *ad-infinitum* at intervals of ω_s. This implies that there is an infinite number of associated pole-zero patterns in the s-plane representation of $x^*(t)$, separated along the $j\omega$-axis by intervals of ω_s. However, the difficulty of dealing with a system with an infinite number of poles and zeros is overcome by using an equivalent z-tranform representation [7].

* If the sampling theorem $\left(\omega_s \geq 2\omega_H\right)$ is violated *aliasing* occurs.

* Compared with the spectrum of $x(t)$, the spectra of $x^*(t)$ are amplitude scaled by a factor equal to $1/T$.

It has been shown that digital filters process sampled signals, and that consequently the existence of the complementary spectra (Figure 7.2) dictates that the input signal must be bandlimited to a maximum value of $\omega_s / 2$ radians per second. This is normally achieved using a suitable analogue lowpass filter (e.g. Butterworth or Chebyshev), which is the anti-aliasing filter.

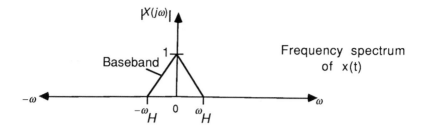

Frequency spectrum
of x(t)

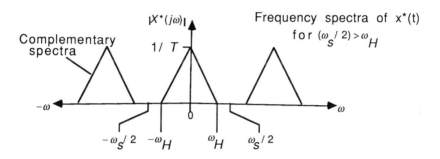

Frequency spectra of x*(t)
for $(\omega_s/2) > \omega_H$

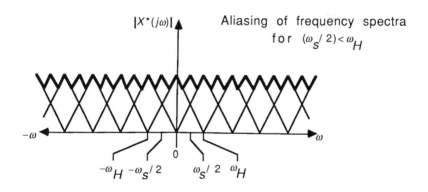

Aliasing of frequency spectra
for $(\omega_s/2) < \omega_H$

Figure 7.2 Significance of the Sampling Theorem

An adequate mathematical description of the sampling process may be obtained using an impulse modulator model, see Figure 7.3 below:

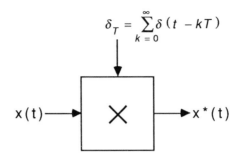

Figure 7.3   Impulse Modulator

The value of $x(t)$ is only known for $t = kT$, therefore we may write

$$x^*(t) = \sum_{k=0}^{\infty} x(kT) \times \delta(t - kT)$$

The corresponding Laplace transform is

$$\mathcal{L}[x^*(t)] = \sum_{k=0}^{\infty} x(kT) \exp(- ksT)$$

but for the standard z-transform, $z = \exp(sT)$, therefore $z^{-k} = \exp(-ksT)$, which corresponds to a delay of k sampling periods.

The z-transform of $x^*(t)$ is therefore given by

$$X(z) = \sum_{k=0}^{\infty} x(kT)\, z^{-k} \tag{7.2}$$

Equation (7.2) simply implies that the z-transform is a method of converting a sequence of numbers (sample values) into a function of the complex variable z. Note that in some works of reference (including this book) this sequence of numbers is denoted as x_k, and consequently equation (7.2) may be written in an alternative form:

$$X(z) = \sum_{k=0}^{\infty} x_k\, z^{-k} \tag{7.3}$$

The input and output signals of a digital filter are related through the *convolution-summation* process, which is defined as

$$y_k = \sum_{i=0}^{\infty} g_i \times x_{(k-i)} \qquad (7.4)$$

where y_k is the filter output sequence and g_i is the filter impulse response sequence.

Using the standard z-transform (equation (7.3)) on equation (7.4) we obtain

$$Y(z) = \sum_{k=0}^{\infty} \left[g_0 x_k + g_1 x_{(k-1)} + g_2 x_{(k-2)} + \cdots \right] z^{-k}$$

$$= g_0 \sum_{k=0}^{\infty} x_k z^{-k} + g_1 \sum_{k=0}^{\infty} x_{(k-1)} z^{-k} + g_2 \sum_{k=0}^{\infty} x_{(k-2)} z^{-k} + \cdots$$

$$= g_0 X(z) + g_1 X(z) z^{-1} + g_2 X(z) z^{-2} + \cdots$$

$$= \left[g_0 + g_1 z^{-1} + g_2 z^{-2} + \cdots \right] X(z)$$

$$\therefore \quad Y(z) = G(z) \times X(z) \qquad (7.5)$$

The ratio $Y(z) / X(z)$, equal to $G(z)$, is the transfer function of the digital filter.

7.2 DIGITAL FILTER TRANSFER FUNCTION

The digital filter transfer function is sometimes derived by z-transforming the transfer function of a known analogue filter, $G(s)$, that is

$$G(z) = \mathbf{Z}[G(s)]$$

In general, $$G(z) = \frac{a_0 + a_1 z^{-1} + a_2 z^{-2} + \cdots + a_p z^{-p}}{1 + b_1 z^{-1} + b_2 z^{-2} + \cdots + b_q z^{-q}} = \frac{Y(z)}{X(z)}$$

where $a_i \ (0 \le i \le p)$ and $b_j \ (0 \le j \le q)$ are the digital filter coefficients.

It follows that

$$X(z)\left[a_0 + a_1 z^{-1} + a_2 z^{-2} + \cdots a_p z^{-p}\right]$$

$$= Y(z)\left[1 + b_1 z^{-1} + b_2 z^{-2} + \cdots b_q z^{-q}\right]$$

i.e. $a_0 X(z) + a_1 X(z)z^{-1} + a_2 X(z)z^{-2} + \cdots + a_p X(z)z^{-p}$

$$= Y(z) + b_1 Y(z)z^{-1} b_2 Y(z)z^{-2} + \cdots + a_q Y(z)z^{-q} \quad (7.6)$$

but z^{-k} corresponds to a delay equal to k sampling periods, consequently equation (7.6) may be written in a **linear difference equation** form:

$$a_0 x_k + a_1 x_{(k-1)} + a_2 x_{(k-2)} + \cdots + a_p x_{(k-p)}$$

$$= y_k + b_1 y_{(k-1)} + b_2 y_{(k-2)} + \cdots + b_q y_{(k-q)}$$

$$\therefore y_k = a_0 x_k + a_1 x_{(k-1)} + a_2 x_{(k-2)} + \cdots + a_p x_{(k-p)}$$

$$- b_1 y_{(k-1)} - b_2 y_{(k-2)} - \cdots - b_q {(k-q)} \quad (7.7)$$

Equation (7.7) is **recursive**, whereby the present output sample value, y_k, is computed using a scaled version of the present input sample, x_k, and scaled versions of previous input and output samples. This form corresponds to an *Infinite Impulse Response (IIR)* digital filter.

Sometimes it is useful to represent the linear difference equation in a block diagram form, as shown in Figure 7.4.

The form of representation shown in Figure 7.4 is often referred to as the **Signal Flow Diagram** of the **direct form** of implementation. It is seen that a feedback path exists, and therefore it is appropriate to study whether this has any effect on the stability of $G(z)$. The following simple example illustrates this consideration.

Let $G(z) = \dfrac{z}{z - \alpha}$, then $Z^{-1}[G(z)] = g_i = \alpha^i$

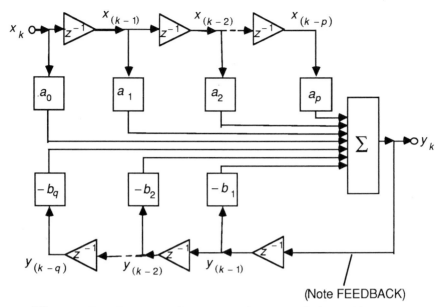

Figure 7.4 Block Diagram of IIR Digital Filter

For $i > 0$, α^i is convergent for $\alpha < 1$ and α^i is divergent for $\alpha > 1$. The former condition corresponds to a stable situation, whereas the latter condition corresponds to an unstable situation. The value of α determines the position of the pole of $G(z)$, and if the pole lies inside the unit-circle $(\alpha < 1)$ in the z-plane, $G(z)$ is stable, however, if the pole lies outside the unit-circle $(\alpha > 1)$ in the z-plane, $G(z)$ is unstable. The stability boundary is the circumference of the unit-circle in the z-plane. In general, for $G(z)$ to be stable, all poles must be within the unit-circle in the z-plane.

The poles and zeros of $G(z)$ may be determined by factorising the digital transfer function numerator and denominator polynomials to yield:

$$G(z) = \frac{f(z - z_1)(z - z_2) \cdots (z - z_p)}{(z - p_1)(z - p_2) \cdots (z - p_q)} \tag{7.8}$$

The multiplying factor, f is a real constant, and z_i $(0 \leq i \leq p)$ and p_j $(0 \leq j \leq q)$ are the zeros and poles respectively. The poles and zeros are either real or exist as complex conjugate pairs.

It is worth noting that in general as $T \to 0$ the poles of $G(z)$ migrate towards the $(1+j0)$-point in the z-plane, thereby making $G(z)$ approach a marginally stable condition.

The issue of stability is eliminated if the feedback (Figure 7.4) is removed. This is the case when the b_j coefficients are zero-valued (no poles in $G(z)$), corresponding to

$$G(z) = \frac{a_0 + a_1 z^{-1} + a_2 z^{-2} + \cdots + a_p z^{-p}}{1} = \frac{Y(z)}{X(z)}$$

and it follows that

$$y_k = a_0 x_k + a_1 x_{(k-1)} + a_2 x_{(k-2)} + \cdots + a_p x_{(k-p)} \qquad (7.9)$$

Equation (7.9) is **non-recursive**, whereby the present output sample value is computed using a scaled version of the present input sample, and scaled versions of previous input samples. This form corresponds to an **Finite Impulse Response (FIR)** digital filter. This form of digital filter is commonly known as a **transversal filter**. The corresponding block diagram representation is shown in Figure 7.5.

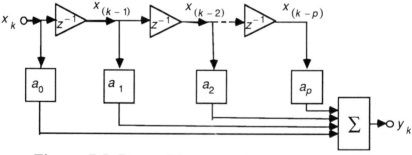

Figure 7.5 Block Diagram of FIR Digital Filter

The steady-state sinusoidal response of $G(z)$ may be obtained by substituting $\exp(j\omega T) = \cos \omega T + j \sin \omega T$ for z in $G(z)$, yielding $G(\exp(j\omega T))$, or it may be estimated by geometrical construction [2]. It is this steady-state sinusoidal response that is usually the main performance specification, and it often influences the choice of design method employed. This is demonstrated in Chapters 8 and 9, which respectively describe a number of FIR and IIR digital filter design methods.

7.3 REFERENCES

[1] A. Cadzow, *Discrete Time Systems: An Introduction with Interdisciplinary Applications* (Prentice-Hall, Englewood Cliffs, NJ, 1973).

[2] T. J. Terrell, *Introduction to Digital Filters*, Second Edition (Macmillan Education Ltd, London, 1988).

[3] D. Quarmby, *Signal Processor Chips* (Granada Publishing, London, 1984).

[4] *TMS320C25 Digital Signal Processor - Product Description*, Texas Instruments, Bedford (1986).

[5] *DSP56000 Product Description*, Motorola, Glasgow (1986).

[6] T. J. Terrell, *Microprocessor-based Signal Filters*, Microprocessors and Microsystems, Vol.11, No.3, April 1987.

[7] E. I. Jury, *Theory and Application of the z-Transform Method* (Wiley, New York, 1974).

Chapter 8

FIR filter design methods

T. J. Terrell and E. T. Powner

8.1 INTRODUCTION

Non-recursive digital filters have a finite impulse response (FIR) sequence, and they are inherently stable. Furthermore, a digital filter with a symmetrical impulse response has a linear phase characteristic, and therefore in this case there is no phase distotion imposed by the filter.

Compared with an IIR filter, the FIR counterpart will generally use more memory and arithmetic for its implementation. The technique of frequency denormalisation (Chapter 9) is not generally suitable for FIR filter design because a recursive filter will normally be produced. The one exception is the lowpass to highpass transformation, in which the linear phase characteristic is preserved.

Generally, for many applications, the advantages significantly outweigh the disadvantages, and consequently FIR filters are often used in practice.

In this chapter FIR filter design will be illustrated by considering the moving-averager filter, the frequency sampling method of design, and frequency-domain filter design using window functions.

8.2 THE MOVING AVERAGER

The output signal of a moving averager [1] digital filter is the arithmetic average of the current input sample value and one or more delayed input sample values. This averaging action is a lowpass filtering process, and smooths fluctuations in the filter input signal. However, it should be noted that there is a characteristic start up transient before the smoothing action is effective. For the N-term moving averager, the start up transient exists between $t = 0$ and $t = NT$, which is the duration of its impulse response sequence.

We will assume that the FIR filter coefficients are all equal to unity, and correspondingly the impulse response sequence of the N-term moving averager is therefore

$$g_i = \left\{ \frac{1}{N}, \frac{1}{N}, \cdots \cdots \frac{1}{N}, 0, 0, 0 \cdots \right\}$$
$$\leftarrow N - \text{terms} \rightarrow$$

and the transfer function of the N-term moving averager is therefore

$$G(z) = \frac{1}{N}\left[1 + z^{-1} + z^{-2} + \cdots + z^{-(N-1)}\right]$$

Expressing the transfer function in *closed form* we obtain

$$G(z) = \frac{1}{N}\left[\frac{1 - z^{-N}}{1 - z^{-1}}\right]$$

The corresponding frequency response function is

$$G(\exp(j\omega T)) = \frac{1}{N}\left[\frac{1 - z^{-j\omega NT}}{1 - z^{-j\omega T}}\right] \qquad (8.1)$$

The magnitude/frequency characteristics for three unity coefficient moving averagers are shown in Figure 8.1. It may be noted that in each case the first null occurs at a frequency $\omega = \frac{2\pi}{NT}$, which applies in general to N-term moving averagers having equal-value filter coefficients. Therefore for a filter with a known sampling frequency, the design of a moving averager simply involves specifying the frequency value for the first null, which in turn determines the number of terms used in implementing the averaging process.

ILLUSTRATIVE EXAMPLE

A lowpass filter based on an N-term moving averager is to be used such that the first null occurs at a frequency of 4 kHz, determine the linear difference equation of the filter. The sampling frequency for the filter is 24 kHz.

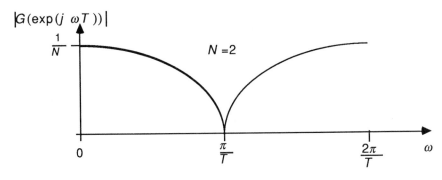

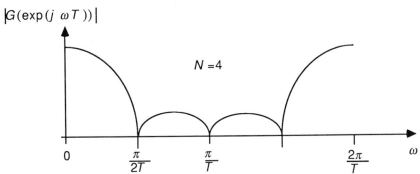

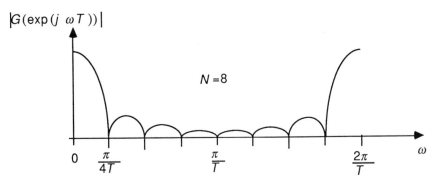

Figure 8.1 Magnitude/Frequency Characteristics for three Moving Averagers

Solution:

$$\omega = \frac{2\pi}{NT} \qquad \therefore N = \frac{f_s}{f} = \frac{24 \times 10^3}{4 \times 10^3} = 6$$

The linear difference equation for the required six-term moving averager is

$$y_k = \frac{x_k + x_{(k-1)} + x_{(k-2)} + x_{(k-3)} + x_{(k-4)} + x_{(k-5)}}{6}$$

The moving averager example presented above has unity-value coefficients, and was easily designed. Furthermore, its frequency response will have the general form shown in Figure 8.1. However, it is possible to introduce unequal coefficient values, which correspondingly modifies the frequncy response of the filter. Unfortunately this complicates the design process, involving the solution of difficult simultaneous equations, and consequently the following method is often preferred.

8.3 FREQUENCY SAMPLING METHOD OF DESIGN

If a digital filter has a finite impulse response sequence, g_i , it will also have an equivalent discrete Fourier transform (DFT) of G_n, where G_n is a sampled version of the required filter frequency response, $G(\exp(j\omega T))$.

The z-transform of g_i , for $i = 0, 1, 2, \cdots, (N-1)$, is

$$G(z) = \sum_{i=0}^{N-1} g_i \, z^{-i}$$

Using the inverse discrete Fourier transform (IDFT) relationship we obtain

$$G(z) = \sum_{i=0}^{N-1} \left\{ \frac{1}{N} \sum_{n=0}^{N-1} G_n \exp\left(\frac{j \, 2\pi \, i \, n}{N} \right) \right\} z^{-i}$$

$$= \sum_{n=0}^{N-1} \frac{G_n}{N} \sum_{i=0}^{N-1} \exp\left(\frac{j \, 2\pi \, i \, n}{N} \right) z^{-i}$$

$$G(z) = \sum_{n=0}^{N-1} \left[\frac{G_n}{N} \times \frac{1 - z^{-N}}{1 - \exp\left(\frac{j\,2\pi\,n}{N}\right) z^{-1}} \right] \tag{8.2}$$

Equation (8.2) implies that to approximate the desired continuous frequency response, we *frequency sample* at a set of N points uniformly spaced on the circumference of the unit-circle in the z-plane, and evaluate the continuous frequency response by interpolation of the sampled frequncy response. In this context, suppose that the desired frequncy response, $G_d(\exp(j\omega T))$ is defined for $-\pi \leq \omega \leq \pi$, i.e. defined for all ω. The design of the desired filter simply involves letting the DFT of G_n be a uniformly spaced N-point representation of the required response at the specified intervals. Therefore we have

$$G_n = G_d(\exp(j\omega T))\Big|_{\omega = \frac{2\pi\,n}{N}} \qquad n = 0, 1, 2, \cdots, (N-1)$$

This G_n sequence substituted in equation (8.2) ensures that the FIR digital filter transfer function gives a frequency response which coincides with the required frequncy response at the sampling points of $\omega = \frac{2\pi\,n}{N}$ for $n = 0, 1, 2, \cdots, (N-1)$

ILLUSTRATIVE EXAMPLE

Specification:

Design an FIR filter based on the 8-point sampling of the ideal frequency response shown in Figure 8.2:

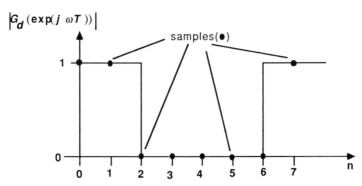

Figure 8.2 Desired Frequency Response

Solution:

In this case $N = 8$, and substituting this value in equation (8.2) gives

$$G(z) = \sum_{n=0}^{7} \left[\frac{G_n}{8} \times \frac{1 - z^{-8}}{1 - \exp\left(\frac{j\,\pi\,n}{4}\right) z^{-1}} \right]$$

$$= \left(\frac{1 - z^{-8}}{8} \right) \sum_{n=0}^{7} \left[\frac{G_n}{1 - \exp\left(\frac{j\,\pi\,n}{4}\right) z^{-1}} \right]$$

but $G_0 = G_1 = G_7 = 1$ and $G_2 = G_3 = G_4 = G_5 = G_6 = 0$

$$\therefore G(z) = \left(\frac{1 - z^{-8}}{8} \right) \left[\frac{1}{1 - z^{-1}} + \frac{1}{1 - \exp\left(\frac{j\,\pi}{4}\right) z^{-1}} + \frac{1}{1 - \exp\left(\frac{j\,7\pi}{4}\right) z^{-1}} \right]$$

This reduces to

$$G(z) = \left(\frac{1 - z^{-8}}{8} \right) \left[\frac{1}{1 - z^{-1}} + \frac{2 - \sqrt{2}\,z^{-1}}{1 - \sqrt{2}\,z^{-1} + z^{-2}} \right]$$

8.4 FREQUENCY-DOMAIN DESIGN USING WINDOW FUNCTIONS

This design method [2,3] is basically a frequency sampling technique, involving four main steps. The design process is:

(1) The desired ideal frequency response is specified.

(2) The specified frequency response is sampled at regular intervals along the frequency axis, and the corresponding impulse response sequence is computed.

(3) The impulse response sequence is modified by a window function sequence (a cosine-like sequence).

(4) The transfer function of the FIR filter is derived from the modified impulse response sequence.

Each step is described in more detail via the following illustrative example.

ILLUSTRATIVE EXAMPLE

(1) The required ideal frequency response is a brickwall lowpass characteristic, with a passband extending to $\omega_s / 4$. An initial zero-phase characteristic may be assumed, but a resulting linear phase is required.

(2) The frequency response of the specified filter is shown in Figure 8.3:

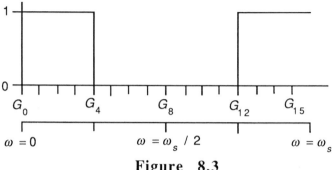

Figure 8.3

The ideal frequency response is sampled at intervals of $1/NT$ Hz, and the impulse response sequence is related to the frequency samples by the inverse discrete Fourier transform (IDFT), which is

$$g_i = \frac{1}{N} \sum_{n=0}^{N-1} G_n W^{-in}$$
(8.3)

where $W = \exp(-j\,2\pi / N)$ and $n = 0, 1, 2, \cdots, (N-1)$.

When a sampling point coincides with a discontinuity, the corresponding sample value may be taken to be the average of the two sample values on either side of the discontinuity.

Referring to Figure 8.3 the sampled values $(G_n$ values$)$ are

$$\{1, 1, 1, 1, 0.5, 0, 0, 0, 0, 0, 0, 0, 0.5, 1, 1, 1\}$$

The corresponding g_i values are evaluated using equation (8.3), namely

$$\{0.5, 0.314, 0, -0.094, 0, \cdots\cdots\}$$

(3) The modifications of the calculated g_i values are required to reduce *oscillations* known as Gibb's phenomenon, which is caused by truncating the Fourier series to N values. The application of the window results in a gradual tapering of the coefficients of the series such that the middle term, g_0, is undisturbed, and the coefficients at the extreme ends of the series are negligible.

A number of window functions exist [4], but for this example we will use the Hamming window defined by

$$W_n = 0.54 + 0.46\cos(n\pi / i)$$

where i is the number of term to be included on either side of g_0. For this example it is convenient to use $i = 4$, resulting in W_n values of
$$\{1, 0.86, 0.54, 0.22, 0.08\}$$

The modification to the impulse response sequence values by the window values, is achieved by forming the product $g_i \times W_n$, resulting in the sequence

$$\{0.5, 0.27, 0, -0.02, 0\}$$

(4) For a linear phase characteristic the modified impulse response sequence must be symmetrical, i.e.

$$\{0, -0.02, 0, 0.27, 0.5, 0.27, 0, -0.02, 0\}$$

The corresponding z-transform is

$$G(z)^\Delta = -0.02z^3 + 0.27z^1 + 0.5z^0 + 0.27z^{-1} - 0.02z^{-3}$$

However, z raised to a positive power is, from an implementation point of view, totally impractical, thus in this example it is neccessary to introduce the appropriate z^{-3} time delay. Therefore the FIR filter transfer function is

$$G(z) = -0.02 + 0.27z^{-2} + 0.5z^{-3} + 0.27z^{-4} - 0.02z^{-6}$$

The corresponding linear difference equation is

$$y_k = -0.02x_k + 0.27x_{(k-2)} + 0.5x_{(k-3)} + 0.27x_{(k-4)} - 0.02x_{(k-6)}$$

The magnitude/frequency and phase/frequency characteristics for this filter are shown in Figure 8.4.

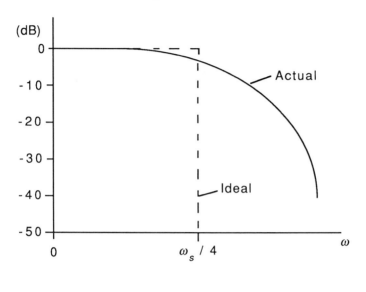

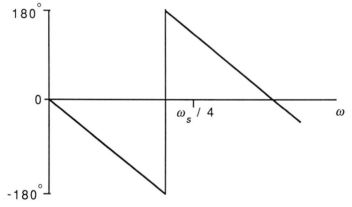

Figure 8.4   Magnitude/Frequency and Phase/Frequency Characteristics for Design Example

A number of comments can be made concerning the design of FIR filters using window functions.

(a) Stopband attenuation can be improved by incresing the window width (increase of i), or by using a different type of window, or by changing both.

(b) Making the filter realisable by shifting the symmetrical impulse response sequence along the time axis, has no effect on the magnitude/frequency characteristic, but it does change a zero phase filter to a linear phase filter.

(c) It is desireable that N is an integral power of 2 so that it will be compatible with the Fast Fourier Transform (FFT) - a technique used to expedite the evaluation of the IDFT. Also N must be at least equal to the width of the window ($N \geq i$). In many practical designs N is chosen so that the interval between sampling points is a fraction (typically ≤ 0.1) of the widths of the transition bands in the final filter design.

Other methods of FIR filter design are sometimes used, such as the equiripple approximation method [5] or the analytical technique [6], but their design is relatively more difficult and complex.

8.5 REFERENCES

[1]   P. A. Lynn, *An Introduction to the Analysis and Processing of Signals - Second Edition*, (Macmillan, London, 1982), Chapter 9.

[2]   B. Gold and C. M. Rader, *Digital Processing of Signals*, (McGraw-Hill, New York, 1969).

[3]   L. R. Rabiner, *Techniques for Designing Finite-Duration Impulse-Response Digital Filters*, Trans. Comm. Tech., IEEE, **19** (1971) pp 188-95.

[4]   T. J. Terrell, *Introduction to Digital Filters - Second Edition*, (Macmillan Education Ltd, Basingstoke and London, 1988).

[5]   O. Herrmann, *On the Design of Nonrecursive Digital Filters with Linear Phase*, IEE Electronic Letters, **6** (1970) pp 328-9.

[6]   J. Attikiouzel and R. Bennett, *Analytic Techniques for Designing Digital Nonrecursive Filters*, Int. J. Electrical Engineering Education, **14** (1977), pp 251-67.

Chapter 9

IIR filter design methods

T. J. Terrell and E. T. Powner

9.1 INTRODUCTION

Infinite impulse response (IIR) digital filters are *recursive* in form (defined by equation (7.7) and shown in block diagram form in Figure 7.4), and their design may be achieved using two main approaches. One method is the *indirect approach*, which transforms a prototype analogue filter transfer function, $G(s)$, to an equivalent digital filter transfer function, $G(z)$. In this context, the s to z mappings used in this chapter are: (a) the standard z-transform (impulse invariant), (b) the bilinear z-transform, and (c) the matched z-transform. The other method is the *direct approach* to the derivation of $G(z)$, which is applicable to the design of frequency sampling filters and filters based on squared magnitude functions.

9.2 DESIGNS BASED ON PROTOTYPE ANALOGUE FILTERS

The main steps involved in this design method are:

Analogue filter specification
↓
(1) derive a realisable transfer function $G(s)$
↓
(2) z – transform $G(s)$ to $G(z)$

Step (1) involves the design of the prototype analogue filter, and since Butterworth and Chebyshev analogue filters are commonly used for this purpose, then the design of both types will be summarised herein.

9.2.1 Butterworth Normalised Lowpass Filter

The magnitude-squared response of a Butterworth normalised

lowpass filter [1] of order n is defined as

$$|G(j\omega)|^2 = \frac{1}{1 + \left(\frac{\omega}{\omega_c}\right)^{2n}} \text{ , where } \omega_c \text{ is the radian cut-off frequncy.}$$

Attenuation (-X dB) =

$$10\log_{10}|G(j\omega)|^2 = 10\log_{10}1 - 10\log_{10}\left[1 + \left(\frac{\omega}{\omega_c}\right)^{2n}\right]$$

$$\therefore X \, dB = 10\log_{10}\left[1 + \left(\frac{\omega}{\omega_c}\right)^{2n}\right] \tag{9.1}$$

Thus if X dB attenuation is specified at a particular value of ω, and if ω_c is given (or assumed to be 1), then the order of filter complexity, n, may be calculated.

Once the value of n is known, then the positions of the s-plane Butterworth poles may be determined. For $1 \le n \le 6$, the pole locations may be obtained from Table 9.1:

n = 1	n = 2	n = 3	n = 4	n = 5	n = 6
-1.0000000	-0.7071068 ± j0.7071068	-1.0000000	-0.3826834 ± j0.9238795	-1.0000000	-0.2588190 ± j0.9659258
		-0.5000000 ± j0.8660254	-0.9238795 ± j0.3826834	-0.3090170 ± j0.9510565	-0.7071068 ± j0.7071068
	Butterworth Pole Locations			-0.8090170 ± j0.5877852	-0.9659258 ± j0.2588190

Table 9.1

In general, if the angle of the k_{th} pole is denoted by ϕ_k, where $k = 0, 1, 2, \cdots, (2n-1)$, and located on the circumference of the unit-circle in the s-plane, then

$$\phi_k = \left.\frac{k\pi}{n}\right|_{n \text{ odd}} \quad or \quad \phi_k = \left.\frac{\left(k + \frac{1}{2}\right)\pi}{n}\right|_{n \text{ even}}$$

However, only the poles in the left-hand half of the s-plane are used in deriving $G(s)$ (to preserve stability). Having determined the pole locations, the transfer function, $G(s)$, may be written down in the form

$$G(s) = \frac{1}{(s + p_1)(s + p_2)(s + p_3) \cdot \cdot \cdot \cdot (s + p_n)}$$

$$\therefore G(s) = \frac{1}{a_n s^n + a_{n-1} s^{n-1} + \cdot \cdot \cdot + a_1 s + a_0} \tag{9.2}$$

For $1 \le n \le 6$, the coefficients of the normalised Butterworth denominator polynomial may be obtained from Table 9.2:

n	a_1	a_2	a_3	a_4	a_5	a_6
1	1.0000000					
2	1.4142136	1.0000000	$a_0 = 1$ for all n			
3	2.0000000	2.0000000	1.0000000			
4	2.6131259	3.4142136	2.6131259	1.0000000		
5	3.2360680	5.2360680	5.2360680	3.2360680	1.0000000	
6	3.8637033	7.4641016	9.1416202	7.4641016	3.8637033	1.0000000

Table 9.2

A practical filter will generally have a cutoff frequency other than 1 rad/s, consequently we must be able to convert a normalised transfer function into one with the required frequency characteristic. This is achieved via a technique known as **frequency denormalisation**, see Table 9.3. Referring to Table 9.3, ω_c is the desired radian cutoff frequency of the lowpass or highpass denormalised filter; ω_{cl} is the desired radian lower transition frequency of the bandstop or bandpass denormalised filter; ω_{cu} is the desired radian upper transition frequency of the bandstop or bandpass denormalised filter.

To transform from normalised lowpass to	Substitute for s in $G(s)$
Lowpass	$\dfrac{s}{\omega_c}$
Highpass	$\dfrac{\omega_c}{s}$
Bandstop	$\dfrac{s\left(\omega_{cu} - \omega_{cl}\right)}{\left[s^2 + \left(\omega_{cu} \times \omega_{cl}\right)\right]}$
Bandpass	$\dfrac{\left[s^2 + \left(\omega_{cu} \times \omega_{cl}\right)\right]}{s\left(\omega_{cu} - \omega_{cl}\right)}$

Table 9.3

ILLUSTRATIVE DESIGN EXAMPLE

Specification:

Derive the transfer function for a Butterworth highpass filter with a magnitude/frequency response of (i) -3 dB attenuation at a frequency of 2 kHz, and (ii) -15 dB attenuation at a frequency of 1 kHz.

Solution:

Firstly, the design is based on a normalised Butterworth lowpass prototype, and therefore the above specification must be translated to the corresponding lowpass case. That is, since the highpass filter stopband attenuation (-15dB) is specified at a frequency (1 kHz) equal to half the cut-off frequency value , then this frequency ratio (1:2) is reversed in the translation to the lowpass prototype (2:1). Thus the magnitude/frequency response of the lowpass prototype becomes (i) -3 dB attenuation at a frequency of 1 kHz, and (ii) -15 dB attenuation at a frequency of 2 kHz.

The next step is to determine the required order of filter complexity.

Substituting the specified values in equation (9.1) gives

$$15 = 10 \, \log_{10}\left[1 + \left(\frac{2}{1}\right)^{2n}\right]$$

$$n = \frac{\log_{10}\{(\text{antilog}_{10}1.5) - 1\}}{2 \, \log_{10}2} \cong 2.47$$

$$\therefore \text{ use } n = 3$$

Now using Table 9.2 the transfer function of the required 3rd-order normalised Butterworth lowpass filter is determined as:

$$G(s)_N = \frac{1}{s^3 + 2s^2 + 2s + 1}$$

This transfer function is now denormalised (denoted by subscript DN) using the transform given in Table 9.3, i.e. $s \rightarrow \frac{\omega_c}{s}$, as follows:

$$\therefore G(s)_{DN} = \frac{1}{\left(\frac{\omega_c}{s}\right)^3 + 2\left(\frac{\omega_c}{s}\right)^2 + 2\left(\frac{\omega_c}{s}\right) + 1}$$

i.e.
$$G(s)_{DN} = \frac{s^3}{s^3 + 2\omega_c s^2 + 2\omega_c^2 s + \omega_c^3}$$

but $\omega_c = 2\pi f_c = 2\pi \times 2 \times 10^3 = 4\pi \times 10^3$ rad/s

$$\therefore G(s)_{DN} = \frac{s^3}{s^3 + 8\pi \times 10^3 s^2 + 32\pi^2 \times 10^6 s + 64\pi^3 \times 10^9}$$

The magnitude/frequncy response may be checked as follows:

$$G(j\omega)_{DN} = \frac{-j\omega^3}{-j\omega^3 - 8\pi \times 10^3 \omega^2 + j32\pi^2 \times 10^6 \omega + 64\pi^3 \times 10^9}$$

$$= \frac{\omega^3}{\left(\omega^3 - 32\pi^2 \times 10^6 \omega\right) + j\left(64\pi^3 \times 10^9 - 8\pi \times 10^3 \omega^2\right)}$$

but $\omega = 2\pi \times f$, and if f is expressed in kHz it follows that $G(j\omega)_{DN}$ is

$$= \frac{8\pi^3 \times 10^9 \times f^3}{\left[\left(8\pi^3 \times 10^9 \times f^3\right) - \left(64\pi^3 \times 10^9 \times f\right)\right] + j\left[\left(64\pi^3 \times 10^9 - 32\pi^3 \times 10^9 \times f^2\right)\right]}$$

$$= \frac{f^3}{\left(f^3 - 8f\right) + j\left(8 - 4f^2\right)}$$

$$\therefore \left|G(j\omega)_{DN}\right| = \frac{f^3}{\sqrt{f^6 + 64}}$$

The corresponding magnitude/frequency characteristic is shown in Figure 9.1:

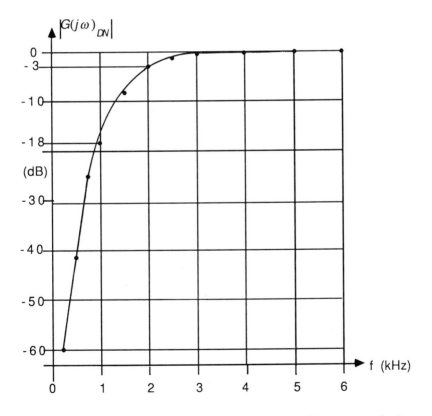

Figure 9.1 Magnitude/Frequency Characteristic

9.2.2   Chebyshev Normalised Lowpass Filter

The magnitude-squared response of a Chebyshev normalised lowpass filter [1] of order n is defined as

$$|G(j\omega)|^2 = \frac{1}{1 + \varepsilon^2 \left[C_n \left(\frac{\omega}{\omega_c} \right) \right]^2}$$

where ω_c is the cutoff frequency, ε is the ripple factor (which is real and <1), and $C_n\left(\frac{\omega}{\omega_c}\right)$ is an n_{th}-order Chebyshev cosine polynomial.

Taking $\omega_c = 1$ (the usual value for the normalised frequency scale), then the Chebyshev cosine polynomial is defined as

$$C_n(\omega) = \begin{cases} \cos\left(n \cos^{-1} \omega\right)\big|_{\omega \le 1} \\ \\ \cosh\left(n \cosh^{-1} \omega\right)\big|_{\omega > 1} \end{cases}$$

when $n = 0$ we have $C_0(\omega) = 1$, and when $n = 1$ we have $C_1(\omega) = \omega$. Higher order polynomials may be determined using the relationship

$$C_n(\omega) = 2\omega\, C_{n-1}(\omega) - C_{n-2}(\omega)$$

In the stopband a frequency is reached whereby $\varepsilon^2 \left[C_n(\omega) \right]^2 \gg 1$, and in this case we may deduce that

$$|G(j\omega)| \cong \frac{1}{\varepsilon\, C_n(\omega)}\bigg|_{\omega \ge \omega_{sb}}$$

$$\text{Attenuation}\,(-X\,\text{dB}) = 20\log_{10}|G(j\omega)|$$

$$= 20\log_{10}\left[\varepsilon\, C_n(\omega)\right]^{-1}$$

$$\therefore X\,\text{dB} = 20\log_{10}\varepsilon + 20\log_{10}C_n(\omega)$$

For large values of ω (in the stopband) $C_n(\omega) \cong 2^{n-1} \times \omega^n$,

$$\therefore X \text{ dB} \cong 20 \log_{10} \varepsilon + 20 \log_{10} \left(2^{n-1} \times \omega^n\right)$$

$$= 20 \log_{10} \varepsilon + (n-1)\, 20 \log_{10} 2 + 20n \, \log_{10} \omega$$

$$i.\, e.\; X \text{ dB} \cong 20 \log_{10} \varepsilon + 6\,(n-1) + 20 \log_{10} \omega \tag{9.3}$$

Thus if the maximum passband ripple is specified, the corresponding value of ε may be calculated, and if X dB is specified at a particular stopband frequency, ω, then the order of filter complexity, n, may be calculated.

Once the value of n is known, then the positions of the s-plane Chebyshev poles may be determined. A straightforward way of achieving this is to firstly determine the position of the n_{th}-order Butterworth poles, and then translate them to the n_{th}-order Chebyshev poles. This is possible because there is a geometrical relationship between the Butterworth pole positions and the Chebyshev pole positions in the s-plane.

A design parameter is defined as $A = \dfrac{1}{n} \sinh^{-1}\left(\dfrac{1}{\varepsilon}\right)$ and is used as follows.

$$\text{Re}\left[\text{Chebyshev}_1\right] = \text{Re}\left[\text{Butterworth}_N\right] \times \tanh A \tag{9.4}$$

$$\text{Im}\left[\text{Chebyshev}_1\right] = \text{Im}\left[\text{Butterworth}_N\right] \tag{9.5}$$

$$\text{Chebshev}_2 = \text{Chebyshev}_1 \times \cosh A \tag{9.6}$$

The translational relationships defined by equations (9.4), (9.5) and (9.6) are illustrated in Figure 9.2. The translation P1 to P2 corresponds to using equations (9.4) and (9.5); the translation from P2 to P3 is achieved using equation (9.6). The result is an overall translation of the normalised Butterworth lowpass filter pole at P1 to a normalised Chebyshev lowpass filter pole at P3.

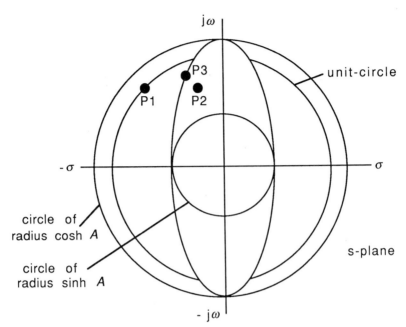

Figure 9.2 Translation of Butterworth Pole Position (P1) to Chebyshev Pole Position (P3)

ILLUSTRATIVE DESIGN EXAMPLE

Specification:

Derive the transfer function of a denormalised Chebyshev lowpass filter with a magnitude/frequency response conforming to (i) 0.5 dB passband ripple, (ii) stopband attenuation of at least 20 dB at 10 kHz, and (iii) a cutoff frequency equal to 2 kHz.

Solution:

At $\omega = 1$, $\left| G(j1) \right| = -0.5 \, dB = 20 \log_{10} \dfrac{1}{\sqrt{1 + \varepsilon^2}}$

$\therefore \quad -0.5 = 20 \log_{10} 1 - 20 \log_{10} \sqrt{1 + \varepsilon^2}$

$i.e. \quad 0.5 = 10 \log_{10} \left(1 + \varepsilon^2 \right)$

$\therefore \quad \varepsilon = \sqrt{10^{0.05} - 1} = 0.3493$

Substituting this value and the specified values in equation (9.3) gives

$$20 = 20 \log_{10} 0.3493 + 6(n-1) + 20n \log_{10} 5$$

$$\therefore n \cong 1.76, \text{ i.e. use } n = 2$$

$$A = \frac{1}{2} \sinh^{-1} \left(\frac{1}{0.3493} \right) = 0.8871$$

$$\therefore \tanh A = 0.71 \text{ and } \cosh A = 1.42$$

Butterworth poles for $n = 2$:

$$k = 0, 1, 2 \text{ and } 3, \therefore \phi_0 = \frac{\pi}{4}, \quad \phi_1 = \frac{3\pi}{4}, \quad \phi_2 = \frac{5\pi}{4} \text{ and } \phi_3 = \frac{7\pi}{4}$$

These normalised Butterworth lowpass filter pole positions are shown in Figure 9.3.

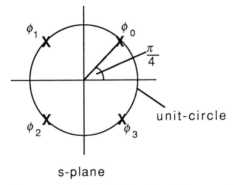

s-plane

Figure 9.3 Butterworth Pole Positions for $n = 2$

For a stable transfer function we only use ϕ_1 and ϕ_2 :

ϕ_1 is located at $-\dfrac{1}{\sqrt{2}} + j\dfrac{1}{\sqrt{2}}$, and

ϕ_2 is located at $-\dfrac{1}{\sqrt{2}} - j\dfrac{1}{\sqrt{2}}$

$$\text{Re}\left[\text{Chebyshev}_1\right] = -\frac{1}{\sqrt{2}} \times 0.71 = -\frac{1}{2}$$

$$\text{Im}\left[\text{Chebyshev}_1\right] = \pm j\,\frac{1}{\sqrt{2}}$$

$$\text{Re}\left[\text{Chebyshev}_2\right] = -\frac{1}{2} \times 1.42 = -\frac{1}{\sqrt{2}}$$

$$\text{Im}\left[\text{Chebyshev}_2\right] = \pm j\,\frac{1}{\sqrt{2}} \times 1.42 = \pm j1$$

$$\therefore \; G(s)_N = \frac{1}{\left(s + \frac{1}{\sqrt{2}} + j1\right)\left(s + \frac{1}{\sqrt{2}} - j1\right)} = \frac{1}{s^2 + \sqrt{2}s + \frac{3}{2}}$$

Denormalising $s \rightarrow \dfrac{s}{\omega_c}$

i.e. $G(s)_{DN} = \dfrac{1}{\dfrac{s^2}{\omega_c^2} + \dfrac{\sqrt{2}\,s}{\omega_c} + \dfrac{3}{2}}$

$$= \frac{\omega_c^2}{s^2 + \sqrt{2}\,\omega_c s + \frac{3}{2}\omega_c^2}$$

but $\omega_c = 2\pi f_c = 2\pi \times 2 \times 10^3 = 4\pi \times 10^3$ rad/s

$$\therefore \; G(s)_{DN} = \frac{16\pi^2 \times 10^6}{s^2 + 4\sqrt{2}\,\pi \times 10^3 s + 24\pi^2 \times 10^6} \tag{9.7}$$

The magnitude response may be checked as follows:

$$G(j\omega)_{DN} = \frac{16\pi^2 \times 10^6}{\left(24\pi^2 \times 10^6 - \omega^2\right) + j\left(4\sqrt{2}\,\pi \times 10^3 \omega\right)}$$

$\omega = 2\pi \times f$, and if f is expressed in kHz we may write

$$G(j\omega)_{DN} = \frac{16\pi^2 \times 10^6}{\left(24\pi^2 \times 10^6 - 4\pi^2 \times 10^6 \times f^2\right) + j\left(4\sqrt{2}\,\pi \times 10^3 \times 2\pi \times 10^3 \times f\right)}$$

$$= \frac{4}{\left(6 - f^2\right) + j(2\sqrt{2} \times f)}$$

$$\therefore \left|G(j\omega)_{DN}\right| = \frac{4}{\sqrt{f^4 - f^2 + 36}}$$

at $f = 0$, for $n = 2$, $\left|G(j0)_{DN}\right| = 0.944 \ (-0.5 \ \text{dB})$

$$\therefore \quad \frac{4k}{\sqrt{36}} = 0.944$$

i.e. $k = 1.416$

$$\therefore \left|G(j\omega)_{DN}\right| = \frac{4 \times 1.416}{\sqrt{f^4 - 4f^2 + 36}}$$

The corresponding magnitude/frequency characteristic is shown in Figure 9.4, and it is seen that the derived transfer function has the desired magnitude response characteristic.

Other forms of prototype filter are available, and in some cases the Cauer filter [1] or the Bessel filter [1] could be used as an appropriate denormalised transfer function, $G(s)_{DN}$.

Having derived the required $G(s)_{DN}$, the corresponding $G(z)$ may be determined using one of the s-plane to z-plane transformation methods described in the following sections of this chapter.

9.3 IMPULSE-INVARIANT DESIGN METHOD

The prototype filter, with transfer function $G(s)_{DN}$, has an impulse response, $g(t)$, defined by $L^{-1}\left[G(s)_{DN}\right]$, and if we consider $G(s)_{DN}$ to have m simple poles, then it follows that

$$g(t) = \mathbf{L}^{-1}\left[\sum_{i=1}^{m}\frac{c_i}{(s+p_i)}\right] = \sum_{i=1}^{m} c_i \exp\left(-p_i t\right)$$

For the impulse-invariant design method [2,3,4], the digital filter is required, at the sampling instants, to have impulse response values equal to the prototype filter's impulse response, i.e. $g_k = g(t)$ for $t = kT$, where T is the sampling period, therefore

$$g_k = \sum_{i=1}^{m} c_i \exp\left(-p_i kT\right) \tag{9.8}$$

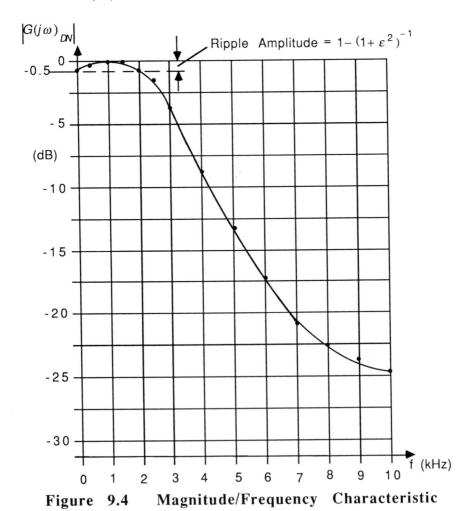

Figure 9.4   Magnitude/Frequency Characteristic

Obtaining the standard z-transform of equation (9.7) and multiplying the result by T yields

$$G(z)^\nabla = T \times G(z) = T \times \sum_{i=1}^{m} c_i \left[\frac{z}{z - \exp(-p_i T)} \right] \qquad (9.9)$$

Multiplying the standard z-transform, $G(z)$, by T ensures that the frequency spectra of the sampled signal, $x^*(t)$, are amplitude scaled to adequately represent the amplitudes of frequency components in the original baseband spectrum of $x(t)$, see Figure 7.2.

ILLUSTRATIVE EXAMPLE

The transfer function for the second-order Chebyshev lowpass filter defined by equation (9.7), derived in the previous illustrative design example, may be re-written in the form

$$G(s)_{DN} = 4\pi \times 10^3 \left[\frac{4\pi \times 10^3}{\left(s + 2\sqrt{2}\,\pi 10^3\right)^2 + \left(4\pi \times 10^3\right)^2} \right] \qquad (9.10)$$

We may now apply the standard z-transform to equation (9.10) to obtain the corresponding $G(z)$. The z-transform used in this case is

$$\frac{\beta}{(s + \alpha)^2 + \beta^2} \rightarrow \frac{z \exp(-\alpha T) \sin \beta T}{z^2 - 2z \exp(-\alpha T) \cos \beta T + \exp(-2\alpha T)}$$

Therefore the transfer function of the equivalent impulse-invariant digital filter is

$$G(z)^\nabla = G(z) \times T$$

$$= T \times \left[\frac{4\pi 10^3 \times z \exp\left(-2\sqrt{2}\,\pi 10^3 T\right) \sin 4\,\pi 10^3 T}{z^2 - 2z \exp\left(-2\sqrt{2}\,\pi 10^3 T\right) \cos 4\pi 10^3 T + \exp\left(-4\sqrt{2}\,\pi 10^3 T\right)} \right]$$
$$(9.11)$$

9.4 BILINEAR z-TRANSFORM DESIGN METHOD

The bilinear z-transform [2,3,4] is a straightforward design method, which ensures that:

(a) A stable $G(s)$ transforms to a stable $G(z)$.

(b) A wideband sharp cutoff $G(s)$ characteristic transforms to a wideband sharp cutoff $G(z)$ characteristic.

(c) The match between the frequency response function $\left|G\left(e^{j\omega T}\right)\right|$ and the frequency response function $|G(j\omega)|$ for breakpoints and zero frequency, is good.

The bilinear z-transform compresses the response at $\omega_a = \infty$ to $\omega_d = \dfrac{\pi}{T} = \dfrac{\omega_s}{2}$. Also in transforming $G(s)$ to $G(z)$ via the bilinear z-transform the impulse response and phase response are not preserved.

The relationship between the analogue filter radian frequency scale and the digital filter radian frequency scale is

$$\omega_a = \frac{2}{T}\tan\left(\frac{\omega_d T}{2}\right)$$

(9.12)

However, this relationship between ω_a and ω_d is non-linear due to the tan function. Consequently if a particular value of ω_d is required (say a digital filter cut-off frequency) then firstly it is **pre-warped** (converted) to a corresponding analogue frequency value, ω_a, using equation (9.12) for a given (known) value of T. The transfer function, $G(s)$, is then transformed to the required $G(z)$ by substituting $\dfrac{2}{T}\left[\dfrac{z-1}{z+1}\right]$ for s in $G(s)$.

ILLUSTRATIVE EXAMPLE

This example shows how the bilinear z-transform method is used to design a lowpass digital filter having a cut-off frequency, f_{cd}, equal to 100 Hz. It will be assumed that the sampling frequency is 625 Hz, and that the transfer function of the prototype denormalised lowpass analogue filter is

$$G(s) = \frac{\omega_{ca}^{2}}{s^2 + \sqrt{2}\,\omega_{ca}s + \omega_{ca}^{2}} \cdot$$

Solution:

$f_{cd} = 100$ Hz (given), $\therefore \omega_{cd} = 2\pi \times 100$ rad/ s

Now prewarping using equation (9.12):

$$\omega_{ca} = \frac{2}{T} \tan\left(200\pi \times \frac{T}{2}\right), \text{ but } T = \frac{1}{f_s} = \frac{1}{625} = 1.6 \text{ ms}$$

$$\therefore \omega_{ca} = \frac{2}{1.6 \times 10^{-3}} \tan\left(100\pi \times 1.6 \times 10^{-3}\right) = 687.2 \text{ rad/ s}$$

$$\therefore G(s)_{pw} = \frac{(687.2)^2}{s^2 + \sqrt{2} \times 687.2 \, s + (687.2)^2}$$

For the bilinear z-transform

$$s = \frac{2}{T}\left[\frac{z-1}{z+1}\right] = \frac{2}{1.6 \times 10^{-3}}\left[\frac{z-1}{z+1}\right] = 1250\left[\frac{z-1}{z+1}\right]$$

$$\therefore s^2 = 1250^2\left[\frac{(z-1)^2}{(z+1)^2}\right]$$

Now substituting for s and s^2 in $G(s)_{pw}$ gives

$$G(z) = \frac{(687.2)^2}{1250^2\left[\dfrac{(z-1)^2}{(z+1)^2}\right] + \sqrt{2} \times 687.2 \times 1250\left[\dfrac{z-1}{z+1}\right] + (687.2)^2}$$

$$\therefore G(z) = \frac{z^2 + 2z + 1}{6.88z^2 - 4.62z + 1.74}$$

9.5 MATCHED z-TRANSFORM DESIGN METHOD

The matched z-transform method [5,6] directly maps the s-plane poles and zeros of $G(s)$ to corresponding poles and zeros in the z-plane.

Real s-plane poles/zeros are mapped using the following relationship

$$(s + \alpha) \quad \rightarrow \quad \{1 - [\exp(-\alpha T)z^{-1}]\}$$

Complex s-plane poles/zeros are mapped using

$$(s + \alpha)^2 + \beta^2 \quad \rightarrow \quad \{1 - [2\exp(-\alpha T)\cos\beta T]z^{-1} + \exp(-2\alpha T)z^{-2}\}$$

ILLUSTRATIVE EXAMPLE

The transfer function of the second-order Chebyshev filter defined by equation (9.10) simply transforms to

$$G(z) = 4\pi\,10^3 \times \left[\frac{4\pi10^3 \times z^2}{z^2 - 2z\exp\left(-2\sqrt{2}\pi10^3 T\right)\cos 4\pi10^3 T + \exp\left(-4\sqrt{2}\pi10^3 T\right)} \right]$$

This particular transfer function may be usefully compared to that derived for the impulse-invariant design method, see equation (9.11). The poles are identical, but the multipliction factor is different and the order of complexity of zeros is incresed by one.

9.6 DESIGN BASED ON SQUARED MAGNITUDE FUNCTIONS

Recursive digital filters may be designed using a *direct approach* using *mirror image polynomials*. [2,7] The associated trigonometric functions produce half of the poles within the unit-circle in the z-plane, the other half lying outside. Clearly to ensure that the transfer function of the digital filter is stable, only the poles inside the unit-circle are used to form $G(z)$.

Consider the squared magnitude function defined as

$$\left| G[\exp(j\omega T)] \right|^2 = \frac{1}{1 + \left[F_n(\omega T) \right]^2}$$

We require suitable trigonmetric functions for $\left[F_n(\omega T) \right]^2$, which for the substitution $z = \exp(j\omega T)$, produces a mirror image polynomial in the z-plane. For example, the following relationship satisfies the lowpass filter requirement:

$$\sin^2\left(\frac{\omega T}{2}\right) = -\frac{(z-1)^2}{4z}$$

Taking this approach further, consider the following squared magnitude function

$$|G[\exp(j\,\omega T)]|^2 = \cfrac{1}{1+\left[\cfrac{\sin^2(\omega T/2)}{\sin^2(\omega_c T/2)}\right]^n} = \cfrac{\left[\sin^2(\omega_c T/2)\right]^n}{\left[\sin^2(\omega_c T/2)\right]^n + \left[\sin^2(\omega T/2)\right]^n}$$

where ω_c is the desired angular cut-off frequency.

Now substituting z for $\exp(j\,\omega T)$ we obtain

$$|G(z)|^2 = \frac{q^n}{q^n + p^n}, \quad \text{where}$$

$$q = \sin^2(\omega_c T/2) \quad \text{and} \quad p = -(z-1)^2/4z.$$

Consequently roots in the p-plane lie on a circle of radius q, thus

$$p_k = q\,\exp(j\,\phi_k), \quad k = 0, 1, 2, \cdots, (n-1) \quad \text{and}$$

$$\phi_k = \begin{cases} \dfrac{(2k+1)\pi}{n} & \text{for } n \text{ even} \\[2mm] \dfrac{2k\,\pi}{n} & \text{for } n \text{ odd} \end{cases}$$

Having solved for p_k the corresponding z-plane factors z_k are determined by solving $p_k = -(z_k-1)^2/4z_k$. That is

$$z_k = (1-2p_k) \pm \sqrt{4p_k(p_k-1)} \qquad (9.13)$$

<u>**ILLUSTRATIVE EXAMPLE**</u>

Specification

A digital lowpass filter is required with a cut-off frequency equal to 1 kHz. Attenuation to be at least 15dB at 3 kHz. The sampling frequency is 10 kHz.

Solution

$$q = \sin^2(\omega_c T / 2) = \sin^2\left(2\pi \times 10^3 / 10^4\right) \cong 0.096$$

$$p = \sin^2(\omega T / 2) = \sin^2\left(2\pi \times 3 \times 10^3 / 10^4\right) \cong 0.655$$

$$|G(z)|^2 = \frac{q^n}{q^n + p^n} = \frac{1}{1 + \left(\dfrac{p}{q}\right)^n} = \frac{1}{1 + \left(\dfrac{0.655}{0.096}\right)^n} = \frac{1}{1 + (6.823)^n}$$

$$15 = 10 \log_{10}\left[1 + (6.823)^n\right]$$

Solving for n we obtain $n = 1.782$, and taking the next higher integer value, we use $n = 2$. Thus

$$k = 0 \quad and \quad 1, \quad \therefore \quad \phi_0 = \frac{\pi}{2} \quad and \quad \phi_1 = \frac{3\pi}{2}$$

Therefore

$$p_0 = 0.096 \angle \pi / 2 \quad and \quad p_1 = 0.096 \angle 3\pi / 2$$

Using equation (9.13) the corresponding z-plane factors are

$$z_0 = 0.58 + j\, 0.27 \quad \text{(inside unit–circle)}$$

$$or \quad z_0 = 1.42 - j\, 0.65 \quad \text{(outside unit– circle)}$$

Similarly

$$z_1 = 0.58 - j\, 0.27 \quad \text{(inside unit–circle)}$$

$$or \quad z_1 = 1.42 + j\, 0.65 \quad \text{(outside unit– circle)}$$

Now selecting the two poles that lie inside the unit-circle we obtain

$$G(z) = \frac{f}{[z - (0.58 - j\, 0.27)]\,[z - (0.58 + j\, 0.27)]}$$

Taking $|G \exp(j\,\omega T)| = 1$ at $\omega = 0,$ then $z = \exp(j\,\omega T) = 1,$ and therefore

$$f = 1[1 - (0.58 - j\,0.27)][1 - (0.58 + j\,0.27)] = 0.25$$

$$\therefore\; G(z) = \frac{0.25}{[z - (0.58 - j\,0.27)][z - (0.58 + j\,0.27)]}$$

i.e $\;G(z) = \dfrac{0.25}{z^2 + 1.16z + 0.41}$

Note that it is possible to *frequency transform* a derived lowpass digital filter to (i) a lowpass digital filter (shift of cut-off frequency), (ii) a highpass digital filter, (iii) a bandpass digital filter, or (iv) a bandstop digital filter. Table 9.4 summarises the frequency transforms used for this purpose. For this table β (rad/s) is the cut-off frequency of the prototype digital filter, ω_c (rad/s) is the desired cut-off frequency, ω_1 and ω_2 (rad/s) are the lower and upper cut-off frequencies respectively, and T is the sampling period (s).

Filter	*Substitute for* z^{-1}	*Design Formulae*
Lowpass	$\dfrac{1 - az}{z - a}$	$a = \dfrac{\sin\left[(\beta - \omega_c)T\,/\,2\right]}{\sin\left[(\beta + \omega_c)T\,/\,2\right]}$
Highpass	$-\left[\dfrac{1 + az}{z + a}\right]$	$a = -\dfrac{\cos\left[(\beta - \omega_c)T\,/\,2\right]}{\cos\left[(\beta + \omega_c)T\,/\,2\right]}$
Bandpass	$-\left[\dfrac{1 - \dfrac{2abz}{(b+1)} + \dfrac{z^2(b-1)}{(b+1)}}{\left(\dfrac{b-1}{b+1}\right) - \dfrac{2abz}{(b+1)} + z^2}\right]$	$a = \dfrac{\cos\left[(\omega_2 + \omega_1)T\,/\,2\right]}{\cos\left[(\omega_2 - \omega_1)T\,/\,2\right]}$ $b = \dfrac{\cot\left[(\omega_2 - \omega_1)T\,/\,2\right]}{1\,/\,[\tan \beta T\,/\,2]}$
Bandstop	$\dfrac{1 - \dfrac{2az}{(b+1)} + \dfrac{z^2(1-b)}{(b+1)}}{\left(\dfrac{1-b}{b+1}\right) - \dfrac{2az}{(b+1)} + z^2}$	$a = \dfrac{\cos\left[(\omega_2 + \omega_1)T\,/\,2\right]}{\cos\left[(\omega_2 - \omega_1)T\,/\,2\right]}$ $b = \dfrac{\tan\left[(\omega_2 - \omega_1)T\,/\,2\right]}{1\,/\,[\tan \beta T\,/\,2]}$

Table 9.4

9.7 CONCLUDING REMARKS

The methods presented in this chapter are suitable for the design of recursive digital filters, and they readily implement signal filtering processes in a variety of practical applications. However, other design methods exist, each generally having merits for particular situations. Further examples and details can be found in References [8-11].

9.10 REFERENCES

[1]   M. E. Van Valkenburg, *Analog Filter Design* (Holt, Rinehart and Winston, New York, 1982).

[2]   C. M. Rader and B. Gold, *Digital Filter Design in the Frequency Domain*, Proc. IEEE, **55** (1967) 149-71.

[3]   H.J. Blinchikoff and A. I. Zverev, *Filtering in the Time and Frequency Domains*, (Wiley, New York and London, 1976), Chapter 9.

[4]   L. R. Rabiner and B. Gold, *Theory and Applications of Digital Signal Processing*, (Prentice-Hall, Englewood Cliffs, N.J., 1975).

[5]   S.A. Tretter, *Introduction to Discrete-time Signal Processing*, (Wiley, New York and London, 1976).

[6]   J. R. Mick, *Digital Signal Processing Handbook*, (Advanced Micro Devices, California, 1976).

[7]   M. H. Acroyd, *Digital Filters* (Butterworths, London, 1973), Chapter 2.

[8]   T. J. Terrell, *Introduction to Digital Filters - Second Edition*, (Macmillan Education Ltd, Basingstoke and London, 1988).

[9]   A.V. Oppenheim and R.W. Schafer, *Digital Signal Processing*, (Prentice-Hall, Englewood Cliffs, N.J., 1975).

[10]   A. Antonious, *Digital Filter Analysis and Design*, (McGraw-Hill, New York, 1983).

[11]   F.J. Taylor, *Digital Filter Design Handbook*, (Marcel Decker, New York, 1983).

Chapter 10

Quantisation and rounding problems in digital filters

T. J. Terrell and E. T. Powner

10.1 INTRODUCTION

All operations in digital filters are performed with discrete digital representations, and consequently they are, by definition, approximate procedures. If the approximations are too coarse or of a particular form, the errors involved can lead to instability (limit cycles and overflow oscillations) as well as inaccuracies.

The main sources of error are:

(a) Quantisation due to resolving the input sequence, x_k, into a set of discrete levels.

(b) Finite word length, giving rise to modification of filter parameters (e.g. coefficients) due to the limited number of bits used in their representation.

(c) Arithmetic involving number truncation or round-off.

The design of digital filters can be undertaken without detailed consideration of the above sources of error (see Chapters 8 and 9). However, for effective implementation these are important practical considerations that should not be ignored. In this context, an understanding of these errors enables the filter designer to estimate the minimum number of bits required to form a word-length in the filter implementation. Additionally, with regard to implementation, this knowledge enables the designer to make informed decisions concerning trade-offs between filter hardware costs and digital signal processing precision.

10.2 QUANTISATION OF THE INPUT SIGNAL

The continuous input signal is sampled using an A/D converter (see Chapter 7), which normally produces fixed-point binary number representations of the input samples - a process commonly referred to as quantising the input signal.

Figure 10.1 shows the transfer characteristic (for the *rounding* case) of an eight-level quantiser. Each level is assigned a particular two's complement 3-bit binary number. The value of Q (the quantisation interval) depends on the input signal peak-to-peak range, and, in general, if an n-bit binary number is used, then

$$\left(2^n - 1\right) \times Q = \text{peak-to-peak range}$$

The corresponding maximum error in an input sample is $\pm Q/2$. For example, if the peak-to-peak voltage range of the input signal is ± 3.5.V, then the quantisation interval for the quantiser having the transfer characteristic shown in Figure 10.1 is 1 V, and the maximum rounding error is ± 0.5V.

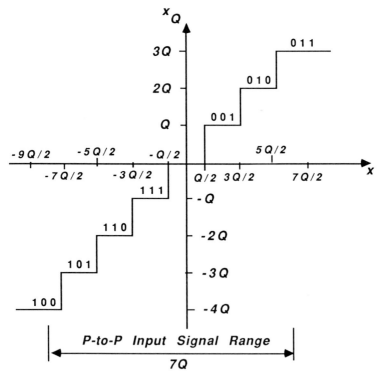

Figure 10.1 Eight-level Quantiser Transfer Characteristic (Rounding Case)

The probability density function for round-off quantisation error is generally assumed to be uniform, and therefore in this case it may be shown that the mean-square error (variance or average noise power) is

$$\sigma_i^2 = \frac{Q^2}{12} = \frac{2^{-2n}}{12} \qquad (10.1)$$

where n is the number of bits representing the magnitude (i.e. excludes sign bit) of the sample value produced by the A/D converter.

The input to the digital filter is convieniently modelled by the sum of two component signals, namely, a noiseless input x_k and a noise input e_k. Thus the quantised input signal is expressed as

$$x_k\Big|_Q = x_k + e_k$$

However, in practice, if the amplitude of the input signal exceeds the dynamic number range of the A/D converter, the signal amplitude should be reduced to eliminate clipping. Consequently the model of the quantisation process accommodates this by including a scaling factor, F, such that

$$x_k\Big|_Q = Fx_k + e_k$$

where $0 < F < 1$. In this case, and assuming that the quantisation errors are due to rounding, the signal-to-noise ratio (SNR) is

$$SNR = 10\log_{10}\left[\frac{F^2\sigma_x^2}{\sigma_i^2}\right] \text{ dB}$$

where $F^2\sigma_x^2$ is the variance of Fx_k, and σ_i^2 is the variance of the quantisation noise. As a rule-of thumb, for negligible clipping, F is generally set equal to $1/(5\sigma_x)$, and for this case it follows that

$$SNR = 10\log_{10}\left[\frac{1}{25\sigma_i^2}\right] \text{ dB} \qquad (10.2)$$

Substituting equation (10.1) in equation (10.2) gives

$$\text{SNR} \cong (6n - 3.2)\,\text{dB} \tag{10.3}$$

The commonly used 12-bit A/D converter has a SNR $\cong 68.8$ dB.

From equation (10.3) it is seen that the signal-to-noise ratio increases by 6 dB for each additional bit added to the A/D converter word-length.

The variance at the digital filter output, due to quantisation of the input signal, may be obtained using Parseval's theorem for discrete systems [1], that is

$$\sigma_o^2 = \frac{Q^2}{12} \sum_{k=0}^{\infty} \left[g_k \right]^2 = \frac{Q^2}{12} \times \frac{1}{j\,2\pi} \oint G(z) G^*(z) \frac{dz}{z} \tag{10.4}$$

The significance of equation (10.4) may be demonstrated using the following simple example.

ILLUSTRATIVE EXAMPLE

Suppose that a digital filter has the transfer function $G(z) = z / (z - \alpha)$, where $|\alpha| < 1$. It follows that

$$\sigma_o^2 = \frac{Q^2}{12} \sum_{k=0}^{\infty} \left[\alpha^k \right]^2 = \frac{Q^2}{12}(1 + \alpha^2 + \alpha^4 + \alpha^6 + \cdots) = \frac{Q^2}{12} \times \frac{1}{(1 - \alpha^2)}$$

Consequently as $\alpha^2 \rightarrow 1$ then $\sigma_o^2 \rightarrow \infty$, and as $\alpha^2 \rightarrow 0$ then $\sigma_o^2 \rightarrow \frac{Q^2}{12}$. Thus it is seen that the pole location (filter structure) has a significant influence on the value of σ_o^2. Indeed, in this respect, this form of round-off error analysis in digital filter implementations may be undertaken to enable the optimum form of realisation structure to be selected [2-5].

10.3 EFFECT OF FINITE WORD LENGTH ON STABILITY AND FREQUENCY RESPONSE

If one or more poles of the digital filter are located close to the circumference of the unit-circle in the z-plane, making $G(z)$

marginally stable, it is then desireable to investigate how a small change in one of the denominator coefficients, b_j $(0 \leq j \leq q)$ - see Section 7.2, Chapter 7, may result in one or more poles moving outside the unit-circle, rendering the filter unstable. This small change is often related to inaccurate coefficient representation using finite fixed-point processor word lengths.

The pole positions of the ideal digital filter may be computed from its characteristic equation, i.e. from

$$1 + \sum_{j=1}^{q} b_j \, z^{-j} = 0$$

If one of the filter coefficients, b_h , is changed to a new value equal to $b_h + \Delta b_h$, the characteristic equation becomes

$$1 + \sum_{j=1}^{q} b_j \, z^{-j} + \Delta b_h \, z^{-h} = 0 \qquad (10.5)$$

Assuming a stable $G(z)$, and that the sampling frequency is relatively high, then all poles will be inside the unit-circle close to $z = 1$, therefore we seek the value of Δb_h which takes a pole outside the unit-circle. Substituting this condition of $z = 1$ in equation (10.5) gives

$$\left| \Delta b_h \right| = 1 + \sum_{j=1}^{q} b_j \qquad (10.6)$$

By comparing $\left| \Delta b_h \right|$ with the largest b_j coefficient the required resolution may be determined [6], and thus the required processor word length is determined.

ILLUSTRATIVE EXAMPLE

Specification:

A digital filter designed by the impulse invariant design method (Chapter 9), which is based on a prototype second-order Butterworth lowpass filter with a cut-off frequency of 1 radian/second and a sampling frequency of 30.2 radians/second, has the transfer function $G(z) = z / \left(26.8z^2 - 45.6z + 20\right)$. Calculate the minimum word length to maintain stability, assuming that filter coefficients are rounded.

Solution:

$$G(z) = z / \left(26.8z^2 - 45.6z + 20\right) \cong 0.04z / \left(z^2 - 1.7z + 0.75\right)$$

Using equation (10.6) we obtain $\Delta b_h = 1 - 1.7 + 0.75 = 0.05$, and the corresponding quantisation interval is $2 \times \Delta b_h = 2 \times 0.05 = 0.1$. Now comparing this with the largest coefficient value we obtain

$$\frac{1.7}{0.1} \le 2^{(n-1)}$$

Thus the minimum word length, n, to maintain stability is $n = 6$. Note, however, that if coefficients are truncated rather than rounded the calculated value is increased by one, to $n = 7$ in this example.

An alternative method of calculating the processor word length to maintain stability has been published by Kou and Kaiser [6]. They assumed truncation rather than rounding, and showed that for an N th-order digital filter having distinct poles at $\left(\cos \omega_k T - j \sin \omega_k T\right), k = 1,2,3, \cdots, N$, the number of processor bits must be

$$n = \text{smallest integer exceeding} \left\{ -\log_2 \left[\frac{5\sqrt{N}}{2^{N+2}}\right] \prod_{k=1}^{N} (\omega_k T) \right\}$$

$$(10.7)$$

ILLUSTRATIVE EXAMPLE

Specification:

Verify the result of the previous illustrative example.

Solution:

The z-plane poles are located at $z = 0.85 \pm j\,0.15$, and $G(z)$ has a second-order denominator, therefore $N = 2$. Since the poles are a complex conjugate pair, then

$$\omega_1 T = \omega_2 T = \left(\tan^{-1}\frac{0.15}{0.85}\right) = 0.175\,\text{rad}$$

Substituting $\omega_1 T = \omega_2 T = 0.175\,\text{rad}$, and $N = 2$ in equation (10.7) we obtain

$$n = \text{smallest integer exceeding}\left\{-\log_2 [0.014]\right\}$$

that is, $2^n \geq \dfrac{1}{0.014} = 71.43$, $\therefore n = 7$.

As a rule-of thumb, to provide the required frequency response precision, three or four bits are added to the word length value calculated for maintaining stability. However, actual changes in the frequency response due to coefficient rounding may be determined by calculating the corresponding movement of the filter's poles and zeros. Taking the poles of $G(z)$ to be located in the z-plane at position p_l, $l = 1, 2, 3, \cdots q$, then it has been shown by Kaiser [7] that the poles of the quantised filter move to $p_l + \Delta p_l$, where

$$\Delta p_l = \sum_{i=1}^{q} \frac{p_l^{(i+1)}}{\displaystyle\prod_{\substack{n=1 \\ n \neq i}}^{q} (1 - p_l / p_n)} \Delta b_i$$

, and where Δb_i is the change in the b_i coefficient. Similar results can be obtained for the movement in the filter zeros.

10.4 ARITHMETIC ERRORS

Arithmetic errors arise due to arithmetic-logic-unit (ALU) operations. For example, the multiplication of two n-bit binary numbers produces a $2n$-bit product, which is normally truncated to n most-significant bits for storage and further arithmetic. Another example is that of adding two n-bit fixed-point binary numbers which produces a result exceeding n bits, thereby creating a number overflow condition. When such quantities are used in successive or subsequent calculations, the effect is cumulative and large overall errors may build up. Consequently arthimetic errors, together with other sources of error, may lead to stability problems, limit cycle oscillations or unacceptable inaccuracy.

ILLUSTRATIVE EXAMPLE

A digital filter has the transfer function

$$G(z) = \frac{1 - z^{-2}}{1 + 1.86z^{-1} + 0.98z^{-2}}$$

which has two zeros at $z = \pm 1$ and a pair of complex conjugate poles at $z = 0.9302957 \pm j\,0.3385999$. The poles are thus located quite close to the circumference of the unit-circle, and any pole movement due to coefficient quantisation, plus cumulative arithmetic errors, may place the poles on or outside the unit-circle, leading to instability.

10.5 LIMIT CYCLE OSCILLATIONS

When the input signal to a digital filter is constant or zero, the fixed-point finite word length arithmetic rounding errors cannot be assumed to be uncorrelated random variables. In fact, in these circumstances, the performance of the filter may exhibit unwanted limit cycle oscillations [8-10]. This implies that the round-off noise is significantly dependent on the input signal, and its effect is easily seen by considering a simple example of a first-order IIR filter.

Suppose that a digital filter has the linear difference equation:

$$y_k = x_k + C\,y_{(k-1)}$$

Assuming that the input signal is zero-valued, and letting coefficient C equal 0.93, then we have

$$y_k = 0.93 \times y_{(k-1)}$$

Furthermore, assuming that the multiplication result, y_k, is rounded to the nearest integer value (interger number resolution), and that the initial condition is $y_{(-1)} = 11$, then the computations are:

$$y_{(0)} = 0.93 \times y_{(-1)} = 0.93 \times (11) = 10.23 \rightarrow 10$$

$$y_{(1)} = 0.93 \times y_{(0)} = 0.93 \times (10) = 9.3 \rightarrow 9$$

$$y_{(2)} = 0.93 \times y_{(1)} = 0.93 \times (9) = 8.37 \rightarrow 8$$

$$y_{(3)} = 0.93 \times y_{(2)} = 0.93 \times (8) = 7.44 \rightarrow 7$$

$$y_{(4)} = 0.93 \times y_{(3)} = 0.93 \times (7) = 6.51 \rightarrow 7$$

$$y_{(5)} = 0.93 \times y_{(4)} = 0.93 \times (7) = 6.51 \rightarrow 7$$

It is seen that the filter output reaches a steady value of 7. This is the **deadband effect** [11]. In fact, in the above case, any initial value in the range $-R \le y_{(-1)} \le R$, where R is the largest integer satisfying $R \le \frac{1}{2} \times (1 - |C|)^{-1}$, will produce the same resulting steady output value.

Let us now consider a change of sign for the coefficient in the above digital filter, i.e.

$$y_k = x_k - C\, y_{(k-1)}$$

Again assuming C is equal to 0.93, and also assuming that the multiplication result, y_k, is rounded to the nearest integer value, and that the initial condition is $y_{(-1)} = 0$, then given $x_k = \{12, 0, 0, 0, \cdots, 0\}$ the computations are:

$$y_{(0)} = x_{(0)} = 12 \rightarrow 12$$

$$y_{(1)} = -0.93 \times 12 = -11.16 \rightarrow -11$$

$$y_{(2)} = -0.93 \times (-11) = 10.23 \rightarrow -10$$

$$y_{(3)} = -0.93 \times 10 = -9.3 \rightarrow -9$$

$$y_{(4)} = -0.93 \times (-9) = 8.37 \rightarrow 8$$

$$y_{(5)} = -0.93 \times 8 = -7.44 \rightarrow -7$$

$$y_{(6)} = -0.93 \times (-7) = 6.51 \rightarrow 7$$

$$y_{(7)} = -0.93 \times 7 = -6.51 \rightarrow -7$$

$$y_{(8)} = -0.93 \times (-7) = 6.51 \rightarrow 7$$

In this particular case it is seen that the filter's impulse response is an oscillatory signal, referred to as a *limit cycle* output, with frequency equal to $\omega_s / 2$ and amplitude ± 7.

10.6 CONCLUDING REMARKS

It has been shown that quantisation and rounding errors sometimes have considerable effects on the performance of digital filters, and therefore it is important that these are taken into account when considering their practical implementation. In this respect, to achieve adequate precision for the signal and number representation, a good estimate of the required processor word length must be made, and, if possible, a *safety factor* should be used so that three or four bits are added to the estimated word length.

10.7 REFERENCES

[1]   R. E. Bogner and A. G. Constantinides, *Introduction to Digital Filtering* (Wiley, New York, 1975) Appendix 10c, pp 179-81.

[2]   B. Liu, *Effect of Finite Word Length on the Accuracy of Digital Filters - A Review*, Trans. Circuit Theory, IEEE, **18** (1971), pp 670-7.

[3]   L. R. Rabiner and B. Gold, *Theory and Application of Digital Signal Processing* (Prentice-Hall, Englewood Cliffs, N. J., 1975), Chapter 5.

[4]   A. V. Oppenheim and R. W. Shafer, *Digital Signal Processing* (Prentice-Hall, Englewood Cliffs, N.J., 1975), Chapter 9.

[5]   L. B. Jackson, *Digital Filters and Signal Processing*, (Kluwer Academic Publishers, Boston, 1986), Chapter 11.

[6]   F. F. Kuo and J. F. Kaiser, *Systems Analysis by Digital Computer*, (Wiley, New York, 1966), Chapter 7.

[7]   J. F. Kaiser, *Some Practical Considerations in the Realization of Linear Digital Filters*, Proceedings of the 3rd Annual Alerton Conference on Circuit and System Theory, 1965.

[8]   L. B. Jackson, *An Analysis of Limit Cycles Due to Multiplication Rounding in Recursive Digital (Sub) Filters*, Proceedings of the 7th Annual Alterton Conference on Circuit and System Theory, 1969.

[9]   L. J. Long and T. N. Trick, *An Absolute Bound on Limit Cycles Due to Roundoff Errors in Digital Filters*, Trans. Audio and Electroacoustics, IEEE, **21** (1973) pp 27-30.

[10]   S.R. Parker and S.F. Hess. *Limit Cycle Oscillations in Digital Filters*, Trans. Circuit Theory, IEEE, **18** (1971) pp 687-97.

[11]   R. B. Blackman, *Linear Data-Smoothing and Prediction in Theory and Practice*, (Addison Wesley, Reading, Mass.), 1965.

Chapter 11

State space control

G. W. Irwin

11.1 INTRODUCTION

Classical design of digital controllers involves the use of frequency domain methods and the root locus. Control is restricted to single–input, single–output systems. The aim of this chapter is to introduce the state space or modern approach to design. Although single–input, single–output systems will mainly be considered, modern control design can readily be extended to cover systems with several inputs and outputs.

The design of state feedback controllers is presented after a brief introduction to state space models including the key ideas of controllability and observability. The selection of feedback gains in order to achieve a desired set of closed–loop poles will be familiar to engineers versed with classical design methods. The implementation of state feedback control laws assumes that all the state variables are known. In practice, the state vector is estimated from the measurements or plant outputs using an observer. Observer design is therefore treated next prior to an analysis of the complete control–estimator system.

The chapter finishes by examining linear quadratic optimal control. Here the state feedback gains are chosen to minimise a scalar quadratic cost or index of performance which is chosen by the designer. This approach removes the need to define the closed–loop pole positions which can be difficult, especially for high order and multi–input, multi–output systems.

11.2 STATE MODELLING

The state variable description of an r–input, m–output continous–time, linear dynamical system consists of a set of first order, ordinary differential equations of the form

$$\dot{\underline{x}} = A_c \underline{x} + B_c \underline{u} \tag{1}$$

$$\underline{y} = C\underline{x} \tag{2}$$

Here $\underline{x}$ is the $(n \times 1)$ state vector, $\underline{u}$ is the $(r \times 1)$ control or input vector and $\underline{y}$ is the $(m \times 1)$ output or measurement vector. Also A_c, B_c and C are the plant, input and measurement matrices respectively which are of appropriate dimensions.

The solution of equation (1), analogous with the scalar case, is given by

$$\underline{x}(t) = e^{A_c(t - t_0)}\underline{x}(t_0) + \int_{t_0}^{t} e^{A(t - \tau)}B_c \underline{u}(\tau)d\tau \tag{3}$$

Equation (3) expresses the familiar idea that the state response $\underline{x}(t)$ has two components, one due to the initial conditions $\underline{x}(t_0)$ and the other from the input $\underline{u}(t)$ via a convolution integral. The matrix exponential $e^{A_c t}$ is defined as

$$e^{A_c t} = I + A_c t + \frac{1}{2!} A_c^2 t^2 + \dots$$

and is produced by extending the notation of the scalar exponential function to matrices.

The corresponding discrete–time plant representation consists of a set of difference equations. Thus consider Figure 1 which illustrates the sampling of the input and output vectors for the plant at intervals of T. If a zero–order hold is employed, the control input $\underline{u}(t)$ is replaced by the staircase function given in Figure 2, where

$$\underline{u}(t) = \underline{u}(t_k), \quad t_k \le t < t_{k+1}$$

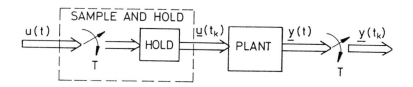

Figure 1. Sampling of multi–input, multi–output plant

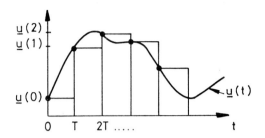

Figure 2. Output from the zero-order hold

Now, in equation (3) set $t_0 = t_k$ and $t = t_{k+1}$ such that

$$\underline{x}(t_{k+1}) = e^{A_c(t_{k+1} - t_k)}\underline{x}(t_k) + \int_{t_k}^{t_{k+1}} e^{A_c(t_{k+1} - \tau)}B_c \underline{u}(t_k)d\tau \qquad (4)$$

Equation (4) can be expressed more simply as

$$\underline{x}_{k+1} = A\underline{x}_k + B\underline{u}_k \qquad (5)$$

where the discrete plant and input matrices are defined by

$$A = e^{A_c T}; \quad B = \int_0^T e^{A_c(T - \tau)}B_c\, d\tau$$

The corresponding discrete measurement equation is

$$\underline{y}_k = C\underline{x}_k \qquad (6)$$

The important concepts of controllability and observability are now introduced which describe structural features of a dynamic system.

Definition: The system in equation (5) is *controllable* provided there exists a sequence of inputs $\underline{u}_0, \underline{u}_1, \underline{u}_2 \ldots \underline{u}_N$ that will translate the system from an initial state $\underline{x}_0$ to any final state $\underline{x}_N$, with N finite.

It can be shown that the system is controllable if the matrix in equation (7) below has rank n i.e. has n independent columns.

$$[B \quad AB \quad A^2B \ \ldots \ A^{n-1}B] \qquad (7)$$

Definition: The system in equations (5) and (6) is *observable* provided that any arbitrary initial state $\underline{x}_0$ can be calculated from the N measurements $y_0, y_1, \ldots y_{N-1}$, with N finite.

It can be shown that the system is observable if the following matrix has rank n i.e. has n independent rows.

$$\begin{bmatrix} C \\ CA \\ CA^2 \\ \vdots \\ CA^{n-1} \end{bmatrix}$$

11.3 STATE FEEDBACK CONTROL

The diagram in Figure 3 illustrates the principle of stable feedback where the control signals are formed from linear combinations of the state variables

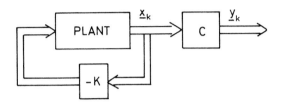

Figure 3. State feedback control

Although the state vector is generally not available for measurement, state feedback control is useful to study for the following reasons. (i) It is instructive to see what can be done in shaping the desired closed–loop response when full state information is available. (ii) Techniques exist for estimating the states from measurements of the plant inputs and outputs, as will be shown later. (iii) Optimal control laws often take a state feedback form so that it is useful to understand the effects of this strategy.

The *open-loop* system is defined in equations (5) and (6) and the control law is given by

$$\underline{u}_k = -K\underline{x}_k \qquad (8)$$

Substituting for $\underline{u}_k$ in equation (5) gives the *closed-loop* system equations

$$\underline{x}_{k+1} = (A - BK)\underline{x}_k \qquad (9)$$

$$y_k = C\underline{x}_k$$

Now, the stability of a discrete linear system is determined by the roots of the *characteristic equation* (or the eigenvalues of the system matrix). The characteristic equations for the open– and closed–loop systems are as follows:

$$\alpha_o(z) = |zI - A|$$

$$\alpha_c(z) = |zI - A + BK|$$

The basic idea behind state feedback is to position the closed–loop poles by choice of K. Suppose it is decided to locate the closed–loop poles at $\lambda_1, \lambda_2, ..., \lambda_n$ in the z–plane. The characteristic equation will then take the form

$$\alpha_c(z) = (z - \lambda_1)(z - \lambda_2)...(z - \lambda_n)$$

$$= |zI - A + BK| \tag{10}$$

Equation (10) can then be solved for the unknown gains by equating coefficients.

The calculation of the control gain matrix K is simplified for a *single-input system* in *controllable canonical form* where the plant and input matrices have the special structure defined in equation (11).

$$\underline{x}_{k+1} = \begin{bmatrix} 0 & 1 & 0 & ... & 0 \\ 0 & 0 & 1 & ... & 0 \\ \vdots & \vdots & \vdots & & \vdots \\ 0 & 0 & 0 & & 1 \\ -b_0 & -b_1 & ... & -b_{n-1} \end{bmatrix} \underline{x}_k + \begin{bmatrix} 0 \\ 0 \\ \vdots \\ 0 \\ 1 \end{bmatrix} u_k \tag{11}$$

This plant has the characteristic equation

$$\alpha_o(z) = z^n + b_{n-1}z^{n-1} + ... + b_1 z + b_0 = 0$$

The term BK in equation (10), for this state model, becomes

$$BK = \begin{bmatrix} 0 \\ 0 \\ \vdots \\ 0 \\ 1 \end{bmatrix} [k_1 k_2 ... k_n] = \begin{bmatrix} 0 & 0 & ... & 0 \\ 0 & 0 & ... & 0 \\ \vdots & \vdots & & \vdots \\ 0 & 0 & ... & 0 \\ k_1 & k_2 & ... & k_n \end{bmatrix}$$

Hence the closed–loop system matrix is

$$A - BK = \begin{bmatrix} 0 & 1 & ... & 0 \\ 0 & 0 & ... & 0 \\ \vdots & \vdots & ... & \vdots \\ 0 & 0 & ... & 1 \\ -(b_0 + k_1) & -(b_1 + k_2) & ... & -(b_{n-1} + k_n) \end{bmatrix}$$

and the characteristic equation becomes

$$z^n + (b_{n-1} + k_n)z^{n-1} + ... + (b_1 + k_2)z + (b_0 + k_1) = 0 \tag{12}$$

Now, if the desired characteristic equation for the closed–loop system is expressed as

$$\alpha_c(z) = z^n + \alpha_{n-1}z^{n-1} + ... + \alpha_1 z + \alpha_0 \tag{13}$$

where the α's are of course all known, then the unknown gains can be determined by equating coefficients in equations (12) and (13) to give

$$k_{i+1} = \alpha_i - b_i, \quad i = 0, 1, ..., n - 1 \tag{14}$$

In practice most single–input state models are not in canonical form. However controllable systems can be transformed to canonical form. The state feedback gains are then calculated using equation (14) and finally transformed for application to the original state model. *Ackermann's formula* is based on this principle. The matrix polynomial $\alpha_c(A)$ is first formed using the coefficients of the desired characteristic equation (13)

$$\alpha_c(A) = A^n + \alpha_{n-1}A^{n-1} + ... + \alpha_0 I$$

Ackermann's formula for the required gain matrix K is then given by

$$K = [0\ 0\ ...\ 0\ 1][B\ AB\ A^2B\ ...\ A^{n-1}B]^{-1}\alpha_c(A) \tag{15}$$

A simple example will now be introduced to illustrate the algebra involved.

Example: Design a state feedback controller for the system below to produce closed–loop poles at $0.888 \pm j\, 0.173$.

$$\underline{x}_{k+1} = \begin{bmatrix} 1 & 0.095 \\ 0 & 0.905 \end{bmatrix} \underline{x}_k + \begin{bmatrix} 0.005 \\ 0.095 \end{bmatrix} u_k$$

$$y_k = [1 \quad 0]\, \underline{x}_k$$

Here the desired characteristic equation is given by

$$\alpha_c(z) = (z - 0.888 - j\, 0.173)(z - 0.888 + j\, 0.173) = z^2 - 1.776z + 0.819$$

Hence,

$$\alpha_c(A) = \begin{bmatrix} 1 & 0.095 \\ 0 & 0.905 \end{bmatrix}^2 - 1.776 \begin{bmatrix} 1 & 0.095 \\ 0 & 0.905 \end{bmatrix} + 0.819 \begin{bmatrix} 1 & 0 \\ 0 & 1 \end{bmatrix}$$

$$= \begin{bmatrix} 0.043 & 0.0123 \\ 0 & 0.0308 \end{bmatrix}$$

Also

$$[B \quad AB]^{-1} = \begin{bmatrix} 0.005 & 0.0139 \\ 0.095 & 0.086 \end{bmatrix}^{-1} = \begin{bmatrix} -95.13 & 15.34 \\ 105.1 & -5.342 \end{bmatrix}$$

Then, using Ackermann's formula in equation (15)

$$K = [0 \quad 1] \begin{bmatrix} -95.13 & 15.34 \\ 105.1 & -5.342 \end{bmatrix} \begin{bmatrix} 0.043 & 0.0123 \\ 0 & 0.0308 \end{bmatrix}$$

$$= [4.52 \quad 1.12]$$

11.4 STATE ESTIMATION

An *observer* or *state estimator* is used to reconstruct the full state vector from the measurements of a smaller number of state variables. Essentially, the observer is a dynamical system, driven by the inputs and outputs of the plant, whose state vector $\hat{\underline{x}}_k$ is an estimate of the unknown state vector $\underline{x}_k$. Figure 4 shows the structure of the observer in block diagram form

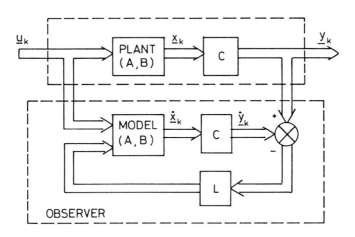

Figure 4. Observer for state estimation

Given a current state estimate $\hat{\underline{x}}_k$ and the control input $\underline{u}_k$, a reasonable *prediction* of the state vector at the next sample instant can be produced from the system model as

$$\hat{\underline{x}}_{k+1} = A\hat{\underline{x}}_k + B\underline{u}_k$$

The quality of this open–loop estimate can then be checked by comparing the observer output $\hat{\underline{y}}_k$ with the actual plant output $\underline{y}_k$. The difference between the two is used as a *correction* term to produce a closed–loop estimator given by

$$\hat{\underline{x}}_{k+1} = A\hat{\underline{x}}_k + B\underline{u}_k + L\{\underline{y}_k - \hat{\underline{y}}_k\}$$

$$= (A - LC)\hat{\underline{x}}_k + B\underline{u}_k + L\underline{y}_k \tag{16}$$

The dynamics of the state estimation error $\underline{e}_k$, defined as

$$\underline{e}_k \, \underline{\Delta} \, \underline{x}_k - \hat{\underline{x}}_k \tag{17}$$

follow from equations (5) and (16). Thus

$$\underline{e}_{k+1} = (A - LC)\underline{e}_k \tag{18}$$

In practice the observer gain matrix L is chosen to stabilise the observer and hence produce satisfactory error convergence.

To select L the same approach is adopted as in designing the state feedback control law. The stability is determined by specifying the desired estimator root locations in the z–plane, typically to make the observer 2 to 4 times faster than the closed–loop control system. Then, *provided* y_k *is scalar and the system is observable*, L is uniquely determined. Two alternatives are available for computing L. The first is to match coefficients in like powers of z on the two sides of the equation below

$$\alpha_e(z) = |zI - A + LC| = (z - \beta_1)(z - \beta_2) \dots (z - \beta_n) \tag{19}$$

In equation (19), the β's are the desired estimator root locations. Ackermann's estimator formula can be employed instead. This is given by

$$L = \alpha_e(A) \begin{bmatrix} C \\ CA \\ CA^2 \\ : \\ CA^{n-1} \end{bmatrix}^{-1} \begin{bmatrix} 0 \\ 0 \\ : \\ 0 \\ 1 \end{bmatrix}$$

It is interesting to note that state feedback design to find K for a single–input system and state estimator design to find L for a single–output system are dual problems i.e. the control design procedure can be used for the "fictitious" closed–loop system $A^T - C^T L^T$ to find L instead of $A - KC$ for K.

11.5 COMBINED CONTROL AND ESTIMATION

The third section of the chapter dealt with the design of a state feedback control law assuming full state knowledge while the last section described estimation of the state vector from the system outputs using an observer. The complete control system is implemented by using this estimated state vector in the control law of equation (8), as shown in Figure 5.

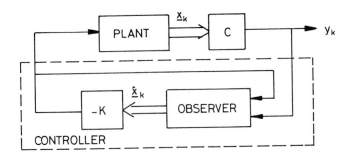

Figure 5. Combined estimation and control

The question arises as to the effect of including an observer in the closed–loop

control system. Will the closed–loop pole positions, used in selecting K, be

altered?

With an observer present, the controlled plant equation (9) becomes

$$\underline{x}_{k+1} = A\underline{x}_k - BK\hat{\underline{x}}_k$$

Substitution for $\hat{\underline{x}}_k$ from equation (17) gives

$$\underline{x}_{k+1} = A\underline{x}_k - BK\{\underline{x}_k - \underline{e}_k\}$$

When this is combined with the estimator error equation (18), the complete system

shown in Figure 5 is described by two coupled equations

$$\begin{bmatrix} \underline{e}_{k+1} \\ \underline{x}_{k+1} \end{bmatrix} = \begin{bmatrix} A - LC & 0 \\ BK & A - BK \end{bmatrix} \begin{bmatrix} \underline{e}_k \\ \underline{x}_k \end{bmatrix}$$

The poles of this composite system follow from the characteristic equation (20)

$$\begin{vmatrix} zI - A + LC & 0 \\ -BK & zI - A + BK \end{vmatrix} = 0 \qquad (20)$$

This can be written as

$$|zI - A + LC|\,|zI - A + BK| = \alpha_e(z)\alpha_c(z) = 0$$

Then the combined control–estimator system includes the *same* poles as those of the

control system above. This illustrates the *Separation Principle* by which control and

estimation schemes can be designed separately, yet used together.

11.6 LINEAR QUADRATIC OPTIMAL CONTROL

So far control design by pole placement has been discussed and only single–input, single–output systems considered. The methods presented can be generalized to multi–input, multi–output systems but there is a loss of uniqueness in the specification of the feedback gains K and L. An alternative way of designing multivariable systems is to use optimal control techniques in which a 'cost function' or 'performance index' is to be minimised by the choice of multivariable feedback control law.

This approach is to be regarded as a useful design procedure and the cost function is chosen to give generally desirable properties to the system and also to be mathematically tractable. The resulting system cannot be regarded as being 'optimal' in any absolute sense.

The most general problem which has been completely solved is that for time–dependent linear systems with quadratic cost function and the treatment will be restricted to an outline of this problem which can be stated as follows.

For the linear discrete system described by

$$\underline{x}_{k+1} = A_k \underline{x}_k + B_k \underline{u}_k \qquad (21)$$

$$y_k = C_k \underline{x}_k,$$

and given the initial state vector $\underline{x}_0$, determine the control

$$\underline{u}_k = \underline{f}_k(\underline{x}_k)$$

that minimises the cost function

$$J_N = \frac{1}{2} \sum_{k=0}^{N} [\underline{x}_k^T Q_k \underline{x}_k + \underline{u}_k^T R_k \underline{u}_k], \qquad (22)$$

where N is finite, Q_k is positive semi–definite and R_k is positive definite. Q and R may be chosen to be symmetric without loss of generality.

The motivation behind this particular choice of cost function is that, in some sense, the states will be driven towards zero whilst at the same time keeping the control effort at a reasonable level.

For the linear–quadratic problem with finite time–horizon, N, it turns out that the optimal control is linear and has the form

$$\underline{u}_k = -K_k \underline{x}_k,$$

where the feedback gains, K_k, are time–varying even if the plant and cost function are time–independent. Furthermore the K_k are independent of the problem initial condition x_0.

The solution to the above problem can be obtained either by treating it as a constrained–minima problem and using Lagrange multipliers[2], (or equivalently by using the Minimum Principle) or by the application of Dynamic Programming[1].

Taking the former course the above constrained problem can be rewritten as the unconstrained problem of minimising

$$J_N' = \sum_{k=0}^{N} [\tfrac{1}{2} x_k^T Q x_k + \tfrac{1}{2} u_k^T R u_k + \lambda_{k+1}^T(-x_{k+1} + A x_k + B u_k)]$$

by suitable choice of x_k, u_k and λ_k.

Necessary conditions for a minimum are

$$\frac{\partial J_N'}{\partial u_k} = R u_k + B^T \lambda_{k+1} = 0 \qquad \text{control equations} \qquad (23)$$

$$\frac{\partial J_N'}{\partial \lambda_{k+1}} = -x_{k+1} + A x_k + B u_k = 0 \qquad \text{state equations} \qquad (24)$$

and

$$\frac{\partial J_N'}{\partial x_k} = Q x_k - \lambda_k + A \lambda_{k+1} = 0 \qquad \text{adjoint equations} \qquad (25)$$

Using (23) to substitute for u_k in (24), these equations provide a set of coupled difference equations which can be solved for the optimal values of x_k and λ_k and hence u_k through (24) again. The initial x_0 is presumed known (in fact it is not required), but not λ_0. However, for minimum J_N, u_N will be zero because it cannot affect x_N, then (23) gives $\lambda_{N+1} = 0$ so that

$$\lambda_N = Q_N x_N. \qquad (26)$$

Thus, finally, it is seen that the problem has been transformed into the solution of two difference equations with the initial x_0 given and the end condition on λ_N

provided by (26). In the above formulae the time–dependence of the matrices A, B, Q and R has been suppressed for notational simplicity.

The solution to this two–point boundary–value problem can be found by assuming

$$\underline{\lambda}_k = P_k \underline{x}_k \tag{27}$$

This assumption arises naturally in the Dynamic Programming method of solution to the original design problem, which also gives a physical meaning to the matrix P_k in terms of the minimum cost

$$J_{N-k} = \tfrac{1}{2} \underline{x}_k^T P_k \underline{x}_k = \tfrac{1}{2} \sum_{r=k}^{N} [\underline{x}_r^T Q \underline{x}_r + \underline{u}_r^T R \underline{u}_r].$$

Substituting (27) into (23) leads to

$$\underline{u}_k = -(R + B^T P_{k+1} B)^{-1} B^T P_{k+1} A \underline{x}_k$$
$$= -S^{-1} B^T P_{k+1} A \underline{x}_k, \quad \text{say.} \tag{28}$$

Using (27) to eliminate $\underline{\lambda}$ from the adjoint equation and combining the result and (28) with the state equations (25) leads to

$$[P_k - A^T P_{k+1} A + A^T P_{k+1} B S^{-1} B^T P_{k+1} A - Q] \underline{x}_k = 0$$

which must be true for arbitrary $\underline{x}_k$, hence the [ ] matrix must be zero. This gives a backward–in–time equation in P_k,

$$P_k = A^T [P_{k+1} - P_{k+1} B S^{-1} B^T P_{k+1}] A + Q \tag{29}$$

Or,

$$P_k = A^T M_{k+1} A + Q$$

where

$$M_{k+1} = P_{k+1} - P_{k+1} B [R + B^T P_{k+1} B]^{-1} B^T P_{k+1}$$

The initial condition for this backward recursion for P_{k+1} is given by (26) and (27), namely

$$P_N = Q_N.$$

Then, from (28)

$$\underline{u}_k = -K_k \underline{x}_k$$

where

$$K_k = [R + B^T P_{k+1} B]^{-1} B^T P_{k+1} A$$

is the optimal time–varying feedback gain.

The optimal feedback gains, K_k, can thus be calculated, independently of the initial state, by the following backward recursion algorithm:

1. Let $P_N \leftarrow Q_N$ and $K_N \leftarrow 0$

2. Let $k \leftarrow N$

3. Let $M_k \leftarrow P_k - P_k B_{k-1} [R_{k-1} + B_{k-1}^T P_k B_{k-1}]^{-1} B_{k-1}^T P_k$

4. Let $K_{k-1} \leftarrow [R_{k-1} + B_{k-1}^T P_k B_{k-1}]^{-1} B_{k-1}^T P_k A_{k-1}$

5. Store K_{k-1}

6. Let $P_{k-1} \leftarrow A_{k-1}^T M_k A_{k-1} + Q_{k-1}$

7. Let $k \leftarrow k-1$

8. Go to step 3.

For any given initial condition $\underline{x}_0$,

$$\underline{x}_{k+1} = A_k \underline{x}_k + B_k \underline{u}_k$$

where, using the stored gains

$$\underline{u}_k = -K_k \underline{x}_k.$$

It is not difficult to show that the resulting minimum cost is

$$J_N = \tfrac{1}{2} \underline{x}_0^T P_0 \underline{x}_0.$$

An example[2] of the time–varying behaviour of the control gains caused by a finite horizon, N, even when the system and cost matrices (A, B, Q, R) are constant, is provided by the system with transfer function $1/s^2$ and $Q = \text{diag}[1, 0]$, $R = 1$, $K = [k_1, k_2]$:

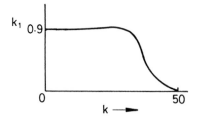
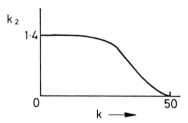

Figure 6.   Example of time–varying feedback gains

As shown in Figure 6, the example demonstrates typical behaviour for values of K_k with distant horizon (i.e. $k \ll N$) provided that the system and cost are time independent. This leads to the concept of 'steady–state optimal gains' which can be obtained by letting $N \to \infty$ in the time–varying problem. Then $P_{k+1} = P_k = P$ and $K_k = K$ and (29) becomes

$$P = A^TPA + Q - A^TPB[B^TPB + R]^{-1}B^TPA, \tag{30}$$

the *Algebraic Riccati equation*.

Various methods exist for solving this equation for P and hence obtaining K. However, perhaps the simplest approach is to set N to a large value in the time–dependent algorithm and calculate the values of the P_k matrix until the matrix elements become constant. This will give the solution to (30) and the corresponding steady–state value of K.

The solution of this optimal steady–state problem thus provides an alternative design method for obtaining constant feedback gains for multi–input, multi–output systems. However, since the process is now assumed to be an infinite–stage one (i.e. $N \to \infty$ in (22)) it is important that the resulting system should be asymptotically stable and this is ensured[3] if the matrices A and Q are such that

$$\text{rank}[Q_1, A^TQ_1^T, ..., (A^T)^{n-1}Q_1^T] = n,$$

where Q_1 is defined by $Q_1^TQ_1 = Q$. Otherwise it would be possible to minimise J_∞ even though $\underline{x}_\infty \neq 0$.

11.7 THE LINEAR QUADRATIC GAUSSIAN OPTIMAL CONTROL PROBLEM

So far only deterministic control problems have been considered in which it is assumed that the output, y_k, can be measured exactly and, once the initial state is known, then the state at any later time is pre–ordained. However, in practice the output can rarely be measured exactly and the identification of the system parameters is seldom precise. To take account of these uncertainties the system is often modelled by the following equations:

$$\underline{x}_{k+1} = A\underline{x}_k + B\underline{u}_k + \underline{v}_k \tag{31}$$

$$y_k = C\underline{x}_k + \underline{w}_k,$$

where $\underline{v}_k$, $\underline{w}_k$ are 'disturbance' vectors which are only known in some statistical sense.

Specification of zero–mean Gaussian disturbances leads to the *LQG optimal control problem* in which the system is described by equations (31) and the control $\underline{u}_k$ as a function of $\{y_{k-1}, y_{k-2}, ..., \underline{u}_{k-1}, \underline{u}_{k-2}, ...\}$ is to be chosen to minimise the cost function

$$J_N = E\{\underline{x}_N^T Q_N \underline{x}_N + \sum_{k=0}^{N-1} (\underline{x}_k^T Q_k \underline{x}_k + \underline{u}_k^T R_k \underline{u}_k)\},$$

where E denotes expectation.

Solution of this optimal control problem can be obtained using Dynamic Programming. It turns out that the *Separation Principle* applies again which means that the design procedure can be separated into two parts; namely, the design of a state estimator which gives the best estimates of the states from the observed outputs, for this purpose a *Kalman filter* (see elsewhere) is used, and the choice of a linear–feedback law using the estimated states. The optimal feedback controller is the same as that used when there are no disturbances acting on the system.

BIBLIOGRAPHY

1. C L Phillips and H T Nagle, "Digital Control System Analysis and Design", Prentice–Hall, Inc., 1984.

2. G F Franklin and J D Powell, "Digital Control of Dynamic Systems", Addison Wesley, 1980.

3. K Ogata, "Discrete–Time Control Systems", Prentice–Hall, Inc., 1987.

Chapter 12

Kalman filters

P. Hargrave

12.1 INTRODUCTION

In what follows we present an elementary introduction to Kalman filtering, starting from the simplest of all estimation problems, namely that of estimating a time independent scalar quantity from a number of noisy measurements. From this we move on to consider the case when the quantity to be estimated is a function of time, and then generalise the results to the estimation of a time dependent vector. Finally we indicate how the resulting Kalman filter equations can be applied to an elementary but nevertheless real problem in navigation.

12.2 ESTIMATING A TIME INVARIANT SCALAR

Let x be a scalar quantity that is constant in time. If n measurements of x are available which are corrupted by noise drawn from uncorrelated zero mean Gaussian distributions of equal variance, the best estimate of x is the mean of all the measurements. Denoting the ith measurement by $\hat{y}_i$, and the best estimate after n measurements by $\hat{x}_n$, we thus have that

$$\hat{x}_n = \frac{1}{n} \sum_{i=1}^{i=n} \hat{y}_i.$$

If a further measurement, denoted by $\hat{y}_{n+1}$ becomes available the new best estimate obviously follows as

$$\hat{x}_{n+1} = \frac{1}{n+1} \sum_{i=1}^{i=n+1} \hat{y}_i.$$

One can express this new estimate in terms of the previous one, to obtain the following update equation

$$\hat{x}_{n+1} = \frac{\left[\dfrac{n}{n+1}\right] \hat{x}_n + \hat{y}_{n+1}}{n+1}$$

This equation can be rearranged to yield

$$\hat{x}_{n+1} = \hat{x}_n + \frac{1}{n+1} [\hat{y}_{n+1} - \hat{x}_n].$$

We may interpret this equation as follows. If we have an estimate $\hat{x}_n$ of a quantity x we may use a new estimate of x to update the estimate $\hat{x}_n$ to an improved estimate, $\hat{x}_{n+1}$. The update is achieved by adding to $\hat{x}_n$ the difference between the new measurement and the previous estimate scaled by a weighting factor. This idea of updating an estimate in the light of a new measurement to form an improved estimate is central to the concept of Kalman filtering.

12.3 THE TIME VARYING CASE

As before, we consider a scalar quantity x, which is observed at discrete points in time. Rather than being time independent we now assume that the change in x from the time of the nth to $(n+1)$th measurement can be modelled by an equation of the form

$$x_{n+1} = \Phi_n \, x_n + w_n$$

where w_n is drawn from a zero mean distribution. The w_n effectively allows for the uncertainty in our model of the time variation of x as expressed by Φ_n, and is known as process noise.

If we have an estimate of x_n denoted by $\hat{x}_n$ (+), the best estimate of x_{n+1}, which we will denote by $\hat{x}_{n+1}$ (-) is obviously given by

$$\hat{x}_{n+1} \ (-) = \Phi_n \, \hat{x}_n \ (+) \ .$$

If $\hat{x}_n$ (+) is characterized by a variance P_n (+) $= E \{(\hat{x}_n \ (+) - x_n)^2\}$, where $E \{...\}$ denotes the expectation operator, the variance of $\hat{x}_{n+1}$ (-) will have a contribution of $\Phi^2_n \, P_n$ (+) from the original uncertainty in $\hat{x}_n$ (+). It will have a further contribution of $Q_n = E \{w_n^2\}$ from the unknown w_n. Denoting the variance of $\hat{x}_{n+1}$ (-) by P_{n+1} (-), we thus have that

$$P_{n+1} \ (-) = \Phi^2_n \, P_n \ (+) + Q_n.$$

Suppose a measurement is made of the scalar y_{n+1} which is linearly related to x_{n+1} by

$$y_{n+1} = H_{n+1} \, x_{n+1}.$$

Let the actual measurement be

$$\hat{y}_{n+1} = H_{n+1} \, x_{n+1} + v_{n+1},$$

where v_{n+1} is drawn from a zero mean distribution and represents measurement noise. If we denote the variance of this measurement by

$$R_{n+1} = E \{v_{n+1}^2\}$$

we may, as in the time invariant case, seek an updated estimate of $\hat{x}_{n+1}$ (-), which we denote by $\hat{x}_{n+1}$ (+), of the form

$$\hat{x}_{n+1} \ (+) = \hat{x}_{n+1} \ (-) + K_{n+1} \ [\hat{y}_{n+1} - H_{n+1} \, \hat{x}_{n+1} \ (-)] \ .$$

We seek the Kalman Gain K_{n+1} which minimizes the variance of $\hat{x}_{n+1}$ (+) which we will denote by P_{n+1} (+), and hence gives rise to an updated estimate with the minimum possible uncertainty.

If we use the notation Δ to represent a change in a variable, we have that

$$\Delta \hat{x}_{n+1} (+) = (1 - K_{n+1} H_{n+1}) \Delta \hat{x}_{n+1} (-) + K_{n+1} \Delta \hat{y}_{n+1}.$$

Hence, assuming that the error associated with $\hat{x}_{n+1}(-)$ is independent of the error associated with $\hat{y}_{n+1}$, we have that

$$P_{n+1} (+) = (1 - K_{n+1} H_{n+1})^2 P_{n+1} (-) + K^2_{n+1} R_{n+1}.$$

By differentiating this expression with respect to K_{n+1} and setting the result equal to 0, we may solve for the value of K_{n+1} that minimizes $P_{n+1} (+)$.

We find

$$K_{n+1} = \frac{P_{n+1} (-) H_{n+1}}{H^2_{n+1} P_{n+1} (-) + R_{n+1}}$$

It then follows that, with this value of K_{n+1},

$$P_{n+1} (+) = (1 - K_{n+1} H_{n+1}) P_{n+1} (-).$$

The complete update procedure is summarized in Fig. 1. The equations presented there are those of the discrete Kalman filter for the case of scalar variables.

12.4 THE VECTOR KALMAN FILTER

We may generalise the variables in the last section from scalars to vectors and matrices. We are now seeking to estimate all of the components of a column vector $\underline{x}$, which is known as the state vector. This is assumed to vary between the nth and (n+1)th measurement according to the propagation equation.

$$\underline{x}_{n+1} = \underline{\Phi}_n \underline{x}_n + \underline{w}_n,$$

where $\underline{\Phi}_n$ is known as the transition matrix, and $\underline{w}_n$ is a process noise vector.

The measurement state relationship is now assumed to be of the form

$$\underline{y}_{n+1} = \underline{H}_{n+1} \underline{x}_{n+1}.$$

Here $\underline{H}_{n+1}$ is the measurement matrix and $\underline{y}_{n+1}$ a column vector of measurements. The Kalman gain is now also a matrix, and all variances become covariance matrices. For example

$$\underline{P}_n (+) = E \{ (\underline{\hat{x}}_n (+) - \underline{x}_n) (\underline{\hat{x}}_n (+) - \underline{x}_n)^T \},$$

where ^{T} denotes the operation of transpose (interchange rows for columns). Such matrices are always symmetric and contain mean square errors along the diagonal. Off diagonal elements are measures of the correlations between the errors associated with different components of the state vector estimate.

By analogy with the scalar case we may propagate an estimate $\hat{\underline{x}}_n(+)$ of the state vector to an estimate $\hat{\underline{x}}_{n+1}(-)$ by

$$\hat{\underline{x}}_{n+1}\,(-)\,=\,\underline{\Phi}_n\,\hat{\underline{x}}_n\,(+).$$

Neglecting for the moment the effects of process noise we then have that

$$\Delta\hat{\underline{x}}_{n+1}\,(-)\,=\,\underline{\Phi}_n\,\Delta\hat{\underline{x}}_n\,(+)$$

and hence that

$$\Delta\hat{\underline{x}}_{n+1}\,(-)\,\Delta\hat{\underline{x}}_{n+1}{}^T\,(-)\,=\,\underline{\Phi}_n\,\Delta\hat{\underline{x}}_n\,(+)\,\Delta\hat{\underline{x}}_n{}^T\,(+)\,\underline{\Phi}_n{}^T.$$

From this it follows that

$$\underline{P}_{n+1}\,(-)\,=\,\underline{\Phi}_n\,\underline{P}_n\,(+)\,\underline{\Phi}_n{}^T.$$

Re-introducing the effects of the process noise, it then follows that

$$\underline{P}_{n+1}\,(-)\,=\,\underline{\Phi}_n\,\underline{P}_n\,(+)\,\underline{\Phi}_n{}^T\,+\,\underline{Q}_n.$$

The derivation of the update equation proceeds by analogy with the scalar case. We seek an updated estimate of $\hat{\underline{x}}_{n+1}\,(-)$, which we denote by $\hat{\underline{x}}_{n+1}\,(+)$, of the form

$$\hat{\underline{x}}_{n+1}\,(+)\,=\,\hat{\underline{x}}_{n+1}\,(-)\,+\,\underline{K}_{n+1}\,[\hat{\underline{y}}_{n+1}\,-\,\underline{H}_{n+1}\,\hat{\underline{x}}_{n+1}\,(-)].$$

Here $\underline{K}_{n+1}$ is the Kalman gain matrix, and $\hat{\underline{y}}_{n+1}$ the vector of actual measurements. The Kalman gain is chosen to minimise the sum of the diagonal elements of $\underline{P}_{n+1}\,(+)$, otherwise known as the trace of this matrix and denoted by $\mathrm{tr}\,[\underline{P}_{n+1}\,(+)]$.

Once again using the notation Δ to represent a change in a variable, we have that

$$\Delta\hat{\underline{x}}\,(+)\,=\,\Delta\hat{\underline{x}}\,(-)\,+\,\underline{K}\,[\Delta\hat{\underline{y}}\,-\,\underline{H}\Delta\hat{\underline{x}}\,(-)],$$

where, for compactness of notation, we have dropped the common subscript of $_{n+1}$.

Hence

$$\Delta\hat{\underline{x}}\,(+)\,\Delta\hat{\underline{x}}^T\,(+)\,=\,\{\Delta\hat{\underline{x}}\,(-)\,+\,\underline{K}\,[\Delta\hat{\underline{y}}\,-\,\underline{H}\,\Delta\hat{\underline{x}}\,(-)]\}$$
$$\{\Delta\hat{\underline{x}}^T\,(-)\,+\,[\Delta\hat{\underline{y}}^T\,-\,\Delta\hat{\underline{x}}^T\,(-)\,\underline{H}^T]\,\underline{K}^T\}.$$

Assuming that the errors associated with the components of $\hat{\underline{x}}(-)$ are independent of the measurement errors associated with $\hat{\underline{y}}$, we therefore have that

$$\underline{P}\,(+)\,=\,\underline{P}\,(-)\,-\,\underline{P}\,(-)\,\underline{H}^T\,\underline{K}^T\,+\,\underline{K}\,\underline{R}\,\underline{K}^T\,-\,\underline{K}\,\underline{H}\,\underline{P}\,(-)\,+\,\underline{K}\,\underline{H}\,\underline{P}\,(-)\,\underline{H}^T\,\underline{K}^T.$$

The variation in $\underline{P}$ (+) brought about by a variation in $\underline{K}$ then follows as

$$\Delta\underline{P}\ (+) = -\ \underline{P}\ (-)\ \underline{H}^T\ \Delta\underline{K}^T + \Delta\underline{K}\ \underline{R}\ \underline{K}^T + \underline{K}\ \underline{R}\ \Delta\underline{K}^T$$

$$-\ \Delta\underline{K}\ \underline{H}\ \underline{P}\ (-) + \Delta\underline{K}\ \underline{H}\ \underline{P}\ (-)\ \underline{H}^T\ \underline{K}^T + \underline{K}\ \underline{H}\ \underline{P}\ (-)\ \underline{H}^T\ \Delta\underline{K}^T.$$

Since $\underline{P}^T$ (-) $= \underline{P}$ (-) and $\underline{R}^T = \underline{R}$, this expression can be written in the form

$$\Delta\underline{P}\ (+) = \Delta\ \underline{K}\ [\underline{R}\ \underline{K}^T - \underline{H}\ \underline{P}\ (-) + \underline{H}\ \underline{P}\ (-)\underline{H}^T\ \underline{K}^T]$$

$$+\ (\Delta\underline{K}\ [\underline{R}\ \underline{K}^T - \underline{H}\ \underline{P}\ (-) + \underline{H}\ \underline{P}\ (-)\ \underline{H}^T\ \underline{K}^T])^T.$$

Thus

$$\Delta\text{tr}\ [\underline{P}\ (+)] = \text{tr}\ [\Delta\underline{P}\ (+)] = 2\ \text{tr}\ [\Delta\underline{K}\ (\underline{R}\ \underline{K}^T - \underline{H}\ \underline{P}\ (-) + \underline{H}\ \underline{P}\ (-)\ \underline{H}^T\ \underline{K}^T)].$$

We seek the value of $\underline{K}$ such that $\Delta\ \text{tr}\ [\underline{P}\ (+)] = 0$ for all $\Delta\underline{K}$. That is, the value of $\underline{K}$ such that $\text{tr}\ [\underline{P}\ (+)]$ is minimized. We thus require $\underline{K}$ to be the solution to

$$\underline{R}\ \underline{K}^T - \underline{H}\ \underline{P}\ (-) + \underline{H}\ \underline{P}\ (-)\ \underline{H}^T\ \underline{K}^T = \underline{0}.$$

Or

$$\underline{K} = \underline{P}\ (-)\ \underline{H}^T\ [\underline{R} + \underline{H}\ \underline{P}\ (-)\ \underline{H}^T]^{-1}.$$

It then follows that, with this value of $\underline{K}$

$$\underline{P}\ (+) = [\underline{I} - \underline{K}\ \underline{H}]\ \underline{P}\ (-).$$

The resulting equations are presented in Fig. 2 and will be seen to have a close correspondence with those in Fig. 1 for the scalar case.

12.5 A SIMPLE EXAMPLE

A Global Positioning System (GPS) receiver operates by making measurements of "pseudo-range" to each of four satellites by timing the occurrence of certain states of received signals of known transmission time against the receiver's clock. These measurements are of pseudo-range rather than true range because of the (at present) undetermined clock offset. Similarly, measurements are made of "pseudo-range-rate" by measuring the Doppler shifts of the received radio frequency carriers. These are in error from true range-rates because of the clock's frequency error. These measurements, together with ephemeris data from each satellite which detail its motion, then enable the observer's position and velocity to be determined, together with the time offset and frequency bias of his clock. This determination may be undertaken using a Kalman filter.

The smallest number of elements that can be used for the state vector, $\underline{x}$, of such a filter is eight. These are the three user's position co-ordinates (x, y, z), three user's velocity components $(\dot{x}, \dot{y}, \dot{z})$,

$$\hat{x}_{n+1} (-) = \Phi_n \, \hat{x}_n (+)$$

$$P_{n+1} (-) = \Phi_n^2 \, P_n (+) + Q_n$$

$$K_{n+1} = \frac{P_{n+1} (-) \, H_{n+1}}{H_{n+1}^2 \, P_{n+1}(-) + R_{n+1}}$$

$$\hat{x}_{n+1} (+) = \hat{x}_{n+1}(-) + K_{n+1} \, [\hat{y}_{n+1} - H_{n+1} \, \hat{x}_{n+1} (-)]$$

$$P_{n+1} (+) = (1 - K_{n+1} \, H_{n+1}) \, P_{n+1} (-)$$

Fig.1.   **The Discrete Scalar Kalman Filter**

$$\underline{\hat{x}}_{n+1} (-) = \underline{\Phi}_n \, \underline{\hat{x}}_n (+)$$

$$\underline{P}_{n+1} (-) = \underline{\Phi}_n \, \underline{P}_n (+) \, \Phi_n^T + \underline{Q}_n$$

$$\underline{K}_{n+1} = \underline{P}_{n+1} (-) \, \underline{H}_{n+1}^T \, [\underline{R}_{n+1} + \underline{H}_{n+1} \, \underline{P}_{n+1}(-) \, \underline{H}_{n+1}^T]^{-1}$$

$$\underline{\hat{x}}_{n+1} (+) = \underline{\hat{x}}_{n+1} (-) + \underline{K}_{n+1} \, [\underline{\hat{y}}_{n+1} - \underline{H}_{n+1} \, \underline{\hat{x}}_{n+1} (-)]$$

$$\underline{P}_{n+1} (+) = [\underline{I} - \underline{K}_{n+1} \, \underline{H}_{n+1}] \, \underline{P}_{n+1}(-)$$

Fig.2.   **The Discrete Vector Kalman Filter**

$$\underline{\Phi} = \begin{bmatrix}
1 & 0 & 0 & 0 & t & 0 & 0 & 0 & \frac{t^2}{2} & 0 & 0 \\
0 & 1 & 0 & 0 & 0 & t & 0 & 0 & 0 & \frac{t^2}{2} & 0 \\
0 & 0 & 1 & 0 & 0 & 0 & t & 0 & 0 & 0 & \frac{t^2}{2} \\
0 & 0 & 0 & 1 & 0 & 0 & 0 & t & 0 & 0 & 0 \\
0 & 0 & 0 & 0 & 1 & 0 & 0 & 0 & t & 0 & 0 \\
0 & 0 & 0 & 0 & 0 & 1 & 0 & 0 & 0 & t & 0 \\
0 & 0 & 0 & 0 & 0 & 0 & 1 & 0 & 0 & 0 & t \\
0 & 0 & 0 & 0 & 0 & 0 & 0 & 1 & 0 & 0 & 0 \\
0 & 0 & 0 & 0 & 0 & 0 & 0 & 0 & 1 & 0 & 0 \\
0 & 0 & 0 & 0 & 0 & 0 & 0 & 0 & 0 & 1 & 0 \\
0 & 0 & 0 & 0 & 0 & 0 & 0 & 0 & 0 & 0 & 1
\end{bmatrix}$$

Fig. 3. The Form of the Transition Matrix

(The matrix appropriate to the eight state model is the upper left portion bounded by the dotted lines) the user's clock offset (r) and the fractional frequency bias of the master oscillator $(\Delta f/f)$. The next level of complexity involves an eleven state filter, the three additional states being the user's acceleration components $(\ddot{x}, \ddot{y}, \ddot{z})$. The state vector may therefore be written

$$\underline{x} = (x\ y\ z\ r\ \dot{x}\ \dot{y}\ \dot{z}\ \Delta f/f\ |\ \ddot{x}\ \ddot{y}\ \ddot{z})^{\mathsf{T}},$$

where the difference between the eight and eleven state model is indicated by the vertical line. The simplest model that can be used for the variation of $\underline{x}$ with time is that of linear motion, together with a constant frequency bias for the master oscillator. The transition matrix for an elapsed time t is then as shown in Fig. 3. In this model the user's acceleration components (jerk components in the case of the eleven state filter) and the rate of change of frequency bias have been neglected. Allowance may be made for these unmodelled parameters by an appropriate choice of the process noise covariance matrix $\underline{Q}$.

The measurement vector, $\underline{y}$, has as components observations of pseudo-ranges and pseudo-range-rates to the satellites. An estimate of the associated measurement covariance matrix can be computed from a knowledge of the signal-to-noise ratios pertaining to the measurements.

If we start with an initial estimate for the state vector, together with a measure of its uncertainty as characterised by a covariance matrix, we may update the state vector estimate in response to the measurements using the vector Kalman filter equations discussed above with one modification. This relates to the fact that the measurement vectors are not linearly related to the state vector. That is, rather than

$$\underline{y}_{n+1} = \underline{H}_{n+1}\ \underline{x}_{n+1}$$

we have

$$\underline{y}_{n+1} = \underline{h}\ (\underline{x}_{n+1}).$$

where $\underline{h}\ (\underline{x}_{n+1})$ is a non-linear function of $\underline{x}_{n+1}$. For example, the pseudo-range to a particular satellite is the true range plus a correction due to the clock error. The true range, d, is given by a non-linear expression of the form

$$d = \{(x-x_s)^2 + (y-y_s)^2 + (z-z_s)^2\}^{\frac{1}{2}},$$

where $(x_s,\ y_s,\ z_s)$ is the position of the satellite.

In order to overcome this problem we use what is known as an <u>Extended Kalman Filter.</u> The full non linear relationship is used to compute the measurement prediction, so that the state vector estimate update equation takes the form

$$\underline{\hat{x}}_{n+1}\ (+) = \underline{\hat{x}}_{n+1}\ (-) + \underline{K}_{n+1}\ [\underline{\hat{y}}_{n+1} - \underline{h}\ (\underline{\hat{x}}_{n+1}\ (-))].$$

Rather than an exact equation relating changes in variables we instead consider the following approximate form

$$\Delta \hat{\underline{x}}_{n+1}(+) \approx \Delta \hat{\underline{x}}_{n+1}(-) + \underline{K}_{n+1}\left[\Delta \hat{\underline{y}}_{n+1} - \frac{\partial \underline{h}(\hat{\underline{x}}_{n+1}(-))}{\partial \hat{\underline{x}}_{n+1}(-)} \Delta\hat{\underline{x}}_{n+1}(-)\right]$$

Here the notation

$$\frac{\partial \underline{h}(\hat{\underline{x}})}{\partial \hat{\underline{x}}}$$

is used for a matrix of first order Taylor expansion coefficients.

The element in the ith row and jth column is given by

$$\frac{\partial[\underline{h}(\hat{\underline{x}})]_i}{\partial[\hat{\underline{x}}]_j}$$

where $[\cdot\cdot]_i$ denotes the ith component of a column vector.

This is identical to the form of the exact equation appropriate to the case when the measurement vector is a linear function of the state vector, except that $\underline{H}_{n+1}$ has been replaced by the matrix of Taylor expansion coefficients, $\dfrac{\partial \underline{h}(\hat{x}_{n+1})}{\partial \hat{\underline{x}}_{n+1}}$.

The computation of the optimum Kalman gain therefore proceeds as in the linear case except that the matrix of Taylor expansion coefficients is used in place of $\underline{H}_{n+1}$.

The GPS receiver provides navigational data as a function of time by continuing to update the state vector estimate in the light of subsequent measurements. At any instant the Kalman filter effectively smooths over many noisy measurements to provide the best possible estimates of the user's position and velocity (and, in the case of the eleven state filter, acceleration).

Although we have described one of the simplest versions of a Kalman filter that can be used to process the measurements made by a GPS receiver, it has been shown to yield excellent navigational performance during trials. It therefore represents an elementary example of the application of Kalman filtering to real world problems.

12.6 FURTHER READING

For further information about Kalman filters and example applications the interested reader is referred to the book edited by Gelb[1]. That by Bierman[2] discusses methods for ensuring the stability of implementation of Kalman filters and gives coded examples of key algorithms.

12.7 ACKNOWLEDGEMENT

The author thanks the Directors of STC Technology Ltd (STL) for permission to publish this tutorial paper. Work at STL on GPS user equipment has been supported and sponsored by the Procurement Executive, Ministry of Defence (Royal Aerospace Establishment).

12.8 BIBLIOGRAPHY

[1]   Gelb, A.(Ed), "Applied optimal estimation", MIT Press, 1974.

[2]   Bierman, G.J., "Factorization methods for discrete sequential estimation", Academic Press, 1977.

Chapter 13

Implementation of digital control algorithms

P. A. Witting

13.1 INTRODUCTION

The application of conventional 8 and 16 bit microcomputers to control systems is now well established. Such processors have general purpose (usually Von Neumann) architectures which make them applicable to a wide range of tasks, though not remarkably efficient in any. In control applications such devices may pose problems such as inadequate speed, difficulties with numerical manipulation and relatively high cost for the completed system. This latter being due to both the programming effort and the cost of the peripheral hardware. These problems may be overcome by the design of specially tailored architectures, provided that there is a sufficient volume of production to carry the overheads inherent in this approach. Special purpose I/O processors and signal processors are examples of applications where dedicated design has been successfully applied.

Digital Signal processors are designed to implement an algorithm of the form

$$K \prod_{m=1}^{n} \left[\frac{b_{0m} + b_{1m} z^{-1} + b_{2m}z^{-2}}{1 + a_{1m}z^{-1} + a_{2m}z^{-2}} \right] \qquad 13.1$$

Single loop digital controllers may also be written in this form although they are usually of a much lower order than algorithms used for signal processing. The use of digital signal processors (DSP) for controller realisation overcomes some limitations associated with controller design using microprocessors. Programmable DSPs use architectures and dedicated arithmetic circuits which provide high resolution and high speed arithmetic, making them ideally suited for use as controllers.

Certain signal processors may have particular advantages or limitations. For instance certain devices can only implement non-recursive structures while some will provide considerable opportunity for programming additional functions. Factors which need to be considered in this context include

   (i)   Ability to implement adaption
   (ii)   Ability to linearise measured signals
   (iii)   Ability to communicate with other devices

   This chapter is concerned with the design of a controller transfer function in the form of equation 13.1 and various practical considerations such as sample rate selection, numerical accuracy etc.

13.2   CONTROLLER DESIGN VIA CONTINUOUS DOMAIN PROTOTYPES

   A large body of literature exists which enables a continuous domain controller to be designed to meet various performance specifications. This methodology may be carried over into the discrete-time domain provided that certain restrictions are observed.

   Consider figure 13.1

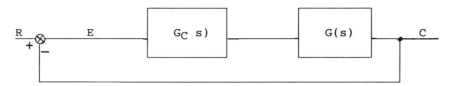

(a)   Continuous time system

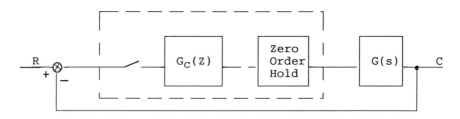

(b)   Discrete-time system

Figure 13.1

   The zero order hold in 13.1(b) reconstructs a continuous (piecewise constant) waveform from the sequence of samples provided by the controller $G_C(z)$. The two systems shown in Figure 13.1 may be considered equivalent if the output of the sections enclosed by a dotted line are identical for identical inputs.

In practice this requires that a fairly high sample rate be used (about ten times the requisite rate) and that $G_c(z)$ be chosen to replicate the action of $G_c(s)$. A number of techniques exist to achieve this and three will be discussed in the following sections, with reference to the example of a lag/lead controller of the form

$$\frac{V}{E}(s) = K \ \frac{1 + s\gamma}{1 + s\alpha\gamma} \qquad\qquad 13.2$$

$$\gamma = \text{time constant}$$

$$\alpha > 1 \text{ for lag compensators}$$

$$\alpha < 1 \text{ for lead compensators}$$

13.2.1 Discrete Simulation of Differential Equations

The controller, 13.2, may be re-written in differential equation form thus

$$v(t) + \alpha\gamma \ \frac{dv(t)}{dt} = Ke(t) + K\gamma \ \frac{de(t)}{dt} \qquad 13.3$$

A simple approximation procedure enables the derivative terms to be computed in terms of the current and previous samples:-

$$\left. \frac{de(t)}{dt} \right|_{t=nT} = \frac{e(nT) - e([n-1]T)}{T} = \frac{e_n - e_{n-1}}{T} \qquad 13.4$$

where T = sampling period

Substituting 13.4 into 13.3 for each of the differential terms (with appropriate changes of variable gives

$$V + \alpha\gamma \left[\frac{V_n - V_{n-1}}{T} \right] = Ke_n + K\gamma \left[\frac{e_n - e_{n-1}}{T} \right]$$

ie

$$V_n \left[1 + \frac{\alpha\gamma}{T} \right] = K \left[1 + \frac{\gamma}{T} \right] e_n - K \frac{\gamma}{T} e_{n-1} + \frac{\alpha\gamma}{T} V_{n-1}$$

$$V_n = e_n K \left[\frac{T + \gamma}{T + \alpha\gamma} \right] - e_{n-1} \left[\frac{K\gamma}{1 + \alpha\gamma} \right] + V_{n-1} \left[\frac{\alpha\gamma}{1 + \alpha\gamma} \right]$$

13.5

This method may readily be extended to compensators with higher order terms in s. For instance

$$s^2 E(s) \equiv \frac{d^2 e(t)}{dt^2}$$

$$\frac{d^2 e(t)}{dt^2} = \frac{\left[\frac{de(t)}{dt} \right]_{t=nT} - \left[\frac{de(t)}{dt} \right]_{t=(n-1)T}}{T}$$

$$= \frac{\left[\frac{e_n - e_{n-1}}{T} \right] - \left[\frac{e_{n-1} - e_{n-2}}{T} \right]}{T}$$

$$= \frac{e_n - 2e_{n-1} + e_{n-2}}{T}$$

13.6

thus $$\frac{d^2 e(t)}{dt^2} = \left[\frac{1 - 2z^{-1} + z^{-2}}{T} \right] E(z)$$

13.7

And the process may be repeated for higher orders

13.2.2 Bilinear Transform Approach

An alternative method for arriving at a Z-transfer function is by simple s -> z substitution.

strictly $z = \exp(sT)$ $s = \frac{1}{T} Ln(z)$

However, this substitution does not lead to the required polynomial form of z transfer function. To achieve this an approximation to the natural logarithm is used:

$$Ln(z) \approx 2 \left[\frac{1 - z^{-1}}{1 + z^{-1}} \right]$$

hence

$$s \approx \frac{2}{T} \left[\frac{1 - z^{-1}}{1 + z^{-1}} \right] \qquad 13.8$$

and the substitution is used directly, so that the simple compensator is transformed:-

$$\frac{V}{E}(s) = K \left[\frac{1 + s\gamma}{1 + s\alpha\gamma} \right]$$

13.9

$$V(z) = E(z) \frac{1 + 2 \frac{\gamma}{T} \left[\frac{1 - z^{-1}}{1 + z^{-1}} \right]}{1 + 2 \frac{\alpha\gamma}{T} \left[\frac{1 - z^{-1}}{1 + z^{-1}} \right]} K$$

$$= E(z) \frac{\left[1 + 2 \frac{\gamma}{T} \right] + \left[1 - 2 \frac{\gamma}{T} \right] z^{-1}}{\left[1 + 2 \frac{\alpha\gamma}{T} \right] + \left[1 - 2 \frac{\alpha\gamma}{T} \right] z^{-1}} K \qquad 13.10$$

ie $\quad V(z) \left[1 + 2 \frac{\alpha\gamma}{T} \right] + \left[1 - 2 \frac{\alpha\gamma}{T} \right] z^{-1} =$

13.11

$$E(z) \; K \left[\left[1 + 2 \frac{\gamma}{T} \right] + \left[1 - 2 \frac{\gamma}{T} \right] z^{-1} \right]$$

or

$$V_n \left[1 + 2\frac{\alpha\gamma}{T} \right] + V_{n-1} \left[1 - 2\frac{\alpha\gamma}{T} \right] =$$

$$e_n \left[1 + 2\frac{\gamma}{T} \right] K + e_{n-1} \left[1 - 2\frac{\gamma}{T} \right] K$$

13.12

$$V_n = e_n \left[\frac{T + 2\gamma}{T + 2\alpha\gamma} \right] K + e_{n-1} \left[\frac{T - 2\gamma}{T + 2\alpha\gamma} \right] K - V_{n-1} \left[\frac{T - 2\alpha\gamma}{T + 2\alpha\gamma} \right]$$

13.13

This transformation, known as the Bilinear z-Transform, is only strictly valid for sT<<1. In fact if this limitation is violated the frequency axis "warped" according to the formula.

$$w' = \frac{2}{T} \ \mathrm{Tan} \left[\frac{wT}{2} \right]$$

13.14

where w is the "designed" frequency and w'is the "achieved" frequency.

Thus the corner frequencies of compensators which violate the wT<<1 rule have to be pre-warped to account to this.

13.2.3 Hold Invariant Approach

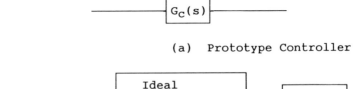

(a)   Prototype Controller

(b) discrete equivalent

Figure 13.2

The two systems shown in Figure 13.2 will produce identical outputs for identical inputs. This occurs because the sampling switch in Figure 13.2b is followed by an <u>ideal</u> data reconstructor. Therefore the input signal to G_C (s) is the same in both cases. However it is possible to find a z transform $G_C'(z)$ for the discrete-time system shown in figure 13.2b and this is the desired discrete controller function.

$$G_C'(z) = \mathcal{Z}[G_H(s)\ G_C(s)] \qquad 13.15$$

Ideal reconstructors do not exist, therefore approximations must be used. In general this means that the reconstructed signal only approximates to the actual signal, and then some signals are reproduced.

Reconstruction	$G_H(s)$	Signal Perfectly Reconstructed
Impulse	1	Pulse
Zero Order Hold	$\dfrac{1 - \exp(-ST)}{S}$	Piecewise Constant
Interpolative first order	$\dfrac{[1 - \exp(-ST)]^2}{TS^2}$	Piecewise Linear

Table 13.1

In practice the zero order hold is used most frequently and gives rise to the following transfer function for a lag/lead controller, using equation 13.15 with equation 13.2

$$\frac{V}{E}(z) = G_C'(z) = \mathcal{Z}\frac{1-\exp(-ST)}{S}\left[\frac{1 + s\ \gamma}{1 + s\ \alpha\gamma}\right] K$$

$$= (1 - z^{-1})\ \frac{K}{\alpha}\ \frac{(1/\gamma + S)}{S(1/\alpha\gamma + S)} \qquad 13.16$$

$$= (1 - z^{-1})\ \frac{K}{\alpha}\left[\frac{\alpha}{s} + \frac{1 - \alpha}{s + 1/\alpha\gamma}\right]$$

$$= (1 - z^{-1})\ \frac{K}{\alpha}\left[\frac{\alpha}{1 - z^{-1}} + \frac{1 - \alpha}{1 - \exp(-T/\alpha\gamma)z^{-1}}\right]$$

$$= \frac{K}{\alpha}\left[\frac{1 + \{\alpha[1 - \exp(-T/\alpha\gamma)] - 1\}z^{-1}}{1 - \exp(-T/\alpha\gamma)z^{-1}}\right] \qquad 13.17$$

or in difference equation terms

$$V_n = \frac{K}{\alpha} e_n + \frac{K}{\alpha} \{\alpha[1 - \exp(-T/\alpha\gamma)] - 1\}e_{n-1} + \exp(-T/\alpha\gamma)V_{n-1} \qquad 13.18$$

It is interesting to note that for small T and $\alpha\gamma >> 1$ equations 13.18, 13.13 and 13.5 become identical.

13.3 DISCRETE DOMAIN DESIGN TECHNIQUES

Designing directly in the z domain removes the restriction which was imposed in Section 13.3, namely that the sampling rate should be sufficiently high to make the system quasi-continuous and the various approximations accurate. Two methods of z domain design of controllers will be presented here

      (i)   Root locus techniques
     (ii)   Direct synthesis

13.3.1 Root Locus Design of Discrete Controllers

In essence this method is the same as that applied in the continuous (s-plane) domain except that the significance of the location of open and closed loop poles is changed. In particular we note the following for the z plane.

    (i)   The stable region is the interior of the unit circle.

    (ii)   Fast modes are characterised by closed loop poles close to the origin.

    (iv)   So far as error performance is concerned the point $z = 1$ takes on the same significance in the z plane as the point $s = 0$ in the continuous domain. In particular, for a unit feedback system the number of open loop poles at $z = 1$ is the system 'type'. Steady-state error performance is governed by 'type' in exactly the same way as in the continuous domain, namely

Zero steady state error to
constant position demand   Type = 1

Zero steady state error to
constant velocity demand   Type = 2

Zero steady state error to
constant acceleration demand   Type = 3

The design procedure thus reduces to the following steps

(a)   Choose the desired z-plane position for the dominant poles to ensure adequate transient performance

(b)   If necessary add open loop (compensator) poles at z = 1 to achieve the desired error performance

(c)   Add open loop (compensator) poles and zeros so that the root locus passes through the location of the desired dominant closed loop poles. As with s-plane design, this amounts to choosing the compensator pole zero positions so that

Σ(angles subtended by closed loop poles at open loop zeros)
-Σ(angles subtended by closed loop poles at open loop poles)

$$= -(2k + 1)\ 180° \qquad 13.19$$

(d)   Calculate the required compensator given to achieve the desired closed loop poles. As with z-plane design this merely amounts to the following graphical calculation

$$K = \frac{\pi(\text{distances from closed poles to open loop poles})}{\pi(\text{distances from closed loop poles to open loop zeros})}$$

13.20

(e)   Verify performance

Steps c, d and e are readily achieved either manually or by computer. It will often be necessary to iterate the five steps above to achieve a final design.

Before proceeding to an example it is necessary to discuss how suitable locations for the closed loop dominant poles may be chosen (stage (b) above), however a satisfactory method of choosing z-plane closed loop pole positions is to chose suitable s-plane closed loop pole positions

$$s_1,\ s_2 = -\ W_n \pm j\ W_d \qquad 13.21$$

$$W_d = W_n\ \sqrt{1 - \zeta^2}$$

using the usual continuous domain criteria, and then map these into the z domain via

$$z = \exp(sT) \qquad 13.22$$

thus $z_1,\ z_2 = \exp(-\zeta W_n T)\ \underline{/\pm\ W_n T} \qquad 13.23$

This turns out to be satisfactory provided that

$$w_d \ T \ < \ 1.2$$

The method may be illustrated by an example :-

A plant has the transfer function

$$G(s) \quad = \quad \frac{\exp(-sT_1)}{(1 + sT_2)} \qquad\qquad 13.24$$

$$T_1 \ = \ 1$$

$$T_2 \ = \ 2$$

A Digital controller D(z) is to be designed to achieve the following specification

(i)   overshoot < 30%

(ii)   5% settling time <9 seconds

(iii)   Zero steady state error for step inputs

The system is shown in figure 13.3

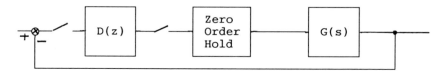

Figure 13.3   Example System

The forward path pulse transfer function is:

$$G(z) \quad = \quad \left[\frac{1 - \exp(-s)}{s} \right] \left[\frac{\exp(-s)}{1 + 2s} \right] \qquad 13.25$$

$$= \quad (1 - z^{-1}) \ z^{-1} \quad \frac{\frac{1}{2}}{s(\frac{1}{2} + s)}$$

$$= \quad \frac{0.4}{z(z- \ 0.6)} \qquad\qquad 13.26$$

The specification requires:

(i)   % overshoot = $100 \exp \left[\dfrac{-\zeta \pi}{\sqrt{1 - \zeta^2}} \right]$ 13.27

$$\geq 0.38$$

(ii)   5% settling time $\approx \dfrac{3}{\zeta W_n}$ 13.28

$$W_n \geq 0.88$$

(iii)   For zero steady state error to step inputs one pole is required at z = 1

Using the equalites we have the desired S-plane continuous poles as

$$S_1, S_2 = -0.33 \pm j\ 0.81$$

and thus the Z-domain poles

$$Z_1, Z_2 = 0.72\ \underline{/\pm\ 46.4°}$$

The original system has no poles at z = 1 thus the compensator must be of the form

$$D(z) \quad = \quad \dfrac{I(z)}{(z - 1)}\ K$$ 13.29

Investigation shows that if I(z) = 1 the required closed loop poles are not on the root locus. Thus we propose a zero at z = a

$$D(z) = K\ \dfrac{z - a}{z - 1}$$ 13.30

and chose a such that the angle-sum criterion (13.19) is satisfied at the desired closed loop poles. This gives the result

$$D(z) \quad = \quad K \left[\dfrac{z - 0.6}{z - 1} \right]$$ 13.31

The gain criterion (13.20) must also be satisfied at the closed loop poles. Graphical (or Computer) calculation leading to the result that the "Root Locus gain" is 0.52.

Hence Root Locus gain = 0.52 = K x 0.4

Hence   K = 1.3

and $D(z) = 1.3 \left[\dfrac{z - 0.6}{z - 1} \right]$

The resulting step response is shown in Figure 13.4

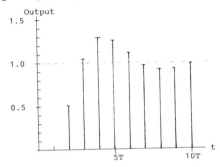

Figure 13.4   Step response of design

13.3.2   Direct Synthesis Approach

This is an algebraic method which assumes that the closed loop transfer function

$\dfrac{C}{n}(z) = K(z)$ can be specified.   (Figure 13.3)

$$K(z) = \dfrac{C}{R}(z) = \dfrac{G(z) \; D(z)}{1 + G(z) \; D(z)} \qquad 13.32$$

where G(z) is the z transfer of the plant-plus-hold

Re-organising 13.32 gives

$$D(z) = \dfrac{1}{G(z)} \cdot \dfrac{K(z)}{1 - K(z)} \qquad 13.33$$

The choice of K(z) is not entirely free.

(a)   It may readily be shown [1] that [1-K(Z)] must be of the form:

$$1- K(Z) = (1-Z^{-1})^{l} \quad (F(Z)$$

if the steady-state output is to equal the input (zero error)

The index, l, governs the kind of input for which the steady-state error will be zero, namely

$$l = 1 \quad \text{Zero error to position inputs}$$
$$l = 2 \quad \text{Zero error to velocity inputs}$$
$$l = 3 \quad \text{Zero error to acceleration inputs}$$

(b) D(z) contains terms which cancel the plant. This is not permissible if G(z) contains zeros outside the unit circle since any attempts at such cancellation by D(z) will be, in practice, inexact and guarantee a segment of the root locus outside the unit circle thus leading to instability. To avoid this F(z) is required to include terms corresponding to zeros of G(z) outside the unit circle

ie $[1 - K(z)] = (1 - z^{-1})l \begin{bmatrix} \text{Zeros of G(z)} \\ \\ \text{Outside unit circle} \end{bmatrix} A(z)$

(c) A(Z) may be of the form

$$\frac{A_n (z)}{A_d (z)}$$

However, if $A_d (z) = 1$ the system has only numerator terms and will settle to its final value within a finite number of samples. The actual number being governed by the order of the $(1 - K(z))$ polynomial.

(d) The expression for K(z) must contain as many delay terms as the plant, G(z). If this is not done the controller D(z) will be non-causal

$$K(z) = z^{-n} I(z)$$

(e) Since D(z) contains terms which cancel the plant it is important that I(Z) contains any poles of G(z) which are outside the unit circle, otherwise instability will be guaranteed, as explained above.

ie $K(z) = z^{-n} \begin{bmatrix} \text{Poles of G(Z)} \\ \\ \text{Outside unit circle} \end{bmatrix} B(Z)$ 13.35

(f) B(z) may be of the form $\frac{B_n(Z)}{B_d(Z)}$

For finite settling $B_d (Z) = 1$

From the above, for finite settling we assume

$$A(z) = a_0 + a_1 z^{-1} - - - - - -$$

$$B(z) = b_0 + b_1 z^{-1} - - - - - -$$

The length of each sequence is chosen so that

(i)   The Polynomials for $K(z)$ and $1 - K(z)$ are of the same order.

(ii)   The number of unknown coefficients a_0, a_1 ..., b_0, b_1... exactly equals the number of equations available.

The total design procedure may be illustrated by an example using the system shown in Figure 13.3 and the plant defined by equation 13.24, as before.

In this instance we chose the following specification:

(i)   Zero error to a step input

(ii)   Finite, minimum settling time

$G(z)$ has no poles, or zeros outside the unit circle, which simplifies matters.

Thus

$$(1 - K(z) = (1 - z^{-1})^1 (a_0 + a_1 z^{-1} ....) \qquad 13.36$$

Also, $G(z)$ may be written

$$\frac{0.4z^{-2}}{1 - 0.6z^{-1}}$$

which shows that the plant has a double unit delay

Hence

$$K(z) = z^{-2} (b_0 + b_1 z^{-1} ...)$$

If we initially minimise the settling time by choosing $a_1 = a_2 = 0$

then $1 - K(z) = (1 - z^{-1})a_0$

and $K(z) = z^{-2}b_0$

Clearly the orders of the polynomials are mismatched and we must add an extra term to $1 - K(z)$

ie $1 - K(z) = (1 - z^{-1})(a_0 - a_1 z^{-1})$ 13.37

$$K(z) = b_0 z^{-2}$$ 13.38

there are 3 equations (equating coefficients) and three unknowns, thus the system is solvable.

$$
\begin{aligned}
1 - a_0 &= 0 \\
a_0 - a_1 &= 0 \\
a_1 &= b_0
\end{aligned}
$$

or

$$
\begin{aligned}
a_0 &= 1 \\
a_1 &= 1 \\
b_0 &= 1
\end{aligned}
$$

Thus $K(z) = z^{-2}$

$$1 - K(z) = (1 - z^{-1})(1 + z^{-1})$$

and

$$D(z) = \frac{1}{G(z)} \frac{K(z)}{1 - K(z)} = \frac{1 - 0.6z^{-1}}{0.4 z^{-2}} \frac{z^{-2}}{(1 - z^{-1})(1 + z^{-1})}$$

$$= 2.5 \frac{(1 - 0.6 z^{-1})}{(1 - z^{-1})(1 + z^{-1})}$$

or $2.5 \dfrac{z(z - 0.6)}{(z - 1)(z + 1)}$

13.4 CONTROLLERS FOR SYSTEMS WITH LARGE TIME DELAYS

A particular problem arises when systems with large time delays are to be controlled. The time delay makes the system prone to instability and severely limits the gain that may be used. Two methods of dealing with this will be described, that due to Dahlin [2] and that due to Smith [3]. In both cases the procedure relies upon a realisation that the system cannot respond faster than the time delay and the controller is configured accordingly. Inevitably the controllers themselves include pure time delay elements and are, therefore, ideal candidates for digital implementation.

Both of the methods presume that the plant transfer function is of the form

$$G(s) = G'(s) \exp(-st)$$ 13.39

where $G'(s)$ is a plant transfer function without any transport lag.

13.4.1 Dahlin's Method

Dahlin's method proposes that the desired closed loop control system should conform to the model

$$K(s) = \frac{\exp(-sT)}{1 + sT_2} \qquad 13.40$$

Where T_1 is chosen to equal the pure time delay present within the plant itself.

The method was originally designed with continuous controllers in mind. In the context of a digital control scheme it is necessary to express the closed loop system as a pulse transfer function. This may, for instance, be achieved using the hold invariant approach (section 13.2.3). The model closed loop transfer function thus becomes

$$K(z) = \frac{1 - \exp(-sT)}{s} \quad \frac{\exp(-sT_1)}{(1 + sT_2)} \qquad 13.41$$

and provided that the time delay in the plant is an integer (N) number of samples we have

$$T_1 = NT$$

$$K(z) = z^{-(N+1)} \left[\frac{1 - \exp(-T/T_2)}{1 - \exp(-T/T_2)z^{-1}} \right] \qquad 13.42$$

The control scheme is exactly as shown in figure 13.3 with G(s) given in equation 13.39.

The required controller is, therefore, specified by the direct synthesis equation (13.33).

The plant is assumed to approximate to the model

$$G(s) = K \frac{\exp(-sT_1)}{(1 + sT_3)} \qquad 13.43$$

and the forward path open-loop pulse transfer function is given by

$$G(z) = \mathcal{Z} \left[\frac{1 - \exp(-sT)}{s} \right] \left[K \frac{\exp(-sT_1)}{1 + sT_3} \right]$$

since $T_1 = NT$

$$G(z) = \frac{z^{-(N + 1)} K[1 - \exp(- T/T_3)]}{1 - \exp(- T/T_3) z^{-1}} \qquad 13.44$$

using 13.44 and 13.42 in the direct synthesis equation 13.33 gives

$$D(z) = \frac{1}{G(z)} \quad \frac{K(z)}{1 - K(z)}$$

$$= \frac{1}{K} \left[\frac{1 - \exp(- T/T_2)}{1 - \exp(- T/T_3)} \right] \left[\frac{1 - \exp(-T/T_3)z^{-1}}{1 - \exp(-T/T_2)z^{-1}} \right.$$

$$\left. - [1 - \exp(- T/T_2)] \; z^{-(N+1)} \right]$$

The philosophy of the Dahlin approach can be extended to more complex plant models and more complex closed loop transfer functions, if required.

13.4.2 Smith Predictor Controllers

The reason why the control of time delay systems is so difficult is that it is physically impossible for the system to respond instantly. Conventionally designed controllers find this slowness difficult to deal with and tend to apply ever more control effort in an attempt to achieve a response. Dahlin's requirement that the closed loop system should incorporate a time delay at least equal to that in the plant is an attempt to sidestep this difficulty.

An alternative approach is to predict what the plant output will be and then use this prediction as feedback to the controller. If the prediction time is equal to the delay time in plant then the controller is, in effect, working with a non-delay system.

If the plant has the form previously outlined

$$G(s) = G'(s) \exp(- sT_1) \qquad 13.45$$

then we can predict the output of the plant T_1 seconds before they actually appear by constructing a model of $G'(s)$ and subjecting it to the same input as the actual plant. The output of this model is fed back and used to form the error signal for the controller. The output of the controller is, of course, the actuation signal and is applied to both the real plant and the model. Figure 13.5 illustrates the arrangement

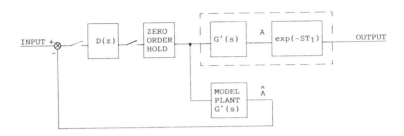

Figure 13.5   Derivation of 'Smith Predictor'

In figure 13.5 there is no arbitrary restriction on the form of G'(s), nor on the kind of control objectives that may be used in the design of D(z) (cf the Dahlin approach).

The system shown in figure 13.5 is open loop so far as the real plant is concerned (but closed loop to the model). Provided the model is exact, and environmental disturbances impact plant and model equally, the signals A or Â remain identical. In practice this ideal is impossible and the model will deviate from reality. To avoid this the loop must be closed around the real plant, although this need not be too 'tight' since it is only trimming out quite small effects. Clearly this deviation between model and system cannot be achieved by comparing A and Â since the former is inaccessible. Similarly comparing Â with the plant is inadmissible since the latter is delayed by T seconds. Therefore the comparison must be made between the plant output and a delayed version of the prediction, as shown in Figure 13.6.

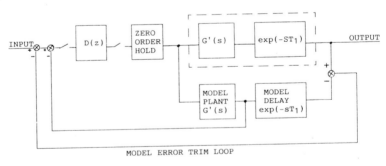

MODEL ERROR TRIM LOOP

Figure 13.6   'Smith Predictor'

In practice the model plant and delay will be realised digitally via digital plant model G'(z) where

$$G'(z) = \left[\frac{1 - \exp(- sT)}{s} \right] G'(s) \qquad\qquad 13.46$$

which gives the system depicted in Figure 13.7.

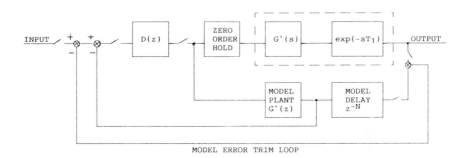

Figure 13.7   Digital Equivalent of Smith Predictor

Figure 13.7 may be re-drawn more neatly as shown in Figure 13.8

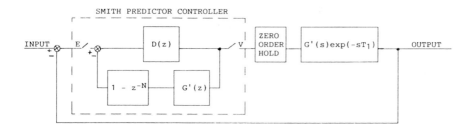

Figure 13.8   Revision of Fig. 13.7

The components filling the box in Figure 13.7 make up the 'Smith Predictor' controller devised by O.J.M. Smith [3].

In summary $D(z)$ is designed to control the undelayed plant $G(s)$ $[G(z)]$ in some desirable way. Then this controller and models of both the plant and delay are combined into a single controller as depicted in Figure 13.7.

The derivation of the Smith Predictor may, alternatively, be obtained mathematically [3] by predicting that the 'ideal' response of the delayed plant is that obtained from an undelayed plant, with the addition of a delay T_1. This leads to a controller equation

$$\frac{V}{E}(z) = \frac{D(z)}{1 + D(z) \, G'(z)(1 - z^{-N})} \qquad 13.47$$

which is exactly the equation of Controller block in Figure 13.7.

Although the Smith Predictor is an attractive proposition stability and performance problems may arise if the model plant and/or delay are in error. Furthermore, it is a relatively complex structure when compared to the other controllers presented in this chapter.

13.5   CONTROLLER ENHANCEMENTS

Where a DSP device is capable of being programmed, additional enhancements may be incorporated. Some possibilities are discussed briefly below.

13.5.1   Linearisation

Most process control systems include significant non-linearities whose presence may cause degraded performance. Typical of such non-linearities are those occurring in flow measurement (square root) and temperature measurement.

By applying an inverse non-linear law to the smoothed, measured input a significant reduction in the effect of the non-linearity may be achieved.

There are a number of ways in which the correction may be applied.

(i)   The correction law may be expressed mathematically using whatever functions are required. This method is often rather wasteful of computer time and is totally impossible if the relevant functions are not available.

(ii)   A lookup table may be constructed. This simply represents the x - y co-ordinates of the inverse function. It is very wasteful of computer memory but is the quickest to execute.

(iii)   A lookup table may be constructed with very few points, intermediate values being determined by interpolation. This represents a compromise solution.

13.5.2 Adaption

A full adaptive control scheme is outside the scope of this chapter. However, simple schemes in which the control algorithm or the control parameters are switched according to the state of one or two easily measured plant parameters are readily implemented with digital controllers.

Such schemes are termed "programmed adaption" and can provide very worthwhile improvements in plant performance. One problem that must be solved in such schemes is that of providing "bumpless" transfer from one control algorithm to the next.

13.6 SAMPLE RATE SELECTION

The sampling rate employed by a digital controller is an important parameter. The choice is governed by a number of considerations.

(i)   An excessively high sampling rate imposes a heavy burden on the computer.

(ii)   A high sampling rate makes the simple design methods more exact.

(iii)   A high sampling rate aggravates the problems caused by derivative action with noisy signals.

(iv)   The sampling rate may, for some design techniques, need to be selected so that the closed loop system poles are in an acceptable position in the z-plane (See Section 13.3.1).

(v)   The sampling rate must satisfy Shannon's law.

(vi)   Excessively high sampling rates aggravate the problems of finite-precision arithmetic.

Practical considerations relating to the computation hardware also need to be considered. An 8-bit microcomputer programmed in interpreted BASIC can achieve a sampling interval of about 140 ms or slower. The same machine using a compiled language would be capable of achieving sampling intervals down to about 30 ms. Modern 16/32 bit processors using a compiled language might achieve a sampling interval as short as 1 ms. To achieve faster sampling rates than these requires special purpose hardware. Typically the Plessey MS2014 processor can achieve sampling periods as low as 16 µS for a 2nd order controller while the Inmos A100 can implement a 16th order controller with sampling periods down to 400ns (See Chapter 14).

Clearly rule (v) imposes a lower limit on the sampling rate, all of the other rules (except ii) provide some "fuzzy" guidance as to the upper limit.

Rule (iii) may be explained thus:

$$\frac{de}{dt} \approx \frac{e_n - e_{n-1}}{T} \qquad\qquad 13.48$$

but the measured values of e are in error due to noise by an amount $\leq \varepsilon$ (say). Thus the calculated value of de/dt is

$$\frac{de}{dt} = \frac{(e_n \pm \varepsilon) - (e_{n-1} \pm \varepsilon)}{T} \qquad\qquad 13.49$$

$$= \text{true derivative} \pm \frac{2\varepsilon}{T}$$

Clearly the error term, $2\varepsilon/T$ gets larger as the sampling period shortens.

Rule (vi) may be illustrated by reference to the difference equation of the low pass filter given by equation 13.50.

$$\frac{e'}{e}(s) = \frac{1}{1 + s\gamma} \qquad\qquad 13.50$$

i.e $\quad e'_n = \dfrac{T}{\gamma} e_n - e'_{n-1} \qquad$ if $T/\gamma \ll 1$

T = sample time

γ = filter time constant

Clearly we have to represent the coefficients one and T/γ. Assuming that we require at least 5% precision on the representation of the relative sizes of the coefficients (which, of course, govern the filter action) this implies a dynamic range in the number systems of

$$1 : \frac{T}{20 \gamma}$$

Hence a careless choice of $T \approx \gamma/100$ leads to a dynamic range of 2000 : 1 requiring 12 bit accuracy (including sign bit). In general rapid sampling implies high arithmetic accuracy, and this matter has been given a more extensive treatment in the literature than is possible in this chapter. [4 - 10]

13.6   COMPUTATIONAL ACCURACY

Both the coefficients and the variables can only be held to a limited accuracy in digital processors. The effects of this limitation are many and varied and have been discussed at length in Chapter 10. It will suffice here to remark that, because controller structures are usually less complex than filters, the problems encountered are less pronounced. Typically 10 to 12 bits will suffice for the realisation of a second order controller [11].

The principal effects which need to be taken account of are

(i)   Inaccurate representation of the input. This can be regarded as equivalent to the addition of 'white' noise to the input of mean square value

$$\frac{E_0{}^2}{12}$$

where E_0 = (Full scale peak - peak range) $\div 2^n$

and n = number of bits in use

(ii)   Each n-bit multiplication results in a 2-n bit answer which is normally immediately truncated or rounded to n-bits for storage. This introduces an equivalent inaccuracy to (i) above and the overall algorithm can be modelled as comprising perfect, multiplications, each in cascade with a 'white' noise source of mean square value

$$\frac{E_0{}^2}{12}$$

the output noise from the system is obviously a complex function of the point of injection and the poles/zeros through which it has to pass to reach the output. The matter has received detailed treatment in the literature [12].

Note that such errors do not occur for cases where the multiplication is by unity or zero.

(iii)   The inaccurate representation of coefficients (coefficient rounding/truncation) will lead to the actual poles/zeros being located in different places to the designed regions. This will inevitably result in a variation from the intended performance and perhaps even instability.

The sensitivity of the ith pole to changes in the Kth coefficient of the transfer function

$$F(z) = \frac{N(z)}{K \sum_{a=1}^{n} b_n z} \qquad \frac{N(z)}{\prod_{p=}^{n} (1 - \frac{z}{z_p})}$$

may be shown to be [13]

$$\frac{\partial b_k}{b_k} \Bigg/ \frac{\partial z_i}{z_i} = \prod_{\substack{p=1 \\ p \neq i}}^{n} (1 - \frac{z_i}{z_p}) \frac{1}{b_k} \frac{1}{(z_i)^k} \qquad 13.51$$

and this may be used to estimate the required coefficient precision.

Other limits to ensure stability alone have been deduced for instance Kaiser [13] gives the following minimum word length

$$\approx 1 + \left[- Ln \left[\frac{5\sqrt{n}}{2^{n+2}} \prod_{i=1}^{n} s_i T \right] \right] \text{bits} \qquad 13.52$$

where s^i are the equivalent continuous domain poles and n is the system order. Clearly a word length 3 or more bits greater than that given by 13.52 would be needed for reasonable performance. It is possible to deduce from 13.52 the following general rules:

Word length α Log $(^1/T)$ $\qquad\qquad$ 13.50a

Word length α n $\qquad\qquad\qquad\qquad$ 13.50b

(iv) As a result of the various inaccuracies it is possible for erroneous results to propagate round an algorithm so that changes in the error in the system do not produce corresponding corrective actions. This leads to a 'dead band' effect which can be very serious in controllers designed to reduce the error to zero. Knowles [14] has derived a bound for the width of such 'dead bands'.

If the algorithm is given by

$$\left[\frac{\displaystyle\sum_{K=0}^{m} b_k \, z^{-k}}{1 + \displaystyle\sum_{1=1}^{n} a_L \, z^{-1}} \right] \qquad 13.51$$

then the dead band is

$$\text{dead band width} \le \pm q \left[(1 + {}^1/_\lambda) + \frac{\mu}{\lambda \displaystyle\sum_{k=0}^{m} b_k} \right] \qquad 13.52$$

when   q   is the maximum size of the roundown
               error

          μ is the number of non-zero, non-unity
               coefficients a_1 , b_k

REFERENCES

(1) SHINNERS S.M.   "Control Systems Design", Chapter 9
                        Section 10 pp401,402, Wiley 1964.

(2) DAHLIN E.B.   "Designing   and   Tuning   Digital
                        Controllers", Instruments & Control
                        Systems, Vol. 42, June 1968, pp77-83.

(3) SMITH, O.J.M.   "Closer Control of Loops with dead-
                        time", Chem. Eng. Prog. Vol. 53 No 5,
                        May 1957, pp217-219.

(4) PERTRAM, J.E.   "The   Effect   of   Quantisation   in
                        Sampled-Feedback   Systems",   1958.
                        Trans. AIEE, 77 pt.2, 177.

(5) SLAUGHTER, J.B.   1964, "Quantisation Errors in Digital
                        Control Systems", Trans IEEE PTGAC,
                        1964, 70.

(6) TSYPKIN, Ya Z.   1960, "An Estimate of the Influence of
                        Amplitude Quantisation of Processes in
                        Digital Automatic Control Systems,
                        "Automatica i Telemkhanika, 1960, 3.

(7) KNOWLES, J.B.   1965, "The Effect of a Finite Word-Length Computer in a Sampled-Data Feedback System", Proc. IEE, 112, 1197.

(8) KNOWLES, J.B.   1965, "Finite Word-Length Effects in
     & EDWARDS, R.   Multi-Rate Direct Digital Control Systems", Proc. IEE, 112, 2376.

(9) KANEKO, T. &   1968, "Round-Off Error of Floating-
     LIU, B.   Point Digital Filters", 6th Annual Allerton Conference.

(10) KNOWLES, J.B.   1966, "Computational Error Effects in a Direct Digital Control Systems", Automatica, 4 7.

(11) WITTING, P.A.   "Digital Controllers for Process Control Applications", in Rees, D., Warwick K (eds) 'Industrial Digital Control Systems (2nd Edition), Peter Peregrinus 1988, pp67-70.

(12) JACKSON, L.B.   "Round off Noise Analysis for Fixed-Point Digital Filters Realised in Cascade or Parallel Form", Trans. I.E.E.E. on Audio and Electro-acoustics, Vol. Au-18, No.2, June 1970, pp. 107-122.

(13) KAISER, J.F.   "Digital Filters", in Kuo, F.F. and Kaiser, J.F., "System Analysis by Digital Computer", Chap. 7, P266, John Wiley and Sons Inc., New York 1966.

(14) KNOWLES, J.B.   1980, "A Simplified Analysis of Computational Errors in a Feedback System Incorporating a Digital Computer" in "The Design and Analysis of Sampled Data Control Systems" - and IERE Seminar held at the Royal Institution 19 Feb. 1980.

Chapter 14

Comparison of DSP devices in control systems applications

P. A. Witting

The use of digital signal processors as controllers offers a number of potential advantages. In particular DSPs incorporate a high degree of parallelism enabling the use of digital control in some very demanding applications. Areas such as robot control, turbogenerators and flight control systems immediately suggest themselves. In this context it is relevant to note that traditional 8-bit microprocessors can execute a 2nd order controller algorithm (including "housekeeping") in about 30 ms while more modern devices are capable of the same computation in 1ms. DSP devices can, by contrast, achieve times as short as 400ns.

Because the architectures are often specifically oriented towards the implementation of appropriate algorithms, the software development time can be significantly reduced and the reduced complexity ought to result in greater reliability. It might be argued that the hardware reliability begins to dominate in such circumstances.

In achieving an applications-specific orientation to the hardware there may well be a reduction in the flexibility available thus reducing the possibility of "value-added" functions such as linearisation and adaption within the controller in the absence of an additional support processor.

The specialisation of the architectures involved may also result in the integration of additional functions into the DSP device thus simplifying the support hardware requirements. Such economics can result in reduced cost, complexity and power requirements for the complete system, with a corresponding improvement in reliability.

Signal processing devices can be divided into two broad groups which we can describe as

   (i)   Specialised Programmable devices

   (ii)   "Silicon Algorithm" devices

The former are characterised by having specialist instruction sets suitable for signal processing. They are more complex to program but retain considerable flexibility. The TMS 320 series and the now obsolete Intel 2920 are examples of this class of device. The latter class typically have a negligible instruction set with the functionality incorporated into the hardware. They are consequently very much simpler to program but have restricted flexibility. Typical examples are the Inmos A100 and the Plessey MS2014 (FAD). The ultimate in this category is, of course, the Application Specific Integrated Circuit (ASIC) class of devices.

The remainder of this chapter contains a survey of four particular devices

Plessey MS2014
Inmos A100
Intel 2920
Texas TMS 320

The first two are examples of "Silicon Algorithm" devices while the latter two are programmable devices. The Intel 2920 is included because, although it is no longer in production, it was the first commercial DSP and is unique in integrating analogue-digital and digital-analogue convertors onto the integrated circuit.

14.1 PLESSEY MS2014

The Plessey MS2014 was originally developed by British Telecom as their Filter and Detect (FAD) circuit for use in digital telephone systems. In contrast to general purpose computing hardware the MS2014 can execute a second order algorithm in between 16µS & 64µS (depending on clock speed). It requires little in the way of support hardware and is capable of operating as an adaptive controller under the supervision of other hardware. Unfortunately, the use of dynamic memory for certain parts of the internal hardware means that the clock speed cannot be reduced. Thus sampling periods between the slowest limit of the MS2014 (64µS) and the fastest limit of simple 8-bit hardware (30mS) cannot be obtained directly. However the addition of some relatively simple circuitry enables the operation of the MS2014 to be extended to arbitrarily low sampling rates.

14.1.1   The Architecture of the Plessey MS2014

The MS2014 [1] [2] is designed to implement the standard second-order digital algorithm:-

$$S \left[\frac{1 + Az^{-1} + Bz^{-2}}{1 - az^{-1} - bz^{-2}} \right] z^{-1} \qquad\qquad 14.1$$

The processing occupies a whole computation cycle thus giving rise to the z^{-1} term.

A Canonic structure is used to implement the processor as shown in Figure 14.1 which represents the logical architecture rather than the detail of the electronics. The coefficients are held with a maximum error of 1.22×10^{-4} and are arranged so that the poles and zeros may be placed anywhere within the unit circle.

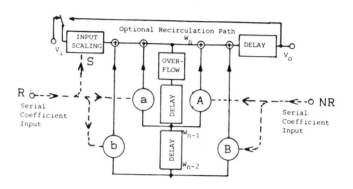

Figure 14.1   Programmer's Model of MS2014

The control structure of the MS2014 permits its output to be recycled to the input so that cascades of second order sections may readily be implemented. The output may be sampled at the same time that it is being fed back and therefore it is possible to implement a tapped cascade of second order sections.

The coefficients of the algorithm are supplied in serial format, as is the data, each time a section is implemented. Although it is usual for the coefficients to remain constant between each instance of a section, it is easy to arrange for these to vary.

Where multiple biquadratic sections are used the intermediate data may be held on-chip (for 8 sections) or off-chip (for other than 8 sections). Because of the serial nature of the data, the storage required is merely a shift register of length 32(N-1) bits where N is the number of stages.

Multiplication and addition within the MS2014 are carried out to full precision before the results are truncated to 16 bits. Thus the multiply-add errors do not build up in the same way as for a typical fixed-point arithmetic implementation in a general purpose computer.

In addition the MS2014 contains special circuitry which detects over-range intermediate results W_n . The variable is immediately zeroed and, while erroneous outputs are produced for a short period, this procedure avoids gross errors entering the feedback paths of the controller algorithm.

Although figure 14.1 shows a "programmer's model" of the MS2014 adequate for general use, the internal architecture used to realise the canonic section is somewhat more complex [3]. Arithmetic is realised with serial units and the results accumulated in shift registers. Shift registers also serve to store the intermediate variables of the section W_{n-1} and W_{n-2} . A recirculation path via a multiplexer permits the MS2014 to capture W_{n-1} at the end of a computation ready for use as W_{n-2} during the next computation. The output of this multiplexer is available at the "DELAY OUT" pin of the MS2014 package, thus giving external circuitry access to it. This fact is utilised to achieve regulation of the sampling rate of the MS2014-based controller. [4]

In practice, the computation of one section occupies three cycles, each of 32 clock periods. During the first cycle the coefficients are loaded in sequence to holding registers for a,b and S via the 'R' (recursive coefficient) input and A & B via the 'NR' (non-recursive coefficient) input. During cycle two, data is entered (or the output re-entered, depending on the setting of the input switch) and the intermediate variable W_n and output are computed. The result is output during cycle three. These operations may proceed independently, thus the coefficients for t = (k+1)T are entered at the same time as the result for t=kT is being calculated and that for t = (k-1)T is being output. In this way, once the system is in operation, each section only occupies one cycle of 32 clock periods.

14.1.2 Implementation Aspects

The practical implementation of MS2014-based controllers requires additional support circuitry such as an ADC, DAC, EPROM and some logic circuitry in the form of SSI and MSI devices (Figure 14.2). This circuitry can for the most part be replaced by a suitable semi-custom VLSI device, and in this way the complexity could be reduced to 4 or 5 packages.

The MS2014 circuit is specified for operation with a clock speed of between 2.0MHz and 0.5MHz. The computation of a single second order section occupies 32 clock periods, giving a sample period of between 16µs and 64µs. The sample period increases proportionally with N, the number of second order sections implemented.

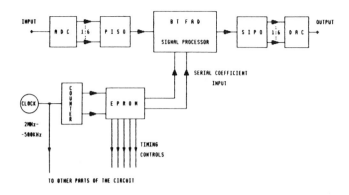

Figure 14.2   MS2014 Support Circuit

It is not possible to reduce the sample frequency simply by lowering the clock frequency since the internal circuitry of the MS2014 employs a considerable amount of dynamic memory, the contents of which would be lost. An alternative has, therefore, to be sought. This may be achieved by forcing the device into a 'hold' state where it repeats the previous calculation continuously. All input data and intermediate results (available at the 'delay out' connection) are held constant. Full details of the method have been reported elsewhere.

It will be observed from Figure 14.2 that the MS2014 requires no additional support processor. However, if adaptive control is to be implemented then some means of both organising and changing the system coefficients will be necessary since the MS2014 does not contain a general purpose processing unit. A conventional 8- or 16-bit microprocessor could be used for this purpose with the EPROM shown in Figure 14.2 becoming shared read/write memory for both the MS2014 and the support processor.

The maximum controller complexity is essentially unlimited although the sampling rate capability falls as the number of second order sections rises, according to the formula

$$fs = \frac{fc}{32N} \qquad 14.2$$

N = Number of second order sections
fc = Clock rate (2MHz to 500 KHz)
fs = Sampling rate < 62.5 KHz

14.2 INMOS A100 PROCESSOR

The A100 processor is a relatively new device which promises to provide extremely high performance enabling control systems with natural frequencies as high as 250KHz to be handled. The device does not implement any general instructions thus a support processor is needed for adaption, however, such a device is required anyway to control and configure the device so this is not an additional burden. The use of static memory means that there is no lower limit to the sampling frequency.

14.2.1 The Architecture of Inmos A100 Processor

The Inmos A100 is a relatively new device which promises to provide extremely high performance. Figure 14.3 shows the internal architecture.

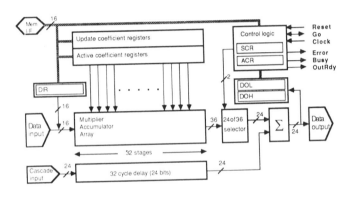

Figure 14.3 Programmer's Model of A100
(Reproduced by kind permission of Inmos Ltd)

The computation section of the device centres around an array of thirty-two multiply-accumulator units with delay units. This section implements a 32-top transversal filter with two banks of coefficient registers, enabling rapid changes of filter characteristic. The output of the array may be added to the delayed data from a "cascade input" port before being output. This feature enables several A100 processors to be cascaded to produce 32N tap (N A100s).

The input and output ports contain logic signals to facilitate the connection of fast ADC and DAC devices.

The highest coefficient accuracy is 3×10^{-5} (16-bit) but a feature of the device is that it may be configured to operate at lower accuracy (8 or 4 bits) but higher speed. With a 20MHz (maximum) clock, at maximum accuracy, the minimum sampling period is 400ns (2.5 MHz). While at

minimum accuracy the sample period can be reduced to 100ns
(10MHz). The coefficient accuracy in use and various other
aspects of the A100 configuration are controlled by a number
of configuration registers. There is no upper limit on the
sampling period achievable by the A100; the sample period
achieved being determined either by the host processor or by
signals applied to the 'Go' terminal.

No overload protection is provided within the A100 and it
is the programmer's responsibility to ensure that scaling is
chosen appropriately. The internal data bus is wide enough
to avoid overload problems and an output selector is
provided to enable the 36-bit results to be appropriately
positioned to provide the 24 bit output.

14.2.2 Implementation Aspects

A typical hardware arrangement for an A100 is shown in
Figure 14.4.

The support microprocessor is always required in order to
permit the coefficient and configuration registers to be
loaded. The availability of this support processor means
that adaptive control is readily implemented without further
hardware by simply adjusting the active coefficient
registers after pre-loading the update registers.

Recursive operation to provide controllers with both
poles and zeros is not possible directly with the
transversal filter structure built into the A100. To
achieve a recursive structure as required by most
controllers two approaches are possible. The most direct

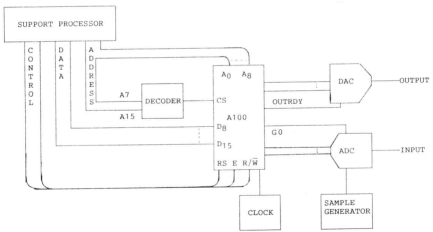

Figure 14.4   Support Hardware for A100

approach is to use two A100s, one as the forward path and
one as the feedback, as shown in figure 14.5. This is

relatively expensive but retains the full speed inherent in
the device. Alternatively the support microprocessor can
have the ADC and DAC connnected directly to it. The support
processor then reads the input and passes it to the A100's
data input register (DIR) via the bus, recording the result
via the data output registers (DOH/DOL). The coefficents
are then changed to those appropriate for the feedback path
and the data passed through the A100 a second time. The

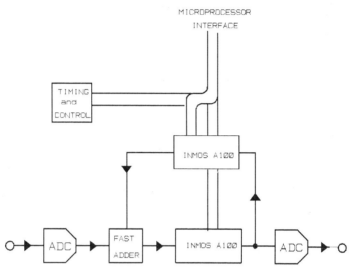

Figure 14.5   Implementation of a recursive
structure with A100 Hardware

combining of the feedback and input signals is handled by
the support processor as is data output via the DAC. The
A100 has a "bank-swopping" mode to facilitate such
operations. This second approach is inevitably slower than
the first as data transfer is limited by the speed of the
support processor.

14.3   INTEL 2920

The Intel 2920 was the first purpose-designed digital
signal processor introduced almost a decade ago as an
"analogue microprocessor". Apart from being the first such
processor the device was unique in integrating both ADC and
DAC onto the integrated circuit. It also contained
programmable instructions enabling logic and general
arithmetic to be carried out.

14.3.1   The Architecture of the INTEL 2920

The programmer's model of the 2920 is shown in Figure
14.6.

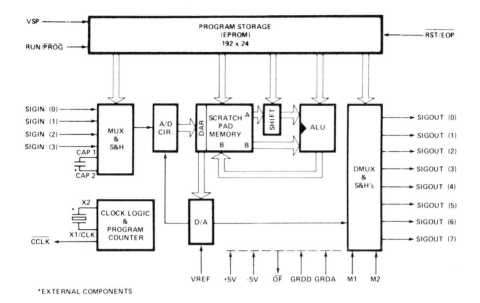

*EXTERNAL COMPONENTS

Figure 14.6   Block Diagram of Intel 2920
(Courtesy of Intel)

The device incorporates a Digital-Analog convertor and an Analog-Digital convertor with 4 multiplexed input and eight multiplexed outputs. All input and output data is passed via a register (DAR) whch forms part of the read-into memory (scratch pad) comprising an array of registers. Two operands, A/B, may be addressed and sent to the ALU for processing. The A operand may be scaled by the programmable shifter and the result is returned to the 'B' address. The arithemtic unit can respond to the following instructions (A' is the output of the scaler).

$$A' + B \qquad -> \qquad B$$
$$B - A' \qquad -> \qquad B$$
$$A' \qquad -> \qquad B$$
$$A' \oplus B \qquad -> \qquad B$$
$$A' . B \qquad -> \qquad B$$
$$|A'| + B \qquad -> \qquad B$$
$$|A'| \qquad -> \qquad B$$

plus a number of conditional versions of the above which are dependent on the contents of DAR or "carry". No multiply instruction is included, this being achieved by a series of shift and add instructions.

Constants are stored and appear as the contents of certain specified "read only" registers. Data is held to an accuracy of 6×10^{-8} .

The program memory consists of 192 24-bit words of EPROM which is incorporated into the integrated circuit. The program is not permitted to have any jumps (no such instructions are provided) and executes until the "End of Program" instruction (EOP) or the last address is executed. The sampling period is, therefore, dependent on program length (which may be padded with NOP instructions). At the maximum clock rate of 10MHz the sample rate will be 13KHz for a full program.

14.3.2   Implementation Aspects

The 2920 is a highly integrated device and requires minimal support apart from power supplies, clocks and signal filters.

As indicated above there is only limited provision for sample rate control. However, this is an important consideration for controller design in order to ensure an adequate response and reasonable controller coefficients. Despite the difficulties Rees [5] has devised a novel method of adjusting the sampling rate of the 2920, a technique which he terms "submultiple sampling".

A program consists of a series of 2920 instructions which are executed sequentially at a rate fixed by the clock frequency. The process allows no internal jumps and the program is of fixed length and execution time, and therefore it is not possible to implement a delay using a subroutine. The sample interval is determined by the time taken for one pass through the program which is given by

$$T = 4 \times \frac{\text{Number of instructions}}{\text{Clock Frequency}} = \frac{4N}{f} \qquad 14.3$$

The factor '4' is due to the program counter which increments its count value every four clock cycles and the maximum number of instructions is 192. The system has a maximum clock frequency of 10MHz, which gives, for one program pass, a maximum period of 0.31 milliseconds. This can be extended using submultiple sampling, a techique which makes use of the conditional load operation. By this method the calculation of the controller equations is made conditional on the sign of a constant, say SUB in the Digital-Analog Register. SUB is decremented by a small constant, say DEL, during each pass of the program and when it becomes negative, the controller equations are evaluated. Using this procedure the effective sampling period is given by

$$T = 4(n+1) \, \frac{N}{f} \qquad 14.4$$

where n = SUB/DEL. For a sampling interval of 8 seconds n is 52082.

One consequence of sub-multiple sampling is that the program length has to be increased. This mainly arises from the fact that instructions which are conditionally executed cannot have I/O operations simultaneously occurring because both operations use the "analogue" field of the instructions. This increase in program length constrains the complexity possible in the remainder of the controller.

As mentioned above, the programme length is severely constrained in the 2920 and this limits controller complexity. In practice it has been found just possible to implement a Smith predictor based on a PID controller and a second order plant model.

14.4 TMS 320

The TMS 320 is probably the most widely used signal processing device currently available. For this reason it is widely discussed in other chapters of this book.

The TMS 320 is available in a number of speeds and in a variety of variants targetted to different cost and performance applications. It is blessed with a number of support chips which simplify hardware design and interfacing.

14.4.1 The Architecture of the TMS 320

There are a number of variations in the Internal Architecture of the TMS 320 family and Figure 14.7 shows the basic "generation 1" - The TMS 32010.

From the diagram it is immediately obvious that the program memory and control unit are entirely separated from the data store and manipulation hardware. Thus address set up etc does not affect the signal processing calculations. There are variations in the capacity of the program memory and its realisation as ROM/EPROM amongst the various members of the family. Program memory can also be expanded "off chip".

The main calculation unit comprises a fairly conventional Arithmetic-Logic unit supported by a 16x16 hardware multiplier and a pair of shifters. This echoes, to some extent The Intel 2920 except that the latter did not have a hardware multiplier. The calculation unit of the TMS 320 is, therefore, well equipped for the repeated multiply-add instructions so commonly used in signal processing.

Data is stored in a separate read-write memory which varies in size across the family of devices. Additionally second generation TMS 320s incorporate an arithmetic unit dedicated to data address calculation. Other versions include serial ports and companding hardware, timers etc.

The instruction set of the TMS 320 resembles that of a typical microprocessor. The 32010 implements 61 instructions and later versions have increased this to 133. The majority of instructions take only one cycle which is 200ns on the basic device and which varies between 225ns and 100ns for other versions. Programming is largely carried out in assembly language and is well supported by various development tools in both hardware and software. A 'C' compiler is available.

The first two generations of the TMS 320 used Integer Arithmetic and required the user to exercise care in the choice of block floating point format since no inherent overload hardware was included. Later versions are available with floating point arithmetic.

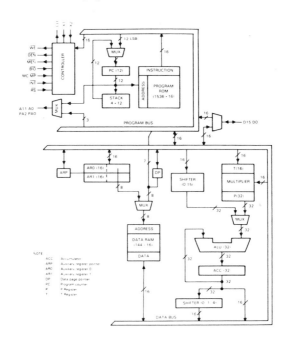

Figure 14.7   Architecture of TMS 32010
(Courtesy of Texas Instruments Ltd)

14.4.2   Implementation Aspects

Since the TMS 320 implements a general instruction set it is possible to programme general routines to carry out adaption etc. However, since programming is likely to take place in assembly language considerable development effort will be required. Certain versions have the facility to interact with support processors via a 'coprocessor port' should this be thought desirable.

Fixed-point arithmetic is implemented in the first two generations with data stored to 16-bit accuracy giving a coefficient and data accuracy of 3×10^{-5}. Obviously multi-precision arithmetic may be used to improve this at the cost of reduced processing speed.

Hardware support for a TMS 320 is fairly simple as shown in Figure 14.8.

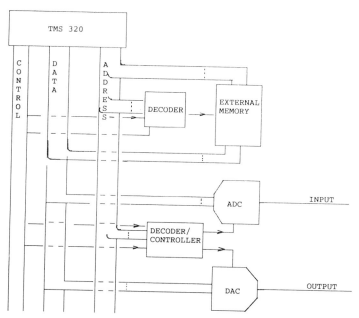

Figure 14.8 TMS 320 Support Hardware

The sampling period achievable by the TMS 320 depends on the cycle time of the device in use and the complexity of the algorithm. For a 200ns device a second order controller can be achieved with a sampling period of 404µs. Under program control this can be increased, as required, without limit. There is no practical limit to the complexity of the controllers that can be implemented, although the sampling periods will increase with program length.

14.5 CONCLUSION

The preceding sections have compared four DSP devices in the context of their application to controller applications. All four have various advantages and disadvantages but all could usefully be applied to high speed control applications in robotics, machinery, flight control etc. The key features of the devices are summarised in table 14.1

CRITERION	Intel 2920	MS2014	Inmos A100	TMS 32010
Hardware	Single Chip & Clock	MS2014 & 11 chips (MSI)	2 x A100 & 11 chips (MSI)	1xTMS32010 & external ADC/DAC
Cost	£105 (one off)	£60 (one off)	£1000 (one off)	£130 (one off)
Maximum Sample Rate Controller	130 Hz – 13 KHz (basic)	4 Hz – 16 KHz depending on clock rate	2.5 MHz (16-bit coeffic-ients)	227 KHz
Lower Limit on sampling rate	No lower limit with "submult-iple sampling"	No lower limit with continuous recalcula-tion mode	No lower limit	No lower limit
Complexity Limit	PID/Smith Predictor with 2nd order model	Essenti-ally unlimited	32nd Order recursive with 2 A100 devices	Essentially unlimited with memory expansion
Coefficient Resolution	6×10^{-8}	1.22×10^{-4}	3×10^{-5}	3×10^{-5}
Overload	NONE	Reset stored values to zero	NONE	No inherent Protection available

Table 14.1   A comparison of the Intel 2920, Plessey MS2014, Inmos A100 and Texas TMS 320 Processors for controller realisation.

REFERENCES

1.   ADAMS, P.F. HARBRIDGE, J.R. MACMILLAN, R.H. 'An MOS integrated circuit for digital filtering and level detection', IEEE Journal of Solid State Circuits, vol SC16 No.3, pp 183-190, 1981.

2.   ADAMS, P.F. MACMILLAN, R.H. 'An LSI circuit for digital filtering and level detection (367): functional description and user guide', British Telecommunications Research Laboratories, 1980.

3.   ADAMS, P.F. 'Filter and detect (FAD) circuit: functional description', BTRL internal communication, August 1978.

4.   WITTING, P.A. 'Control Systems Applications of the British Telecom FAD Processor' 12th IMACS World Congress Paris, July 18-22 1988.

5.   REES, D. 'Controller implementation using a monolithic signal processor', 1985, International journal of Microcomputer Applications, Volume 4, No.3.

Chapter 15

Digital communication systems

E. C. Thomas, E. M. Warrington and T. B. Jones

15.1 INTRODUCTION

Modern communication systems increasingly employ digital techniques for the interchange of information. The growth in the use of digital communications can be attributed to several factors including:-

1) The ability to provide virtually error free communication through the use of digital data encoding techniques.
2) The increasing need for inter-computer data transfer in near real-time.
3) Modern integrated solid state electronics technology allows for the implementation of complex data handling systems at low cost.
4) Digital techniques allow for the handling of data from various sources in a uniform and flexible manner.
5) The availability of wide band channels provided by geostationary satellites, optical fibres and coaxial cables.
6) The use of computer systems as tools for communications.
7) The ability to implement powerful encryption algorithms for security sensitive applications.

Whilst all of these factors have played an important role in the trend towards digital communication systems, it is the economic benefits which have provided the main justification for the development of digital switching and signal processing. This is perhaps most apparent in the rapid adoption of digital transmission for telephony.

Communication systems allow the transmission of information from one point to another. The main elements of such a system are illustrated in Figure 15.1. Information from the data source is encoded before being transmitted over a communication channel to some remote location were it is received, decoded and the original data regenerated. The channel may take many forms (eg. electrical conductors, optical fibres or radio links), however the signal is almost certain to be subject to some degree to distortion, interference and noise. The effect of these may be minimised through the application of a range of appropriate techniques.

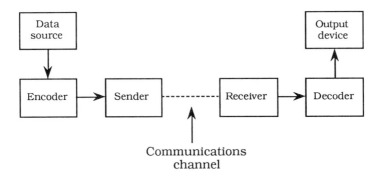

Figure 15.1. Idealised communication system

In digital communications systems the data to be transmitted are represented by a finite set of discrete values. Since the majority of digital electronic equipment utilises the binary number system because of the ease with which the 0 and 1 states can be generated and manipulated, the majority of systems are confined to just two values. This chapter is restricted to such binary systems and in keeping with modern usage the term bit denotes a binary digit.

Digital data may be generated in several ways. Data stored in computers are already held in digital form, eg. text, numbers etc. Other forms of data may readily be represented in digital form (eg. numerical values can be converted to their binary equivalents). Analogue signals can be digitised and transferred as a series of numbers representing the signal level at predetermined intervals provided certain quantisation conditions are met.

15.2 DIGITAL SOURCE CODING

15.2.1 Number codes

Numerical information can readily be expressed in a form suitable for transmission over a digital communication channel since integer numbers can be expressed as a binary code. Non-integer number numbers can be approximated by an integer number and an exponent field, both of which are encoded as binary numbers and transmitted together.

15.2.2 Text codes

The simplest method of encoding a textual message is to assign a single numerical code to each letter of the alphabet The most common code of this type is the American Standard Code for Information

Interchange (ASCII) which represents each symbol, letters (both upper and lower case), numerals and various 'control' codes, as a unique 7 bit code (see Figure 15.2). This fixed length code is employed extensively in both computers and communication systems.

Several other codes of this type have been developed, including a 5 bit code designed by Baudot in the early 1900's for teletypewriters applications. Since only 32 possible combinations are available with this code it can not represent all of the required letters (capitals only), numbers, punctuation and control characters (eg. carriage return). To overcome this problem two code words are reserved which allow a switch between two alternative character sets. By this means, a total of 59 characters are represented. Despite the disadvantage of two character sets, this code is still employed in the Telex network and for Radio Teletype (RTTY) services.

Character	Binary code	Character	Binary Code
Communication control characters	0000000 0011111	A	1000001
		Z	1011010
Space	0100000	[	1011011
!	0100001	Punctuation and symbols	
Punctuation and symbols		\	1100000
/	0101111	a	1100001
0	0110000		
		z	1111010
9	0111001	{	1111011
		Punctuation and symbols	
:	0111010		
Punctuation and symbols		~	1111110
@	1000000	Delete	1111111

Figure 15.2. Summary of the 7-bit ASCII code

ASCII represents a fairly efficient method of transmitting text, however, numerical information can be more efficiently coded (in a character-by-character form) by utilising fewer than 7 bits of information. Each numeral can be uniquely represented by a 4 bit code, leaving six of the possible 16 codes unused to provide additional characters (eg. decimal point, space, etc.).

15.2.3 Variable length codes

In many codes (eg. those described in the previous sections) each symbol is represented by a fixed number of bits. However, in many circumstances such as in English, some source symbols occur more often than others, consequently, a more efficient code cay be devised whereby shorter codes are assigned to the more frequently occurring symbols.

One of the earliest codes which attempts to take account of the different probability of occurrence of source symbols was developed by Morse in 1843. This code is based upon three signalling elements dot, dash and a pause to separate letters and words. The letter e, the most commonly occurring letter in the English language, is represented by a dot, whilst the letter y (less common) is represented by dash, dot, dash, dash. Morse code is still extensively used by radio amateurs and maritime radio operators for long distance short wave radio communication.

Variable length codes suffer from the major disadvantage that it may be difficult to regain synchronism between sender and receiver if some of the data bits become corrupt. This occurs because the receiver cannot determine, merely from the time, the start and end of each character. This is a particularly difficult problem when complete computer control of the link is envisaged.

15.2.4 Information content

A signal source which produces symbols from an alphabet of K possible symbols requires $\log_2 K$ bits to represent the characters. Unless K is a power of 2, the number of bits required, B, will be the next integer larger than $\log_2 K$. In this case, the code efficiency, η, may be given by

$$\eta = \log_2 K / B \tag{15.1}$$

If each source symbol occurs with equal probability p and is not dependent upon any previous symbols, then the number of bits required is

$$\log_2(1/p) = -\log_2 p \tag{15.2}$$

This parameter is related to the quantity of information conveyed by the symbol and is known as the 'information content' of the symbol. If the symbols occur with different probabilities, p_k, where $k=0,1, \ldots K-1$, then the information content of each symbol may be expressed as $-\log_2 p_k$, and the average information content of the message as

$$H = -\sum_{k=0}^{K-1} p_k \log_2 p_k \tag{15.3}$$

This parameter, H, is known as the entropy of the code.

For the case of variable length codes, if each symbol s_k is allocated a code of length b_k, the average length of each code word (L_{av}) is

$$L_{av} = \sum_{k=0}^{K-1} p_k l_k \qquad (15.4)$$

The average information content of a message is related to the optimum average code word length by Shannon's source coding theorem which states:

Given a discrete memoryless source of entropy H, the average code word length L_{av} for any source encoding is bounded as $L_{av} \geq H$.

Thus the minimum average number of bits necessary to encode a discrete memoryless source (ie. one in which the output is independent of the previous outputs) can be determined. The entropy therefore represents a fundamental limit on the average number of bits per source symbol, L_{av}, necessary to represent the source.

15.2.5 Huffman codes

The Huffman code is a source code whose average word length approaches the limit imposed by the entropy of a discrete memoryless source. This code is optimum in that no other uniquely decodable set of code words has a smaller average code word length for a given source. The algorithm required to synthesise the code replaces the prescribed set of source statistics of a discrete source with a simpler one. This reduction process is repeated until only two source statistics remain for which a single bit is the optimum code. Starting at this point and by working backwards it is possible to construct an optimal code for the given source. The Huffman encoding algorithm can be stated as follows:

1) The source symbols are listed in order of decreasing probability. The two source symbols of lowest probability are assigned 0 and 1.
2) These two source symbols are regarded as being combined into a new source symbol with a probability equal to the sum of the two original probabilities, thereby decreasing the list of probabilities by one. The probability of the new symbol is placed in the list in accordance with its value
3) The procedure is repeated until only two source probabilities remain to which 0 and 1 are assigned.

This process is best illustrated by the example in Figure 15.3.

Although the efficiency of the source encoding can be maximised by the adoption of variable length codes, they are poorly suited to situations where the data may be corrupted during transmission. Errors introduced during transmission can therefore cause loss of synchronisation between the sender and the receiver. For fixed length codes, however, errors during transmission may result in corrupt data but synchronisation is maintained. It is therefore more difficult to implement error detection and correction procedures for variable length codes than for fixed length codes. Consequently fixed length codes are more frequently employed for digital communications.

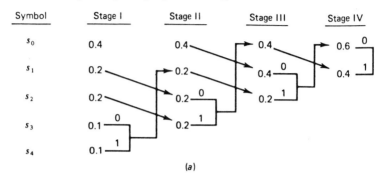

(a)

Symbol	Probability	Code word
s_0	0.4	00
s_1	0.2	10
s_2	0.2	11
s_3	0.1	010
s_4	0.1	011

(b)

Figure 15.3.   (a) Example of Huffman encoding algorithm,
(b) Source code.

15.2.6   Error detection and correction

Communication channels are subject to noise, interference and other defects which may introduce errors into the digital information carried by the channel. Various error detection schemes have been devised which seek to identify corrupt data by introducing additional information into the transmitted data sequence. By adding sufficient redundant information to the signal, it becomes possible not only to detect the presence of an error but to identify its location within the data and to regenerate the original information. This is referred to as error detection and correction (EDC).

15.2.6.1   Parity Checks   The simplest form of error detection is the addition of an extra bit to a data word. This extra bit is referred to as a parity bit and its value is set to make the total number of 1's in the data word and parity bit an even number for even parity, or to an odd number for odd parity. This scheme is often applied to 7 bit ASCII codes to make a total word width, with parity, of 8 bits. The addition of this parity bit allows each received character representation to be checked and any single bit error to be detected. This technique can provide an error free channel provided a reverse channel is available to allow the receiver to request the re-transmission of a data word containing an error.

This idea can be extended by considering a group of character representations as a block of data. The digital representation of the characters within the block can be summed and this value is appended as an additional item of data. Transmission errors will cause this transmitted sum check to differ from that calculated at the receiver, thus a check on the consistency of the data is provided within the block. The data block may be of any length but the receiver must be aware of the block length to ensure that the sum check is calculated at the correct point.

The parity checking scheme may be extended to incorporate parity bits calculated over each bit including the individual word parity bit of a block of data. This row and column parity checking is illustrated in Figure 15.4 in which the block of transmitted data is represented in a table. The location of a single bit error can be detected by identifying both the row and column containing a parity error with the bit common to both being in error. This scheme allows single bit errors to be both be detected and corrected.

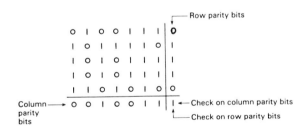

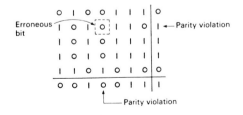

Figure 15.4. Row and column parity checking. (a) No errors; (b) single error detection and correction

Many codes have been developed to provide EDC but they are all derived from the simple parity checking technique. They are broadly divided into two classes:-

1) Block coding in which a additional check information of a fixed length is appended to the data, and
2) Convolution coding in which the check codes are distributed throughout the transmitted data.

15.2.6.2. <u>Hamming code</u>. A commonly used block code is the Hamming code [1] in which a number of information bits are accompanied by a normally smaller number of check bits. We will consider the (7.3) Hamming code in which 4 data bits (I_0 - I_3) are accompanied by 3 check bits (C_0 - C_3). This code provides for the detection and correction of single bit errors. The check bits are calculated from the information bits in the following manner (where $\oplus$ represents modulo-2 addition) :-

$$C_0 = I_0 \oplus I_1 \oplus I_3$$
$$C_1 = I_0 \oplus I_2 \oplus I_3 \qquad (15.5)$$
$$C_2 = I_1 \oplus I_2 \oplus I_3$$

The information bits and the check bits are transmitted together and the receiver calculates the its own version of the check bits by means of the same algorithm from the received information bits. Each of the received check bits are compared with those calculated at the the receiver. If we assign them the value 0 if they are the same, and 1 if they differ then we can readily establish the bit that was received in error as illustrated in Table 15.1 [2].

C_0	C_1	C_2	Digit in error
0	0	0	None
0	0	1	C_0
0	1	0	C_1
0	1	1	I_0
1	0	0	C_2
1	0	1	I_1
1	1	0	I_2
1	1	1	I_3

Table 15.1. (7.3) Hamming code error checking.

The usefulness of the Hamming code lies in this ability to locate the bit in error even if it is one of the check bits. Hamming codes can only detect single bit errors. However, there is a method which will permit Hamming codes to correct burst errors [3]. A sequence of k consecutive code words are arranged as a matrix, one code word per row. Normally the data would be transmitted one code word at a time, from left to right. To correct burst errors, the data are transmitted one column at a time, starting at the leftmost column. When all k bits have been sent the second column is sent, and so on. When the message arrives at the receiver, the matrix is reconstructed, one column at a time. If a burst error of length k occurs, one bit in each of the k code words will have been affected, but the Hamming code can correct one error per code word, so the entire block can be restored. In this method kr check bits make blocks of km data bits immune to a single burst error of length k or less. Where m is the number of message bits and r is the number of check bits. This general concept can be expanded to include much larger blocks of data and for the detection and correction of multiple errors. (eg. Golay codes [4]).

These forward error correcting schemes are confined to the detection of a limited number of errors per block. If the communication channel is subject to impulsive noise which causes bursts of closely spaced errors, but the average error rate is low, then error detection and correction may be more effectively achieved by distributing both data and check digits throughout the transmitting sequence, this is referred to as convolution coding.

15.2.6.3 Automatic Repeat Request system (ARQ). In general the detection of errors is very much simpler than the correction of the errors and an alternative method of providing error free communication can be provided by Automatic Repeat - Request Systems (ARQ). Some form of error detection is required within the receiver which requests re-transmission of blocks of data received in error. In practice a cyclic redundancy code (CRC) is normally employed for this purpose since it provides an effective method of detecting a variety of error types. Some form of reverse channel is required to convey information from the receiver to the transmitter. This reverse channel occupies some part of the bandwidth of the communication channel and itself may be subject to errors. Somewhat complex protocols are therefore often necessary to implement such systems.

It is clear that the choice of the most suitable form of EDC depends upon the the performance of the communication channel over which it is to be implemented. The ability to detect and possibly correct errors relies on the addition of enough redundant information to overcome the limitations of the channel. This process decreases the effective bandwidth of the channel in exchange for the reliable transfer of information.

15.3 ANALOGUE WAVEFORM ENCODING

The advantages of digital representation of data have been established in the previous section, including the possibility of error free transmission by the application of suitable coding schemes. Some data can readily be represented digitally (eg. text may be represented as a sequence of ASCII codes), whereas analogue waveforms do not have an inherent digital equivalent. To enable an analogue signal to be transmitted by digital means, the signal may be represented as a series of sampled values obtained at equally spaced time intervals. Provided that the period between successive samples is sufficiently small, the original signal may be regenerated without error.

15.3.1 Signal sampling

The value of a continuous signal $x(t)$ is sampled at a constant rate $(1/T)$ at the transmitter. The numerical values of the samples are transferred across the communications link and modulate the amplitude of a train of unit impulse functions $p(t)$ with the same constant time separation T (see Figure 15.5a). This modulated waveform $y(t)$ may therefore be considered as the product of the original signal with the train of unit impulse functions $p(t)$.

In the time domain we have

$$y(t) \quad = \quad x(t).p(t) \qquad\qquad (15.6)$$

which may be expressed in the frequency domain as

$$Y(\omega) \quad = \quad \frac{1}{2\pi}\big[X(\omega) \,*\, P(\omega)\big] \qquad (15.7)$$

$$= \quad \int X(\Omega).P(\omega - \Omega)\, d\Omega$$

Provided that the maximum frequency component W of the original signal is less than half of the separation of the shifted spectra, the spectrum of the sampled signal $(Y(\omega))$ comprises an infinite number of replicas of $X(\omega)$ shifted in frequency by $2\pi/T$. This is illustrated in Figure 15.5b. If the sampled signal is passed through an ideal low pass filter to remove all frequency components $> W$, the original waveform will be reproduced.

$$Y(\omega) \quad \xrightarrow[\text{LPF}]{} \quad X(\omega) \qquad\qquad (15.8)$$

If the sampling rate is insufficient, the replicas of $X(\omega)$ will overlap and it will not be possible to recover the original signal $x(t)$ by filtering.

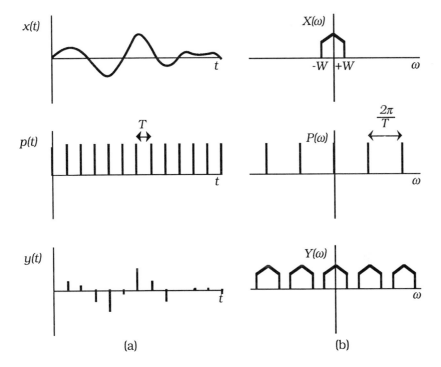

Figure 15.5.   (a) Illustration of combination of a continuous waveform
               x(t) with a sampling function p(t) to produce a series
               of signal samples, y(t).
               (b) Illustration of the signals in the frequency domain.
               Note: only a limited part of the spectrum is shown and
               the signals extend to ∞ in the frequency domain.

In practice, the signal samples modulate the amplitude of a train of
pulses of finite width (each pulse has a constant amplitude - ie. it is flat
topped). This imposes a distortion onto the signal which is negligible
provided that the pulse width is narrow compared to the period of the
original signal. If the pulse width does not satisfy this condition,
corrections must be applied to remove the effect of the distortion.

15.3.2 Signal quantisation

The previous section describes how an analogue signal may be fully
represented by a sequence of instantaneous samples provided that the
sampling rate is sufficiently high. However, in digital communication
systems, the sample values are transmitted in the form of a series of
binary digits. If each sample value is represented by a sequence of n
bits, the code can only represent 2^n distinct signal levels.

Consequently, a range of signal levels are allocated to each binary code (eg. see Figure 15.6). This process is known as quantisation and results in the transmitted signal being only an approximation of the original signal.

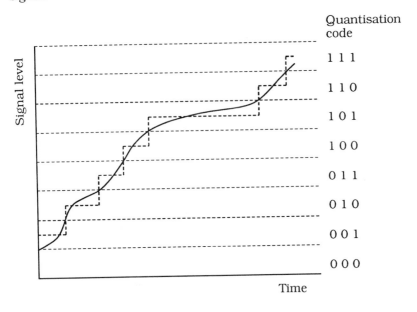

Figure 15.6.   Illustration of the quantisation of a continuous waveform indicated by the solid line. The signal is approximated by that shown by the broken line.

The difference between the continuous and quantised signals may be regarded as a noise component in the transmitted signal and its effect described in terms of a quantisation signal to noise ratio (SNR_Q). This parameter is defined as the ratio of the mean power in the signal to the mean power in the noise (error) component and may be derived as follows. (For a more detailed discussion of quantisation errors see [5].)

If the range of signal voltage corresponding to the full range of 2^n quantisation levels is $2v_p$ (ie. $x(t)$ is in the range $-v_p$ to $+v_p$), the quantisation interval Δ is given by Equation 15.9.

$$\Delta = \frac{v_p}{2^{n-1}} \tag{15.9}$$

The maximum difference between the signal $x(t)$ and its quantised equivalent $q(t)$ is 0.5Δ.   Since there will usually be a random distribution of error voltage (e), between -0.5Δ and 0.5Δ the mean square error voltage (proportional to the quatisation noise power) may be expressed as:

$$\overline{e^2} = \frac{1}{\Delta}\int_{-\frac{\Delta}{2}}^{\frac{\Delta}{2}} v^2\, dv = \frac{\Delta^2}{12} = \frac{v_p^2}{3 \cdot 2^{2n}} \qquad (15.10)$$

SNR_Q is therefore given by Equation 15.11, or in decibels by Equation 15.12. The parameter α relates the peak to the mean power ratio of the signal. From Equation 15.12, it is apparent that the quantisation signal to noise ratio is increase by 6 dB for each additional bit included in the quantisation coding.

$$SNR_Q = \frac{\overline{m^2}}{\overline{e^2}} = 3 \cdot \left(\frac{\overline{m^2}}{v_p^2}\right) \cdot 2^{2n} \qquad (15.11)$$

$$
\begin{aligned}
SNR_{Q(dB)} &= 10 \cdot \log(3) + 20n \cdot \log(2) - \alpha \qquad (15.12)\\
&= 4.77 + 6n - \alpha
\end{aligned}
$$

where :

$$\alpha = 10 \cdot \log\left(\frac{v_p^2}{\overline{m^2}}\right)$$

15.4 DIGITAL MODULATION

15.4.1 Synchronisation

When the binary data stream is received some form of timing signal is necessary to indicate when each data bit is present. This timing signal may be derived in one of two ways:

15.4.1.1 Synchronous data transmission. In this case, a suitable clock signal is transmitted with the data. Synchronous transmissions start with a unique pattern of 8 or 16 bits to establish the initial synchronisation of the clock within the receiver. Subsequently the clock is locked to the timing information derived from the data stream until the message is complete. Synchronous transmission is normally confined to higher data rates where the additional complexity can be justified.

15.4.1.2 Asynchronous data transmission. In this case, data are segmented into groups of 7 or 8 bits, together with 2 (or 3) additional bits to provide timing information. These are known as start and stop bits and are illustrated in Figure 15.7.

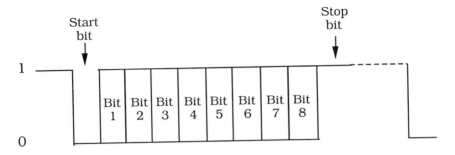

Figure 15.7. Start and stop bits in asynchronous communications.

When no data is being transmitted a continuous 1 state is maintained. Immediately before data transmission, a single 0 bit (start bit) is sent and this indicates to the receiver the beginning of the data. A local clock at approximately the data rate is required to provide timing information for the correct sampling of the subsequent 7 or 8 data bits. After the data has been sent at least one stop bit (1) is transmitted to ensure that the next start bit will be recognised.

The simplicity of the this method has resulted in its widespread adoption for low speed character orientated information.

15.4.2 Baseband

For some applications the binary information can be represented by two signal levels which can be applied directly to a set of wires and carried to the receiver where a simple comparator can regenerate the original data. In this baseband system the bandwidth required extends from DC to some upper limit related to the data rate.

15.4.3 Modulation

The majority of applications require the data to be transferred over a link which can only carry signals at a much higher frequency, (eg. a radio link). In this broadband system a carrier signal is modulated by the baseband data. Modulation also allows frequency diversity multiplexing to permitting multiple channels to be carried over a single link.

Three main types of modulation are currently employed.

15.4.3.1 Amplitude shift keying (ASK).
In this modulation type, the sinusoidal carrier signal of frequency ω_c is either on or off, depending upon the data stream $d(t)$.

$$x(t) = d(t) \cos(\omega_c t) \qquad (15.13)$$

This modulation is illustrated in Figure 15.8 (a).

15.4.3.2 Phase shift keying (PSK). In this case, the phase of the carrier signal is varied, in the simplest case by 180°, in sympathy with the data stream as illustrated in Figure 15.8 (b).

$$x(t) = \cos(\omega_c t + d(t) \phi) \qquad (15.14)$$

To increase the data rate, quadrature phase shift keying is often employed in which each of the four possible phases (0°, 90°, 180° and 270°) represents the value of 2 bits of data. The data rate is therefore twice the the rate at which the phases are switched ie. the binary data rate (bits/s) exceeds the signalling rate, measured in baud: 1 baud corresponding to 1 signalling element per second.

15.4.3.3 Frequency shift keying (FSK). A third alternative is to switch the instantaneous frequency of the carrier as illustrated in Figure 15.9 (c).

$$\omega(t) = \omega_c + d(t)\omega_s \qquad (15.15)$$

This may be considered as ASK (on/off) of two carriers, one turned on for a 0 state and the other turned on for a 1 data bit.

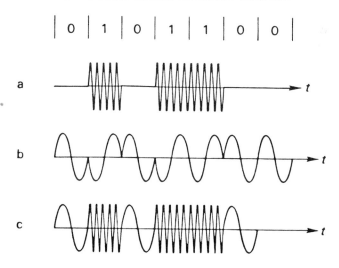

Figure 15.8. Digital modulation schemes. (a) Amplitude shift keying (ASK); (b) phase shift keying (PSK); (c) frequency shift keying (FSK).

15.5 CONCLUDING REMARKS

The design of a digital communication system is a complex problem depending upon many factors, however, the various areas requiring attention are generalised below.

1) The specification of the system in terms of the data to be transmitted, the required data rate, acceptable error rate etc.
2) Source encoding must be selected to determine how the information is to be encoded into binary data.
3) The method by which error correction and detection is to be implemented must be specified. The characteristics of the communication channel will determine the sophistication of the EDC techniques required.
4) The method by which the binary data are to be encoded on to the channel (line coding) must be determined. This will depend upon the type of communication channel, how clocking information will be passed to the receiver and the available bandwidth.
5) If the communication channel is broadband then the type of modulation must be selected.

The special case of digital communications over a high frequency radio link is discussed in Chapter 23.

15.6 REFERENCES

1. R. W. Hanning. Error detecting and error correcting codes. Bell Systems Technical Journal, 29, 1950, 142-160.

2. J. D. Gibson. Principles of Digital and Analogue Communications. Macmillan, 1989, Chapter 14.

3. A. S. Tanenbaum. Computer Networks. Prentice Hall, 1981. Chapter 3.

4   S. Haykin. Digital Communications. John Wiley, New York, 1988, p 390.

5. J.J. O'Reilly. Telecommunication Principles. Van Norstrand Reinhold (UK), 1984, Chapter 6.

Chapter 16

Simulation of DSP devices and systems

D. J. Mayes

16.1 INTRODUCTION

Traditionally, digital signal processors have performed relatively simple, well defined tasks, mainly within the communications industry.

As DSP devices become more complex, they are finding their way into much larger, system level applications requiring inter-processor communications and realtime multitasking operating systems; as lower performance devices become cheaper their use in medium to low cost, high performance industrial control applications is increasing. In most DSP applications, sophisticated tool sets for system design and development are almost essential to allow DSP users to get high performance products to the market place in the shortest possible time. Simulation tools are becoming an increasingly important part of the tool set, improving productivity at both the system design stage and in the development of software.

16.1.1 What is simulation?

Simulation is the process by which the characteristics and conditions of a real-world system or device are either mathematically modelled or reproduced in such a way as to allow the performance of the system or device to be investigated dynamically and experiments carried out independently of the actual system or device.

16.1.2 The benefits of simulation.

The availability of fast, powerful and relatively cheap computing equipment means that simulation tools are now almost universally used alongside more traditional methods in the design of electronic devices, circuits and systems. The ease with which computer simulations can be set up and modified encourages constructive experimentation. Little time is wasted in trying alternative approaches. Compared with traditional bench prototyping methods, computer simulations can offer productivity improvements which can be estimated in orders of magnitude. Simulation allows "computer breadboarding" or "soft prototyping" of circuits and systems. This enables performance to be extensively investigated and operation optimised before the circuit or system is built. Provided that the models used in the simulation are an accurate representation of the actual process or system, simulation results can give a very high degree of confidence in the performance and characteristics of the final hardware.

Some of the many benefits of simulation include :

* Increased productivity.

* "Computer breadboarding" of a system, allowing off-line evaluation and testing.

* Thorough testing with a wide variety of stimuli including noise.

* Monitoring of all system variables including intermediate values which may be difficult or indeed impossible to measure in a real system.

* The "safe" testing of systems to the limits of performance, and under abnormal operating conditions, without danger or risk of damage to sensitive components.

* A high level of confidence that the real system, when built, will have the required performance and characteristics.

* A fast turn-round from concepts and ideas to graphical results.

* The facility to make minor changes and perform a rapid "what-if?" analysis.

* The investigation of the effects of component tolerance, either by a dedicated simulator function or by performing a number of simulations with different component values.

The features mentioned above are not specific to any particular simulation system or language, but are intended to outline some of the benefits of using simulation to aid the design of products.

16.1.3 Simulation of DSP devices and systems.

Within the context of this chapter, the subject of simulation can be divided into two distinct areas:

* The simulation of DSP software using device specific software simulators.

- The simulation of DSP algorithms and systems using both DSP specific and general purpose simulation tools.

The following sections will discuss these two areas in more detail.

16.2 SOFTWARE SIMULATORS.

A DSP software simulator is a software package that fully models the internal registers, memory map, instruction set and operation of a specific digital signal processor or family of digital signal processors allowing cost-effective software development and program verification in non-real-time.

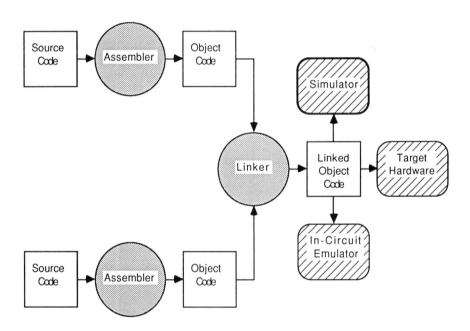

Fig 16.1 - Software development sequence.

Figure 16.1 shows a typical software development cycle. The assembly language code must first be written and assembled. If the software consists of multiple modules they must also be linked together. This linked object code can then be loaded into the simulator. The software simulator is a powerful tool in the software development process, which allows code to be executed and bugs identified and corrected quickly and effectively. Once debugging is complete, the linked object code can be either downloaded directly to the target hardware, or loaded into an in-circuit emulator for checking and monitoring the integration of the target hardware and software.

A good software simulator is in many ways similar to an in-circuit emulator, with two important exceptions. With a simulator, only the software is simulated, allowing code to be debugged and functionally checked with no hardware requirement. In general, software simulators do not operate in realtime. An in-circuit emulator however is a tool which emulates the processor in the target system, allowing both the software and hardware to be exercised in real-time.

For a software simulator to effectively execute a program, there must be facilities available for connecting the simulated processor to the "outside world". To simulate I/O devices connected to the processor, I/O channels are usually assigned to data files, enabling "real" input data to be used and output data to be collected for examination and analysis. Interrupt flags can be set at user-defined intervals, allowing "scheduled" real-time software to be simulated. A clock counter is a useful feature which allows accurate timings to be made for the purpose of code optimisation.

During program execution, the internal registers and memory locations of the simulated processor are updated as each instruction is interpreted. As with an in-circuit emulator, breakpoints can be established, triggered on read or write instructions to specific memory locations. Execution is suspended when a breakpoint is reached, an error is detected or a break command is entered by the user. Once execution is suspended, registers and memory locations can be inspected and/or modified and the simulation restarted.

In summary, the main features of a typical software simulator include :

- Data files associated with I/O channels.

- Breakpoints programmable to trigger on:
  - Memory reads or writes
  - Instruction acquisition
  - Specific data patterns
  - Error conditions

- Timing analysis in terms of clock pulses.

- Interrupt generation at user-defined intervals.

- Trace facilities.

- Display and modification of memory contents.

- Inspection and modification of registers.

16.3 SYSTEM LEVEL SIMULATORS

Applications for digital signal processors can be split into two distinct categories, each of which has very different simulation requirements.

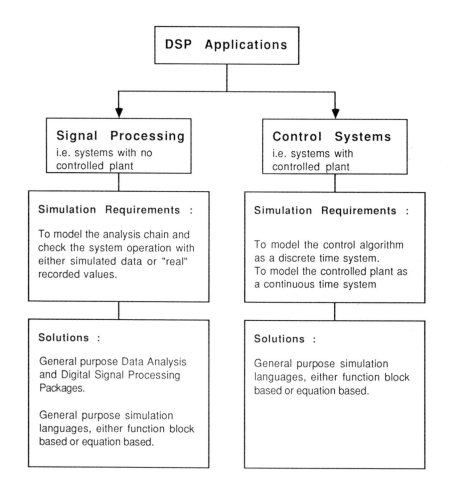

Fig 16.2 - The simulation requirements of DSP systems.

Firstly, there are those applications where the DSP is performing a pure data acquisition and signal processing function, for example a digital filter or a speech processing system. In these cases, the main purpose of simulation is to verify and tune the chosen algorithm. This is achieved by testing a model of the algorithm with either simulated data or "real" values obtained from a data acquisition system.

The second category contains applications where the DSP is performing data acquisition and signal processing as part of a closed loop control system or as part of any system containing controlled plant, for example a self tuning temperature controller or the control of a servo system. For this type of application, the simulation system needs to simulate both the discrete time control algorithm running in the DSP and also the controlled plant. The response and stability of the complete system to changes in operating conditions can then be assessed.

Figure 16.2 outlines the simulation requirements and indicates possible solutions for the two categories of DSP applications.

16.3.1 Data analysis and digital signal processing packages.

Several workstation-based interactive data analysis and digital signal processing packages are commercially available. Such packages can be used for processing data either as a complete solution for non real-time applications or as a design aid and simulation tool for real-time processor based systems. The data can be either "real" values, recorded with a suitable data acquisition system or measuring instrument remotely controlled by the computer over a communications link, or values synthesised from internally generated functions.

A typical data analysis and digital signal processing package which can be used for the design and simulation of DSP systems is DADiSP, produced by DSP Development Corporation. In this system, data is displayed and manipulated in a series of windows, each window containing one step in the analysis chain. The data displayed in any window can be related, by formulae and functions, to data in other windows. Complex analysis chains can therefore be constructed without the need traditional programming skills. Most commonly used signal processing functions are available including :

- Mathematical operations

- Fourier transforms and related functions

- Trigonometric functions

- Digital filter functions

- Statistical functions

- Peak analysis functions

- Data type conversions

In many ways the sequence of windows forming an analysis chain is very similar to a conventional spreadsheet. Once an analysis chain has been created, new data can be collected and processed, each dependent window being automatically recalculated and updated as a new data set is entered. This allows very rapid testing with large amounts of data.

DADiSP provides a comprehensive range of data display formats ranging from tabulated data, through simple line and bar graphs to complex three dimensional plots and waterfall plots. Graphs can be overlaid allowing raw data and the results of analysis to be instantly compared. The digital filter module is particularly comprehensive, allowing FIR (finite impulse response) and IIR (infinite impulse response) filters to be designed with low-pass, high-pass, band-pass or band-reject characteristics. User defined programs may be run from within DADiSP allowing customised analysis routines to be easily integrated into the analysis chain, and automated data acquisition software to be run.

By using recorded "real" data, this type of product allows complex systems to be algorithmically verified, optimally tuned and thoroughly tested independently of any specific DSP chip or hardware configuration, and without the time or effort required to write, debug and then modify large quantities of device specific DSP code.

Other data analysis and signal processing packages include ILS from Signal Technology Inc., Hypersignal-Workstation/Hypersignal-Plus from Hyperception Inc and SPW from Comdisco Systems Inc.

ILS offers many features including waveform display and editing, spectral analysis, pattern classification, data manipulation and speech processing. For specialist applications not covered by the standard features of ILS, the Fortran source code is available allowing customised solutions. ILS is widely used in such areas as speech processing, biomedical research, communications and noise and vibration monitoring.

Hypersignal-Workstation is an integrated DSP environment, supporting a wide variety of analogue data acquisition and DSP boards, allowing real-time data capture and display, powerful waveform analysis and design and simulation of digital filters. For digital filter design, simulation of quantization effects is included as well as automatic code generators for a number of DSP chips.

SPW is a package which allows a DSP or communications algorithm to be graphically captured and edited using a block diagram editor. The algorithm can then be simulated using a built in Simulation Program Builder, and signals analysed with a Signal Display Editor. SPW has GPIB support which allows simulation and analysis with real-world data.

16.3.2 General purpose simulation languages.

The simulation of physical systems is now a standard analysis method used in the evaluation of systems design prior to the construction of the actual hardware. Many different function block-based and equation based simulation languages are currently available. These run on a wide variety of computing platforms including PC's, engineering workstations and mainframes.

A typical general purpose equation based simulation language which is widely used throughout the process and servo control industries is the Advanced Continuous Simulation Language, ACSL, (pronounced "axel"), from Mitchell and Gauthier, Associates Inc.

ACSL was developed for modelling and evaluating the performance of continuous systems which may be described by a set of time-dependent, non-linear differential equations. ACSL provides a simple method of representing these mathematical models on a computer. Starting from either an equation based description of the model or a block diagram, ACSL statements are written to describe the system under investigation. The structure of an ACSL model allows the components of a simulated system to be broken down into their constituent elements. These can be classical control functions such as integrators, comparators etc. or non-linear functions such as saturation, hysteresis, backlash etc. The ability to separate models into function blocks introduces great flexibility into the simulation process.

Statements describing a model need not be explicitly ordered since the ACSL processor contains a sort routine. This automatically orders the equations or statements to ensure that no values are used before they have been calculated. This feature of ACSL contrasts with traditional high level computer programming languages, where program execution is critically dependent on statement order.

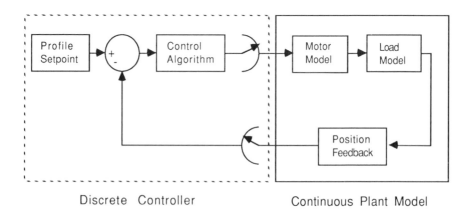

Fig 16.3 Discrete position control of a servo drive system

As an example, consider the discrete position control of a servo drive system, as shown in figure 16.3. The ACSL model of this system is divided into a number of sections. The two main sections are the DERIVATIVE section and the DISCRETE section. The DERIVATIVE section contains the description of the continuous system or controlled plant, that is those parts of figure 16.3 enclosed by the continuous line. The DISCRETE section contains the discrete time control algorithm i.e. a model of the algorithm which will eventually run in the DSP system. In this example the DISCRETE section contains those parts of figure 16.3 enclosed by the dashed line. Figure 16.4 shows the structure of a typical ACSL model program.

PROGRAM   Discrete Controller

```
CONSTANT            - - - - - - - - - - - - - - - - - -
CONSTANT                         Constant  Definitions
CONSTANT            - - - - - - - - - - - - - - - - - -

INITIAL             - - - - - - - - - - - - - - - - - -
                                 Initial  Conditions  and
                                 Parameter  Calculations
END       $"Of initial"   - - - - - - - - - - - - - - - - - -

DERIVATIVE          - - - - - - - - - - - - - - - - -
                                 Continuous  Plant  Model
TERMT $"the  terminate  condition'
END   $"Of  derivative" - - - - - - - - - - - - - - - - -

DISCRETE            - - - - - - - - - - - - - - - - - -
INTERVAL TSAMP = $"The sample period"

                                 Discrete  Controller

END   $"Of  discrete"   - - - - - - - - - - - - - - - - -

END   $"Of  program"
```

Fig 16.4 Typical ACSL model structure for discrete control system

Other general purpose simulation languages and control systems design packages include CYPROS from Camo AS in Norway, ACET from Information and Control Systems, Ctrl-C/Model-c from Systems

Control Technology Inc and MATLAB from The Math Works Inc.

CYPROS is an integrated package for control systems design and data analysis, which incorporates data acquisition, signal processing, statistical analysis, mathematical functions including matrix operations and curve fitting, simulators for complex dynamic systems and linear systems and a variety of tool boxes for advanced control functions such as parameter estimation, extended Kalman filtering and adaptive control.

ACET (Advanced Control Engineering Techniques) is an interactive control engineering tool which integrates the different stages of the design cycle including model building, control system design, system analysis and simulation.

Model-C is a block diagram-driven modelling and simulation program which models continuous, discrete time or mixed systems. Hierarchical block diagram model structures are supported, and Fortran simulation code is automatically generated. Ctrl-C is an interactive language for classical and modern control system design and analysis, signal processing and system identification.

MATLAB is an interactive system for control system design applications. Features include control systems design tools, matrix computations, one and two dimensional signal processing, system identification and multidimensional graphics. Optional toolboxes provide facilities for expressing systems as either transfer functions or in state-space form, with support for both continuous and discrete time systems.

When using any simulation language to investigate and evaluate the performance of control systems and algorithms, it is essential to ensure that the plant model being used is sufficiently accurate. Any inaccuracies in the model description or invalid assumptions used in the modelling process will be reflected in the simulated control action. This will cause discrepancies between the simulated results and the actual response of the real plant. To achieve the best possible correlation between simulated results and the actual system response, the simulation process must be performed in a number of stages. Firstly, the continuous plant model must be verified by applying a known stimulus or set of stimuli to both the model and the real plant and comparing the responses. By iteratively improving the model and repeating the tests, a stage will be reached when the match between the simulated response and the actual response has the required accuracy. At this stage the design of the control algorithm can proceed, with confidence that the simulated response will be accurately reproduced by the real plant.

16.4 COMBINED SIMULATION SYSTEMS AND PROGRAMMING ENVIRONMENTS.

To create complex DSP-based systems quickly and reliably, designers are using a wide variety of advanced tools including system level simulators, high level language compilers and powerful, block diagram driven development environments.

High level languages (the C language being the most popular), can significantly reduce the time taken to develop a program, but have the disadvantage of producing inefficient code. Typically, a well written assembly language program may achieve above 90% of the theoretical device peak processing power. A high level coded program could limit the performance to between 20% and 30% of peak processing power, depending on the degree of optimisation and the efficiency of the particular compiler. In many cases this reduction in code efficiency and therefore processing speed will be unacceptable.

Block diagram driven development environments are systems which combine a graphics editor, an interpreter and a library of basic pre-coded function blocks. Using this type of system the engineer does not need to write device-specific code. Instead he defines the system using a series of simple graphical building blocks, for example a digital filter, a Fourier transform or a data I/O block, and configures each block to give it the required parameters and functionality. On completion of the block diagram entry, the interpreter takes over and automatically translates the block diagram into assembly language modules for a specific DSP chip. Using this method, the time taken to produce software is very much reduced whilst maintaining most of the efficiency of hand written assembly language code.

Since most signal processing and control engineers begin the design process with a block diagram or flow chart of the required system or application, this would seem an ideal entry point for an efficient integrated simulation and programming environment. In diagrammatic form, the integrated simulation and automated code generation development environment is shown in figure 16.5. The concept of generating assembly language code from a block diagram has already been discussed, and such systems are currently available for a number of the more advanced DSP chips. Function block-based simulation languages are widely available, and schematic capture packages have been used for some time as interactive user interfaces to simulation languages.

If the same block diagram forms the input to both a simulation language and an automatic code generator, a very powerful integrated development environment can be envisaged. Algorithms are first checked and optimised by simulation, the optimised design being translated to assembly language code, assembled and downloaded to the target system. In this way a very high degree of confidence that the product or system will be "right first time" can be achieved.

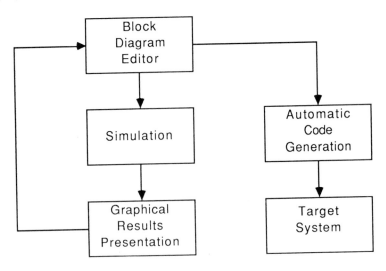

fig 16.5   An integrated simulation and automated code generation development environment.

A system along these lines is currently under development at the Eurotherm International Technology Centre, for in-house use in the design and development of advanced industrial control products using low cost DSP technology.

The structure of the system allows a problem to be defined by a simple graphical block diagram. "Beneath" each graphical symbol contained within the function block library, there is both a dedicated simulator function and one or more pre-coded assembly language modules. The assembly language modules are for specific DSP's, microcontrollers or microprocessors. Diagramatically, this is shown in figure 16.6.

The block diagram simulation language, BLKSIM, is complete and is currently in use as a stand alone system. The libraries of assembly language modules (for the TMS320 family of DSP's in the first instance) are currently being coded in preparation for the development of the automated code generator(s).

The structure of the code generator is more complex than that of the simulation language interpreter since, in addition to the sequencing of the code, it must also handle the real-time issues of a control system including process scheduling and interrupt handling. This will obviously require some form of hardware description language. It is hoped that in the near future limited automated code generation will be achieved, including allocation of variables and sequencing of function blocks, with complete automation including a hardware description language being a longer term goal.

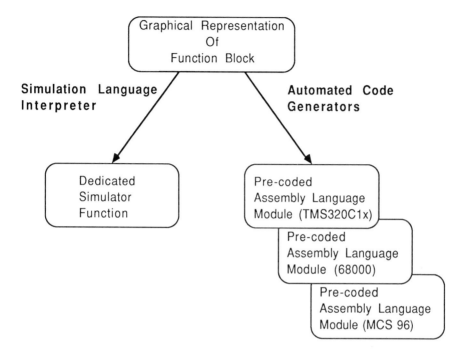

Fig 16.6   Structure of the integrated development environment.

Chapter 17

Review of architectures

R. J. Chance

17.1   INTRODUCTION

There are several approaches to the solution, in real time, of numerically intensive computations of the type used in digital signal processing. The main groups of these are shown in fig 17.1. Those in the largest group can be described as monolithic digital microcomputers, using a sequentially addressed stored program. This group includes most of the best known DSP devices and will be referred to as the general purpose monolithic digital signal processor group (GPMDSP). The operation of these devices can broadly be compared to the familiar von Neumann all purpose microprocessor (vNAPP) in that a program counter addresses a series of instructions from memory, which are decoded and executed. The GPMDSP has evolved into several different subgroups directed, for example, at high performance or high volume production, low cost applications. However there are other quite different approaches to digital signal processing and in some fields, bit slice designs, systolic arrays and array processors have become relevant to practical circuit designers as well as researchers. We shall now look at all these categories in greater detail.

17.2   THE GENERAL PURPOSE MONOLITHIC DSP (GPMDSP)

We have already seen that the realisation of real-time digital signal processing algorithms has produced quite different processor architectures from the traditional von Neumann general purpose designs. Even though there is an identifiable category of devices which could be called general purpose monolithic DSPs, there is not a consensus on a single, optimal approach to their architectures. This puts a heavy responsibility on the system designer to know what is available, especially as the performance of many algorithms in DSP applications is very architecture dependent.

A problem that has been observed among industrial users

DSP TYPE	APPLICATION
GENERAL PURPOSE MONOLITHIC STORED PROGRAM	

STAND ALONE	Digital components
	Controllers
COPROCESSORS	PC peripherals
MULTIPROCESSORS	Numerically intensive
HIGH VOLUME	Telecommunications
	Speech processing
HIGH PERFORMANCE	Aerospace & Military
	Workstations
	Robotics
LOW POWER	Battery operation
FLOATING POINT	Large dynamic range
BIT SLICE	Research prototypes
	Unusual solutions
VECTOR PROCESSORS	Numerically intensive
SYSTOLIC ARRAYS	Particular algorithms

Categories   of   digital   signal   processor

FIG   17.1

of microprocessors results from the desire of many organisations to standardise on a particular general purpose processor. This policy is intended to minimise the range of skills and other infrastructure that needs to be supported within the organisation. However, the performance of the DSP is extraordinarily high in particular areas, compared to conventional processors. The wide variety of DSP applications means that they should be regarded as a completely different class of component from the von Neumann all purpose processor (vNAPP). The inclusion of vNAPP type features in the most recent GPMDSP devices can obscure this fact. One typical example arose in a motor control application, where the GPDSP was able to implement not only the control but also some associated filtering. Although a vNAPP could perform the control, the filtering required substantial additional hardware. One should be sure that the penalty for standardisation is not too great.

17.2.1   What   is   a   DSP?

How can one recognise a digital signal processor? At the present time, the ability to perform a multiply/accumulate operation in one or two machine cycles would probably be a feature of virtually all DSPs and hardly any conventional processors. Current DSP machine cycle times are exceptionally fast, in the region of 80 ns; figures of 35 ns are promised in the near future. Not only are these times short compared to those of the vNAPP, but the GPMDSP often carries out several operations in one cycle which would be performed sequentially by a vNAPP. The single cycle multiplier is at the heart of any GPMDSP. The delivery of data to and from the multiplier has affected the whole architecture of these devices.

Some questions that might usefully be posed by a designer attempting to choose a GPMDSP are:

(a) Is there a particular algorithm that needs to be computed quickly; does it map onto a particular GPMDSP?

(b) Is floating point arithmetic necessary? If not, could 16, 24 or 32 bit integer arithmetic be used?

(c) What general purpose computing must be done in addition to the signal processing?

(d) Should the DSP provide co-processor support for a general purpose computer?

(e) Do DSPs exist with on-chip hardware support for this application?

(f) Is this problem amenable to a multiprocessor solution?

(g) How cost sensitive is the application?

17.3   PARALLELISM

Speed is normally the overwhelmingly dominant requirement in digital signal processing work. Therefore the identification and implementation of operations that may be performed concurrently has been a major influence on GPMDSP design. These can be either within one chip or through interprocessor colaboration.

17.3.1   Intra-processor parallelism

Most (but not all) GPMDSPs use multiple memory spaces, supported by multiple address and data busses. These busses can be used to simultaneously carry an instruction and perhaps several data words. Until quite recently, this has meant that the memory areas to be accessed in this way had to be on the chip, due to a restricted pinout capability. Fig 17.2 shows the memory configurations of some current and announced processors. On-chip memory varies between 0 and 100 percent and the number of address spaces may be one, two or three.

Most DSPs use a Harvard type of structure where the memory used for program storage is separate from that used for data.

DSP	PROGRAM MEMORY		DATA MEMORY	
	WORD LENGTH	ADDRESS SPACE (on chip)	WORD LENGTH	ADDRESS SPACE (on chip)
DSP32C	32	4M (1.5k)	von Neumann	
ADSP2100	24	16k (0)	16	16k (0)
DSP56000	24	64k	24	64k (256 + 256)
uPD7720	23	512	16 13	128 512
uPD77230	32	4k (2k)	32 32	4k 4k (512 + 512 + 1k)
TMS32010	16	4k (0)	16	144 (144)
TMS320C25	16	64k (256)	16	64k (256 + 288)
TMS320C50	16 (!2k +!8k + !256)	64k	16 (!8k + !256 + 288)	4k
	!selectable as program or data			
DSP96002	32	4G (1K)	32 32	4G (512 + 1K) 4G (512 + 1K)

Memory configurations of some general purpose
monolithic DSPs

FIG 17.2

These communicate with the central processor through
separate busses. Thus, as shown in fig 17.3, the instruction
fetch, decode and execute operations can be 'pipelined'. The
three instructions in the pipeline can be at different stages
in their execution at the same time. In this case, this means
that data memory may be accessed concurrently with
program memory. The architecture of most GPMDSPs is not

strictly Harvard, in that it is possible to transfer data in both directions between program and data memory. The establishment of this link means that it has become convenient to use data and program memories of the same word size. Although this may cause the casual observer to think that a common von Neumann memory structure is being used, this is not normally the case. So programs cannot be executed in data memory and data cannot be manipulated to any extent in program memory.

```
FETCH   DECODE   EXECUTE
         FETCH    DECODE   EXECUTE
                  FETCH    DECODE   EXECUTE
                  (BRANCH)

                                    FETCH   DECODE
                                    FETCH
```

```
1        2        3                 4
```

Three   stage   instruction   pipeline   with   branch

FIG 17.3

This pipelining has disadvantages. If the pipeline has to be broken, say for a branch to a new program address or an interrupt, there is a penalty to be paid in refilling the pipeline, see fig 17.3. Thus, for example, the TMS320C25 processor takes one cycle longer to perform a branch than its ancestor, the TMS32020. This is the price paid for the other benefits of a three stage, rather than a two stage pipeline.

Pipelined operations may also affect programming in other situations. The problem is particularly dangerous when the programmer's mnemonic does not imply any pipelining. For example, the AT&T DSP32C pipelines a multiply/ accumulate operation as shown in fig 17.4. The four stages of the multiply/accumulate (fetch, multiply, accumulate, write) can occur simultaneously, but only on sequential data samples. So it is four cycles before the result of a multiply/accumulate operation can be used, although the average throughput is nearly one multiply/accumulate per cycle [1].

Some of the most recently announced digital signal processors, such as the Motorola DSP96002 [2], have increased the parallelism in the movement of data still further. The DSP96002 combines multiple memory spaces with multiple access in one instruction cycle. The ability to house devices in packages with over 200 pins also means that two independent external address and data busses can be accessed simultaneously. Thus each cycle can support two external accesses, two DMA accesses and two read/write memory accessess as well as an instruction fetch. The problem with this architecture is that, for maximum

```
CYCLE          1        2        3        4
1/4  CYCLE     1234     1234     1234     1234
               ^        ^        ^        ^
               FETCH  1
                      MULTIPLY   1
                            ACCUMULATE   1
                                    WRITE   1
                      FETCH   2
                            MULTIPLY   2
                                    ACCUMULATE   2
```

DSP32C   multiply/accumulate   operation

FIG   17.4

efficiency, data has to be ordered so that pairs of operands occupy separate data memory spaces.

17.3.2   The   co-processor   approach

The closely coupled interfacing of a digital signal processor to a general purpose computer is commonly practised. By doing this, the high level language and software support of the general purpose host machine can be

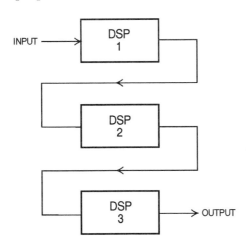

A   series   connected   multiprocessor   system

FIG   17.5

enhanced by the fast arithmetic of the GPMDSP(s), for specific functions only e.g. FFT. Such co-processors also find application as specialised, loosely coupled peripheral devices. Examples of these are the digital filtering of input and output data streams in control systems, such as numerically controlled machine tools and graphics displays.

In most co-processor designs, it is convenient to use RAM

for the DSP program memory. The host can then download a number of different programs. There is a tendency for most modern general purpose DSPs to support stand alone and co-processor systems equally well. Many modern DSPs can be connected directly to the data busses of a host computer, without intervening buffers. An economical multiprocessor system can often be designed where only on-chip DSP memory need be used; the host processor can download programs to on-chip DSP program memory. Thus many GPMDSP devices can form co-processors to a host computer, with very little in the way of additional hardware. Usually, the main problem, in these co-processor systems, is the rapid transfer of data between the host processor and the co-processor. In the past, the sharing of a common memory space has been the usual method. However, processors such as the Texas Instruments TMS320C30, Motorola 96002 and AT&T DSP32 support direct memory access (DMA) controllers on-chip. An important point is that this allows efficient data transfer to and from on-chip as well as off-chip DSP memory.

17.3.3 Multiple processors

Many DSP algorithms can be distributed between several processors, without too much intellectual effort. This may often be done by the connection of processors in series and using a pipeline technique to distribute the computing, as shown in fig 17.5. Most GCMDSP devices support a serial port, primarily for the connection of the CODEC analogue/digital converters used in the telecommunications industry. Such serial ports can normally be used to support quite fast (about 5 Mbaud) serial communication between processors with a low hardware overhead. This can often provide sufficiently fast communication in series connected processors to handle data such as sampled audio.

Where multiple processor systems require fast bidirectional communication, memory to memory transfers between processors may be used. Dual port read/write memory can be used with most devices. The TMS320C25 [3] has hardware support on the chip to allow external data memory to be accessed by more than one processor.

17.3.4 On-chip testing

The in circuit emulator (ICE) is the traditional method for developing microprocessor systems. The ICE normally plugs into the socket intended for the processor and allows the target system program to be executed under the control of a monitor system. In this way, the program developer can set 'break' points to stop program execution under particular conditions and examine the processor status. The very fast clock speeds of modern GPMDSP devices and the large number of pins make the construction of an ICE increasingly difficult. The extra loading on input and output pins can affect performance. Even at slower speeds, it is not uncommmon to find that circuits which operate with the ICE do not perform correctly with the actual processor. A second problem is that when programs are executed in memory on the chip, there is no external bus activity. This is increasingly the normal situation. Therefore the monitoring of this by external ICE hardware, e.g. to set a breakpoint, is impossible. So is the use of logic analysers to monitor bus activity. The use of software breakpoints can also pose certain problems. For example, in the TMS320C25, the program memory where the breakpoint is to be may not exist when program execution is started; it may be configured as data memory. It is now a practical proposition to include a certain amount of program testing capability on the chip, to alleviate these problems.

The Texas Instruments TMS320C30 [4] includes facilities for processor register inspection and modification, breakpoint setting, timing etc. Communication with monitoring and controlling hardware is through a twelve pin connector, leaving the interface pins with only their design loading. The recently announced TMS320C50 [5] and the Motorola 96002 [2] also use a dedicated serial communication link for testing and debugging purposes. The JTAG IEEE 1149.1 standard describes a scheme intended for this purpose. The 96002 provides on-chip hardware to set breakpoints, according to addresses in all three main memory spaces; it also keeps a record of the last five instructions that have been executed.

This type of on-chip testing is becoming essential as projected cycle times reach 35 ns and on-chip memory is used for 100% of the program. It also means that the lower costs and improved printed circuit board layout of unsocketed surface mount packages can be used.

It should be borne in mind that the simulation of a DSP system can be used. This is often a more powerful and convenient environment for developing a system than the real hardware. A simulator can provide and store digital data streams in a more controlled manner than signal generators. Such input and output data streams are essential to the testing of most DSP systems. Some simulators can be expanded [6,7,8,9,10] and can both simulate and communicate with real hardware. Program development and performance evaluation can be carried out before hardware exists at all.

17.3.5 Specialised and low-cost systems

There is a small, but growing group of digital signal processors developed for specialised applications from general purpose architectures. One such is the Motorola DSP56200 finite impulse response filter, which may be adaptive. This is a development of the general purpose DSP56000. The TMS320SA32 encoder/ decoder for telecommunications applications is a mask programmed TMS32010, converting between 64 kbit/s PCM and 32 kbit/s ADPCM (CCITT G721).

These GPMDSPs have simply become components, where the user does not need to be concerned with program development. But the device has the repeatability and functional complexity which is possible with digital processing methods.

The Microchip Technology/ Texas Instruments TMS320C14 is a recent development of the early TMS32010 which is aimed at very cost-conscious mass markets. This processor has on-chip peripheral hardware, designed to reduce external hardware to a minimum. It is particularly aimed at a wide range of digital control applications such as winchester disc drive or hydraulic actuator control. On the chip are 4 counter-timers operating with 6 comparators and 6 'action' registers and it can produce and control six pulse width modulated signals at 80 KHz. There are 4 on-chip first in first out buffers. One timer is normally used as a quite sophisticated watchdog timer and ROM or EPROM versions are available. An interesting detail is that the serial port, incorporated in most GPMDSPs for CODEC interfacing, is able to support a general purpose asynchronous RS232 type of serial communication in the TMS320C14. It should be said that the minimal sized 16 bit data bus of the TMS320 has been an advantage in low cost design, because external memory can be cheaper than in a 24 or 32 bit system. The high performance for very low projected cost of this DSP is bound to open new markets, in spite of the rather old and unfriendly architecture.

17.4 BIT SLICE DESIGNS

Although there seems to be a trend towards putting general purpose processing power into the stored program general purpose DSP, some of the most efficient operations of the GPMDSP are algorithm specific.

Bit slice chip sets, developed in bipolar technology in the nineteen-seventies, allow the development of processors which are fast and specialised for a particular application. Bit slice designs are based on arithmetic logic units (ALU) which form only 4 or 8 bits of a data word, which could be much longer [11]. The ALU has a small vocabulary of operations. The system designer may build an instruction set from sequences of microinstructions. The microinstructions are held in microcode memory. Chips, dedicated to the construction of the remainder of such a design are normally available as part of the system; for example a microsequencer to address the microcode memory. The designer can, of course, incorporate specialised hardware.

Clock rates are normally in the 20 MHz region or higher and manufacturers claim throughputs of about four times current monolithic technology [12]. The main market for such systems is in the construction of minicomputers, high performance graphics generators etc. However, bit slice is ideal for implementing 'no compromise' signal processing computers. Because the architecture and instruction set are so controllable, the bottlenecks which particular algorithms may present to the GPMDSP devices may be avoided. They are also often used to benchmark or investigate experimental architectures, before monolithic construction.

17.4   VECTOR   PROCESSORS

One way of making use of concurrent hardware is to arrange for the same operation to be carried out concurrently on multiple data values. This technique, well known in large 'number crunchers', is well suited to the type of repetition

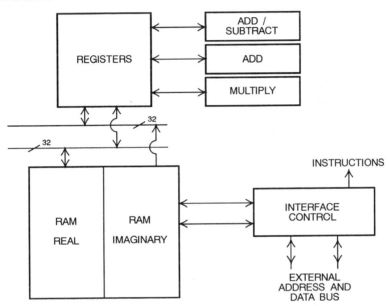

The   Zoran   ZR34325   vector   signal   processor

FIG   17.6

that occurs in signal processing. The Zoran vector signal processors (VSP) are produced in integer (ZR34161) and floating point (ZR34325) forms. These are designed to be used as co-processors, specialising in signal processing algorithms. The integer VSP has a vocabulary of only 23 instructions but they operate at a much higher level than those of a normal assembler language. An example is the ZR34161 instruction 'FFT' which can perform a fast Fourier transform on an array of 128 complex integers. The instruction takes 1856 clock cycles. They appear to the user more like subroutines, but are in fact constructed from specialised hardware and microcoded software. These devices do not use anything approaching the maximum possible degree of concurrency, like a true array processor. All the same, there are several processors operating in parallel. The architecture of the floating point processor is shown in fig 17.6. The dual port on-chip RAM may be used as an array of 64 complex or 128 real words of 32 bits and there are corresponding real and imaginary accumulators. The separate multiplier, adder and subtractor are able to operate not only concurrently, but on different data. Separate processing units allow the data transfer between internal and external memory to be carried out at the same time as instruction execution. These devices look rather like the SPS-41 of the nineteen seventies [13]. They both use parallel arithmetic processors and predefined high-level signal processing instructions. This assists in the programming of a structure with such complex parallelism.

The latest floating point Zoran processor has an extended set of 52 instructions, compared to the earlier integer version. One interesting point about this is that, not only are there are more signal processing functions but there are nine general purpose program control instructions CALL, PUSH, POP etc. This follows the trend in GPMDSP devices towards increased programming versatility as well as increased speed.   ·

17.5 THE INMOS A100 SYSTOLIC MULTIPLIER ARRAY

The transversal filter was one of the first digital signal processing algorithms to be implemented on a stored program computer. The algorithm is based on a series of multiply/ accumulate operations between a series of filter coefficients and a series of data samples. On a GPMDSP, multiply/accumulate operations are carried out sequentially as shown in fig 17.7. It can be seen that, because the multipliers and multiplicands are known before the sequence starts, they could be performed in parallel if more than one multiplier and adder were available. The Inmos A100 is such a system and is used here to illustrate the very high performance that can be achieved by a much more specialised architecture than the digital signal processors mentioned so far.

The architecture of the A100 [14] is shown in fig 17.8. It has an array of 32 multiplier/ accumulator stages. Coefficients are stored in RAM. Two sets of coefficients are stored in order that the inoperative set may be reprogrammed and switched in synchronously. The whole device is designed to be

$$A = A + X(1) * Y(1)$$
$$A = A + X(2) * Y(2)$$
$$A = A + X(3) * Y(3)$$

$$A = A + X(N) * Y(N)$$

Transversal   filter   algorithm

FIG   17.7

interfaced   with   a   general   purpose   computer,   which   can   perform this task. A 16 x 16 bit integer multiply is performed, although coefficient word length can be reduced down to 4 bits, in order to increase speed. The sampling rate can be 2.6 MHz, with 16 bit coefficients, up to over 10 MHz with 4 bits. Even the lowest of these figures is at least an order

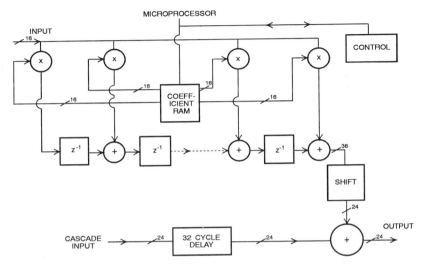

The Inmos A100 systolic multiplier array

FIG 17.8

of magnitude greater than can be achieved with serial operations of the single multiplier in a GPMDSP. The output of the multiplier/ accumulators is 36 bits, which is sufficient to handle the full dynamic range without over-flow. A shifter, provided at this output, allows the user to scale the 36 bit word to the most appropriate 24 bits for the output. In order to allow several A100 devices to be connected in series, a 32 stage, 24 bit wide shifter is pro-vided, able to delay the 24 bit output from a previous stage. A final accumulator adds the two 24 bit words from the present and previous stages.

Because the A100 coefficients are held in read/write memory, the device can be used for adaptive filters. How-ever, it may not be immediately obvious that it can be used to implement other digital signal processing algorithms, such as correlation and the fast Fourier transform (see fig 17.9) faster than most conventional digital signal proces-sors.

17.6 BENCHMARKS

Comparisons of the performance of different computer systems are notoriously unreliable as a guide to the performance of an unwritten program. However reading and quoting them tends to become an addiction. The fast

Fourier transform has become a standard benchmark for digital signal processors. Fig 17.9 shows some figures for a 1024 point FFT, usually as quoted by the manufacturers. What does stand out, is the fact that all the DSPs are much faster than a conventional processor, even with the aid of an arithmetic co-processor. Also that the possession of adequate data memory is an advantage and that the processors too new to be easily purchased are the fastest!

MANUFACTURER	DEVICE	TIME(ms) (fp)=floating point)
Motorola	DSP96002	1 (fp)
Zoran	ZR34325	1.2 (fp)
Inmos	A100	1.87
Zoran	ZR34161	2.4 (complex)
AT & T	DSP32C	3.2 (fp)
Motorola	DSP56000	3.4
Texas Instruments	TMS320C30	3.75 (fp)
Analogue Devices	ADSP2100	4.23
Texas Instruments	TMS320C25	7.1
AT & T	DSP32	7.4 (fp)
NEC	NEC77230	10.75 (fp)
Texas Instruments	TMS32020	14
Texas Instruments	TMS32010	42
NEC	NEC7720	77
Intel	8086+8087 (4.77MHz)	2500 (fp)

Execution times for 1024 point fast Fourier transform

FIG   17.9

The effect of using inappropriate benchmarks may be judged from a comparison of a repeated conventional multiply/accumulate operation i.e.
        a(1) * b(1) + a(2) * b(2)..........

with the multiplication and accumulation of three variables i.e.
        a(1) * b(1) * c(1) + a(2) * b(2) * c(2)..........

While almost all GPMDSP devices can perform the first at a rate of one machine cycle per accumulation, the second may take as long as 11 machine cycles per accumulation on some processors.

17.7   CONCLUSION

This survey of DSP architectures shows how difficult it would be to recommend one processor for all tasks. If economic considerations were included, a choice would become even more difficult. However, increasingly powerful general purpose digital signal processors are being supplemented not only by low cost versions, but by specialised architectures. The optimum processing environment is certainly worth looking for.

REFERENCES

[1] WE DSP32C digital signal processor, AT&T, December 1988.

[2] Motorola semiconductor technical data: DSP96002, 96 bit general purpose IEEE floating point DSP, 1989.

[3] TMS320C25 user's guide, Texas Instruments 1988.

[4] TMS320C30, The third generation of the TMS320 family of digital signal processors rev. 2.1.0 12 Feb. '88.

[5] Texas Instruments TMS320C50 preview bulletin, 1989

[6] Chance R.J. & Jones B.S. ' A combined software/hard ware development tool for the TMS32020 digital signal processor', Journal of Microcomputer Applications Vol. 10, 1988, pp 179-197

[7] Chance R.J., 'Simulation of multiple digital signal processor systems', Journal of Microcomputer Applications, Vol 11, 1988, pages 1-19

[8] Chance R.J., 'A system for the verification of DSP simulation by comparison with the hardware', Microprocessors and Microsystems, Vol 12, No 9 (October 1988) pages 497-503

[9] Motorola sim56000 digital signal processors user's manual 1986

[10] Bier J.C., 'Frigg: a simulation environment for multi-processor DSP system development', Master's Thesis, University of California, Berkeley, 1989

[11] Advanced Micro Devices, Am2900 family data book, 'Bipolar Microprocessor logic and interface', 1983.

[12] Texas Instruments, '32-bit chip set and bit-slice family' product overview, 1986.

[13] Allen, J 'Computer architecture for signal processing', IEEE Proc., 1975, 63,4,624-633

[14] The digital signal processing databook, Inmos/SGS Thomson, 1st edition 1989.

Chapter 18

DSP chips—a comparison

R. J. Chance

18.1 THREE 'TYPICAL' DIGITAL SIGNAL PROCESSORS

The most widely used type of digital signal processor is
the stored program monolithic processor, used for general
purpose signal processing and other applications requiring
fast arithmetic. Three devices: the AT&T DSP32 (DSP32),
the Motorola DSP56000 (56000) and the Texas Instruments
TMS320C25 (TMS320) will be used to illustrate the main
DSP architectures and techniques. In DSP terms, these are
all reasonably mature, dating from 1984 to 1986. They all
have particular, but different, architectural features which
have been exploited in the latest designs of several
manufacturers. Each of these processors belongs to a
family of similar devices; the intention is not to describe a
particular processor but rather to use concrete examples to
compare and contrast the techniques that are used in
current state of DSP technology.

All these processors perform a multiplication in one
machine cycle (about 80 ns in the most recent versions of
all the devices). All support a fast, synchronous serial port,
primarily in recognition of the fact that CODEC analogue/
digital converters, as used by the telecommunications
industry are important peripherals. All the devices can use
on-chip memory for both program and data. However, the
philosophy behind each is quite different.

18.2 The Texas Instruments TMS320C25

The TMS320C25 is derived from two earlier processors,
the TMS32010 and the TMS32020. Perhaps because of the
early introduction (in the TMS32010) of a 'stand-alone'
capability, a reasonably large program memory space and
some general purpose computing capacity, the TMS32010
has spawned the largest family of related digital signal
processors. The TMS320 uses a 16 bit data bus (a minimal
word size for signal processing arithmetic) and supports
only integer arithmetic. As shown in fig 18.1, it uses two
separate memory spaces for program and data. A single
external address and data bus allows both to be expanded
to 65k words. The 16 x 16 multiplier gives a full 31/32 bit
product. The 32 bit accumulator is loaded via a 16 bit

barrel shifter, which allows the 16 bit word to be positioned anywhere in the accumulator. Multiplier inputs normally reside, one in data memory and the other in the T (temporary) register, although one multiplier may be in program and one in data memory (see 18.6.1). The ALU operates on the accumulator and either the P (product) register or data memory, in a similar manner to most conventional processors.

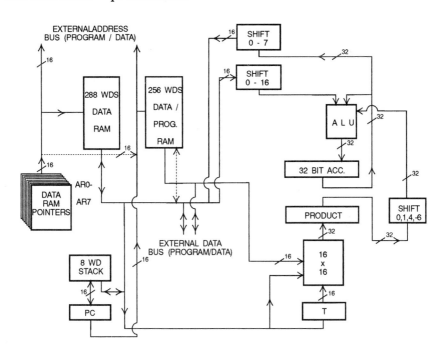

The   Texas   Instruments   TMS320C25

FIG   18.1

18.3   The Motorola DSP56000

Although the 56000 is the most recent of these designs, in many ways, it resembles the multi-chip ECL DSP designs of the nineteen-seventies. One program and two data memory spaces up to 65k words are used, as shown in fig 18.2. A 24 bit program and data word is supported by a 24 bit external data bus. Only integer arithmetic is used. A 24 x 24 bit ALU performs a single cycle unpipelined multiply and accumulate to load one of a pair of 56 bit accumulators. Note that the ALU input data are supplied from four registers, rather than directly from X or Y data memory. These may be loaded from data memory concurrently with an ALU operation. Although a shift can be performed when loading data either into or out of the accumulators, this is only a single bit left or right.

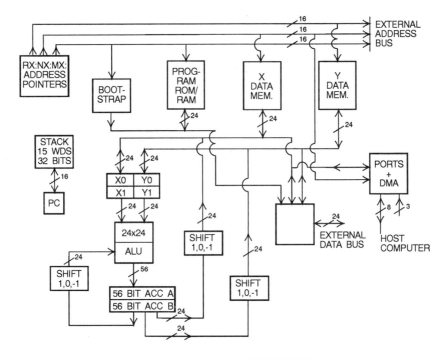

The   Motorola   DSP56000

FIG   18.2

18.4   The   AT&T   DSP32

This   design   was   very   advanced   at   the   time   of   its   introduction   in   1984   and   has   been   enhanced   since.   It   is   the   only   one
of   the   three   processors   to   properly   support   floating   point
arithmetic.   A   very   unusual   feature,   for   a   DSP,   is   the   single
von   Neumann   address   space   for   both   program   and   data.
This   means   that,   unlike   the   other   two   processors,   the   two
inputs   to   a   multiplication   do   not   have   to   be   segregated   into
two   address   spaces   for   the   maximum   efficiency.   The   basic
word   size   is   32   bits,   which   has   now   been   adapted   to   the
IEEE   standard   floating   point   format.

18.5   INTEGER   ARITHMETIC

Integer   arithmetic   is   only   a   viable   proposition   in   the
execution   of   many   DSP   algorithms   if   intermediate   values
can   be   scaled.   Therefore   shifters,   to   perform   this   scaling,
are   an   important   part   of   signal   processors   which   use
integer   arithmetic.   The   need   for   scaling   is   greater   with   the
16   bit   word   of   the   TMS320   than   the   24   bits   of   the   56000.
The   TMS320   and   56000   both   use   an   accumulator   which   is
at   least   twice   as   wide   as   the   data   word.   The   loading   of   a
twos   -   complement   integer   into   an   extended   accumulator

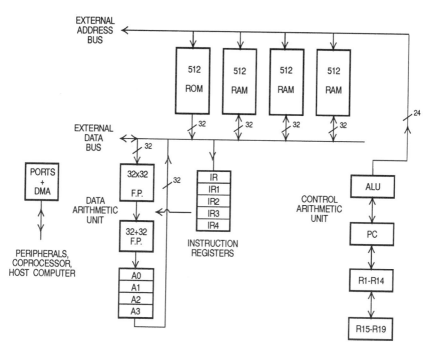

The   AT&T   DSP32

FIG   18.3

requires   that   the   sign   bit   should   be   copied   up   to   the
accumulator   sign   bit.   Although   these   two   processors
perform   integer   arithmetic,   it   is   often   helpful   for   the
programmer   to   regard   register   contents   as   fractional.
Therefore   the   maximum   value   of   a   16   bit   register,   for
example,   would   be   regarded   as   about   0.9999   (32767/
32768)   and   its   minimum   value   as   -1   (-32768/32768).   This,
so   called   Q15   format   is   not   mandatory;   for   example   it   may
be   convenient   to   use   a   Q14   format,   where   +1   can   be   exactly
represented   by   16384   (16384/16384).   The   multiplication   of
two   16   bit   twos   complement   Q15   numbers   gives   a   31   bit
result,   so   that   the   16   bit   result   is   found   in   bits   15   to   30   of
the   32   bit   TMS320   accumulator.   Both   the   TMS320   and   the
56000   are   able   to   perform   a   left   shift   of   one   on   the   product,
without   a   time   penalty,   to   normalise   the   result   of   a
'fractional'   multiplication.

The   TMS320   allows   values   to   be   left   shifted   up   to   16   bits
when   they   are   added,   subtracted   or   loaded   into   the
accumulator.   There   is   no   speed   penalty   for   this.   This
feature   allows   values,   which   are   held   in   memory   with
different   implied   scaling   factors   to   be   used.   Fig   18.4
illustrates   the   use   of   an   extended   accumulator   in   the
addition   of   two   numbers,   stored   to   16   bit   precision,   but

BIT 15 (SIGN)
1011000011110101   (-20235 represents -0.61752 Q15 in memory)

load to the accumulator with a left shift of 12 and
sign extension

```
                        #  BIT 31
                        #####  SIGN BITS
accumulator  =          1111101100001111 0101000000000000
```

0100000000000000   (16384 represents +1.0 Q14 in memory)

add +1.0 adjust scaling with a left shift of 13

```
+1.0 (Q14) represents       ####   SIGN BITS
in accumulator              0000100000000000 0000000000000000

perform addition =          0000001100001111 0101000000000000
```

store with left shift of 3 in Q14 format =

0001100001111010   (6266 represents 0.3824 Q14 in memory)

Using the shift and sign extension for fractional/integer
arithmetic in the TMS320C25

$$0.3826 \ (x2^{14}) = -0.6175 \ (x2^{15}) + 1.0 \ (x2^{14})$$

FIG 18.4

with different scaling factors. Using this technique, accumulator overflow may be avoided even during multiple
arithmetic operations. The 56000 does not perform a
multiple shift in one operation, but as it uses a 24 bit word,
there is not usually so much need to store values with
differing scaling factors.

It is clear from fig 18.4, that the management of the
scaling of intermediate numbers in integer arithmetic is a
non-trivial excercise, as far as the programmer is concerned. As a floating point processor, the situation is
rather different on the DSP32. A 24 bit mantissa is used
with an 8 bit exponent, effectively performing and recording
shift-scaling invisibly in hardware. Four accumulators of
40 bits may be used to hold a 32 bit floating point result.
The floating point arithmetic largly removes responsibility
for overflow control from the programmer.

When performing a multiply/ accumulate operation (fig
18.5) overflow is avoided in different ways in the two integer
processors. The 56000 performs a multiply/ accumulate in
one operation. In this case, a 56 bit accumulator gives 9
bits of headroom when the results of 24 x 24 bit signed
multiplications are accumulated. Thus overflow is avoided
without multi-bit shifts. However the output may need
scaling, when stored to 24 bit precision. The TMS320
performs the accumulation separately from the multiply. A

right shift operation may be performed on the product in the accumulation process, in order to avoid overflow. A further complication, in comparing these two processors, is that the need for such scaling may sometimes be eliminated by scaling down multiplier inputs, e.g. filter coefficients. The longer 24 bit word of the 56000, compared to the TMS320 helps in this.

If an analogue amplifier is overloaded, the output generally saturates at the maximum positive or negative value. While this is undesirable, the effect of an overflow in a conventional digital system is far worse. Adding a small positive number to a large twos-complement value results in a large negative result. For example, in a 16 bit system, 32767 + 1 gives -32768. For this reason, one of the features of the integer processors is saturation arithmetic. This prevents the overflow of a large value to a value of opposite sign. Thus, using saturation arithmetic in a 16 bit system, 32767 + 1 = 32767. It is worth noting that it is a property of twos complement binary arithmetic that a correct result is given even after the overflow of intermediate results, as long as the final result is within range. For example with 8 bit arithmetic:

```
0 1 1 1 1 0 0 0    (120)   +
0 0 0 0 1 0 1 0    (10)    =
1 0 0 0 0 0 1 0    (-126)  (incorrect due to overflow)
0 0 0 0 1 0 1 0    (10)    -
0 1 1 1 1 0 0 0    (120)   (correct 120 + 10 - 10 = 120)
```

This does not apply with saturation arithmetic. Care should also be taken if use is made of the 'roll-over' in conventional binary arithmetic, such as may be used to represent the cyclic nature of an angle.

18.6   MULTIPLIER   CONSIDERATIONS

The most efficient implementation of multiplication requires that the two input operands are delivered simultaneously on two busses, rather than consecutively by means of a single one. The 'traditional' digital signal processing approach to this has been to use two separate address spaces for input data and sometimes a third for the output. The difficulty of providing package pinout for multiple address and data busses means that the use of RAM on the chip is extremely important for supporting an efficient multiplication. A multiply/ accumulate algorithm for a FIR filter is described in fig 18.5. This is repeated for each sample. The x and k arrays represent data and filter coefficients respectively. They must both be delivered in a single machine cycle for optimum performance; ideally the addition and data transfer should also be performed in the same cycle.

The three processors adopt different approaches to this type of multiply/ accumulate operation.

```
i := 1;
a := k[i] * x[i];
i := i + 1;
repeat
        a := a + k[i] * x[i];
        x[i] := x[i + 1];
        i := i + 1;
until i = no_of_taps;
a = a + k[i] * x[i];
x[i] = data_sample;
```

A   FIR   multiply/accumulate   algorithm

FIG   18.5

18.6.1   The   TMS320   FIR   filter

In   the   TMS320,   the   two   multiplier   input   operands   are normally held in a dedicated register (T) and data memory. The   method   used   to   perform   a   single   cycle   multiply/ accumulate   within   this   architecture   is   to   use   program memory as one of the multiplier inputs. Fig 18.6 shows the steps   in   performing   a   FIR   filter   on   the   TMS320.   Coefficients are   in   program   memory   and   data   samples   in   data   memory. A   loop   counter   is   used   to   allow   a   single   instruction   to

```
1              zero   accumulator   &   product   registers
2              load   data   pointer   (highest   address)
3              load   coefficients   pointer   (lowest   address)
4              repeat
               {
                       accumulate   previous   product
                       load   data   to   T   register
                       load   data   to   data   address   + 1
                       multiply   by   coefficient
                       decrement   data   pointer
                       increment   coefficient   pointer
               }
5              accumulate   final   product
```

FIR   multiply/   accumulate   on   the   TMS320C25

FIG   18.6

be   repeated   in   hardware.   This   means   that   the   FIR   multiply/ accumulate   operation   is   carried   out   as   a   specialised instruction   on   the   TMS320.   The   'repeat'   operation   means that   the   instruction   fetch   only   needs   to   be   performed   once, when   the   multiply/   accumulate   instruction   is   first   encoun- tered.   The   multiply   and   accumulation   operations   are pipelined,   but   the   instruction   fetch   is   only   executed   once. The   disadvantage   of   this   approach   is   that   the   instruction   is extremely   specialised   aimed   at   only   this   one   function.   If variable   filter   coefficients   are   to   be   used   e.g.   for   adaptive filters,   program   memory   must   be   RAM.   A   block   of   on   chip memory   may   be   switched   into   either   program   or   data memory   space,   for   the   purpose.   Note   the   two   data   memory accesses per cycle used to move data. This can only be done using on-chip data memory.

18.6.2 Multiply/accumulate on the DSP56000

The 56000 uses two registers to hold the inputs to the multiplier. This processor relies for its efficiency on the pre-loading of these registers in the cycle before the data are needed. The register load operations may be carried out concurrently with a previous arithmetic operation such as multiply/ accumulate. In addition, the 56000 can carry out the multiplication and the accumulation in a single cycle, without pipelining. Fig 18.7 shows the FIR multiply/ accumulate operation on the 56000. Note that the two data memory areas, separate from program memory must hold the data and coefficients; they may then be loaded simultaneously.

```
1   load  data  pointer  with  address  in  X  memory
2   load  coefficient  pointer  with  address  in  Y  memory
3   load  multiplier  input  registers  A  &  B  with  data
    &  coefficient

4   repeat
        {
        multiply  A  by  B  &  accumulate
        load  register  A  from  X  memory
        increment  data  pointer  (modulo  N)
        load  register  B  from  Y  memory
        increment  coefficient  pointer  (modulo  N)
        }

    FIR  multiply/  accumulate  on  the  DSP56000
```

FIG 18.7

The unpipelined multiply/ accumulate has avoided a final accumulation operation. It could also be said that the 56000 FIR filter is performed by less specialised instructions than the TMS320, which could be used for other tasks. Notice that the multiplier input registers are being loaded with the next, rather than the current data. While this is no problem in the simple example given, the need for the programmer to look ahead to keep these registers supplied with data is an undesirable task in more general purpose programming. Modulo addressing (18.9.3) is used to eliminate the need to move data in memory.

18.6.3 The DSP32 FIR filter

The DSP32 is a von Neumann machine in the sense that it uses a single address space. Its speed comes from the multiple memory accesses that are possible during each instruction cycle and used in a pipeline. Each cycle can support one instruction fetch, two operand read operations and one memory write. The DSP32 FIR multiply/ accumulate operation is shown in fig 18.8 in DSP32 assembler language. This is very like C and the algorithm follows fig 18.5 quite well. The algorithm description is certainly more conventional than the 56000 or TMS320. However, the DSP philosophy has resulted in some peculiar, if legal, C. Take the basic multiply/accumulate:

a1 = a1 + (*r3++ = *r4++) * *r2++

Note the parallel data move (*r4 to *r3). In addition, the pipelining results in a 'delayed action'; the result of fig 18.8 is not available for three cycles.

```
/*   r2   points  to   coefficients
     r3  =  r4  points  to  data
     *r5  =  input,  *r6  =  output  */

a1 = *r4++ * *r2++                            /* 1st multiply */
do 0, r1                                      /* r1 = loop counter */
a1 = a1 + (*r3++ = *r4++) * *r2++             /* mult/acc & move*/
a0 = a1 + (*r3 = *r5) + *r2                   /* last m/a & sample */
*r6 = a0 = a0                                 /* output result */
```

A   DSP32   FIR   filter

FIG   18.8

In general purpose programming, which rarely has the symmetry found in DSP algorithms, the ability of the DSP32 to operate on any two values without segregating them into separate multiplier and multiplicand address spaces is surely an advantage.

18.7   ITERATIVE   LOOPS

Iteration is a fundamental operation in all computer programming. It has long been established practice to speed up iterative loops by using 'straight line coding'. In this technique, a repeated operation is coded by making multiple copies of the algorithm. This avoids creating a loop counter, decrementing it, testing its value and performing a conditional branch to the start of the algorithm. DSP algorithms are almost all iterative operations.

We have already seen that the TMS320 can repeat an instruction, by using a dedicated register to hold the number of repeats. In the 56000 and the DSP32, more sophisticated schemes are used to implement a multiple instruction iterative loop in hardware.

The 56000 supports a DO instruction, where the loop address (end of the repeated section) and the loop counter are pushed onto the system stack. Next, the program counter (start of the repeated section) is pushed. From this point onwards, the end of loop address is compared continuously with the program counter. If the end of the loop has been reached, the loop counter is decremented. The program counter is loaded with the start of the loop, if the loop counter is not zero, or the following instruction if the iteration has terminated. The 56000 hardware-implemented DO loop increases the speed of iterations, which involve several instructions. After the loop has been initiated, the iteration control is implemented in hardware at no cost in execution time.

In the DSP32, a similar DO instruction is able to repeat the group of instructions which follow, up to a maximum of 32. Up to 2047 repetitions can be performed, determined either by a constant in the instruction or the value in a register. These fast hardware implementations of generally useful constructs are not, at the present time, implemented without disadvantages and restrictions, such as a limited depth of iteration.

18.8 INTERRUPTS, PIPELINING AND LATENCY

The implementation of interrupt driven input and output systems is rather common in general purpose computing systems. In digital signal processing systems, however, the input/ output rate can often be several orders of magnitude more frequent; a typical figure of one sample every 100, rather than every 10,000 instructions is credible. For this reason, support for interrupts is often limited and inefficient in DSP devices. The implication is that there is rarely sufficient time for interrupt overheads in DSP work.

The simultaneous execution of several different operations, such as exists in a pipeline, creates particular difficulties when the operations have to be interrupted. Obviously, if the pipeline is broken to perform an interrupt, all the parallel operations must be restored to the original state before execution can continue. The TMS320 simply inhibits interrupts during a repeated instruction. This means that the programmer needs to be aware that the interrupt response time may be more than 200 machine cycles. The original DSP32 did not support interrupts at all. Interrupts have been added to the more recent DSP32C. The highly pipelined structure of many DSP32 operations has posed problems. The states of registers used in pipelined operations are saved in 'shadow registers', rather than on a stack. Therefore only a single interrupt level is supported. However, the hardware implemented DO instruction is not interruptable, so that in this case too, a long interrupt response time is possible. The 56000 supports two interrupt types, fast and long. The essence of the system is that a fast interrupt can be provided on the assumption that the processor status does not need to be saved. The purpose of the DSP interrupt is often to cause a minimal overhead response to an unpredictable event, which is part of the program being executed. This is different from the context switch of a conventional interrupt driven system.

The overlapping of the instruction fetch, decode, execute operations means that the instruction pipeline must be broken in all these processors when a branch instruction occurs. The TMS320C25 takes three cycles to execute a branch from off-chip memory. In the DSP32, the instruction immediately following a branch instruction is always executed, irrespective of the branch destination. This is but one example of the the interdependence of consecutive instructions, as a consequence of pipelining. If a pipeline is in operation, it may affect the instruction which is about to be executed. Various delays of up to four cycles (known as latency) occur in the DSP32, before data 'arrives' at its destination. For example

$$a0 = a1 + a2$$
$$a1 = a0 * a2$$

would not give the result $a1 = a2 * (a1 + a2)$, because it takes three cycles to transfer an accumulator to the multiplier input. Unfortunately, this sort of restriction means that although the DSP32 is a von Neumann machine, the algorithm is still architecture dependent.

18.9   ADDRESSING   MODES

18.9.1   Absolute   addressing

On the whole, digital signal processing algorithms avoid absolute addressing modes. Signal processing algorithms do not need to make much use of absolute addressing, as they are orderly and repetitive. On the other hand, general purpose algorithms can usefully access variables in an irregular fashion. This avoidance of absolute addressing is, of course, because reading the complete address from program memory can take extra machine cycles. The DSP32 supports an absolute addressing mode, which incurs no time penalty. This is possible because the 32 bit word size is much larger than the reduced 16 bit address, which is supported. The 56000 also supports an absolute addressing mode, but a second fetch from program memory is necessary to obtain it, with a time penalty. The TMS320 does not support true absolute addressing but uses a scheme which divides data memory into pages of 128 words. In this arrangement, a data page pointer defines the current page. Each page contains 128 words. This 7 bit address within a page can be defined within a single instruction word, as shown in fig 18.9.

16   bit   address   AAAAAAAAABBBBBBB
9   bit   page   pointer   register   AAAAAAAAA
16   bit   instruction   word   XXXXXXXXXBBBBBBB

TMS320   direct   addressing   mode

FIG   18.9

This means that a moderate range of addresses can be accessed randomly, without using a two word instruction. The addresses within a page can be contained within a single 16 bit instruction. The result makes for very fast general purpose programming in assembler language. However it presents major problems for assemblers, compilers and linkers in producing relocatable segments in data memory. It may be regarded, with equal validity as a type of indexed addressing.

18.9.2 Indirect addressing

As with many general purpose processors, the provision of a range of addressing modes is an important contributor to the processing speed of a DSP. DSPs are like RISC (reduced instruction set computer) machines in the sense that multi-cycle instructions tend to be avoided. The provision of an address pointer post-increment or post-decrement is probably universal in the DSP, not only because of the need to 'walk through' memory, but also because it can be more easily implemented in a single machine cycle than pre-increment/decrement. Any pre-increment/ decrement addressing mode would require address arithmetic to be complete before a memory access. The TMS320, 56000 and DSP32 all support post increment and decrement addressing modes.

18.9.3 Modulo addressing

Two adaptations of post-increment addressing, not found in general purpose processors, are commonly used in DSPs. A modulo N arithmetic may be used, where N can be specified by the user. This allows the creation of a circular buffer for the implementation of first-in-first-out (FIFO) or other buffers with zero time penalty, as shown in fig 18.10. The 56000 uses modulo addressing to implement the FIR filter. In this case, data samples can remain in the same memory location, only the address pointer needs to change.

ADDRESS POINTER BEFORE INCREMENT		ADDRESS POINTER AFTER INCREMENT
............111	(BUFFER TOP)	000
............000	(BUFFER BOTTOM)	001

Modulo 8 post-increment addressing
(addresses A3 upwards do not change)

FIG 18.10

This is one of many examples where the DSP design philosophy has produced an algorithm of general usefulness in hardware, rather than software. The circular buffer is very frequently implemented in software on CISC von Neumann microprocessors, with a considerable time penalty for cycling the buffer pointer.

18.9.4   Bit   reversed   addressing

A second type of post-incrementing addressing is directed at the reshuffling of FFT coefficients. The coefficients of a Fast Fourier Transform are deposited in memory in the wrong order. Addressing the coefficients with the address bits reversed restores the correct order. Anybody who has carried out this operation on a conventional computer or microprocessor will particularly appreciate the bit-reversed addressing, which has become a standard DSP addressing mode and is used by the TMS320, 56000 and DSP32. An address pointer is loaded with the lowest address of a buffer full of coefficients. An indexed addressing mode is used, where the content of a second register, containing half the buffer size is successively added to the address pointer. In this case, the carry is in the reverse direction when performing the address pointer arithmetic. The coefficients are thus accessed, by indexing the address pointer, in the bit reversed order, with no time penalty.

18.10   MULTIPLE   DSP   SUPPORT

Many digital signal processing algorithms contain obvious parallelism; the FIR filter is but one of many examples, see fig 18.5, for example. In theory, every multiply/accumulate could be carried out at the same time on a different processor, as is done with the Inmos A100. For this reason, several DSP designs offer some support for interprocessor communication in their architecture.

The TMS320 possesses on-chip support for memory to memory communication. A programmable data memory address decoder provides an external bus request signal to indicate that 'global', i.e. shared, rather than local memory is being accessed. If the shared global memory is free, the access is normal. If it is in use by another processor, the 'ready' input on the TMS320 is made unready by external hardware. This extends the memory access until the shared memory becomes free. The important point is that access to local memory is performed at full speed. Global memory is programmable in software and can occupy from 256 to 32768 words of data memory. This system is ideal where there is a large amount of bidirectional data traffic between processors.

All three of these processors can release external data and address busses, so that other devices may take control of external memory. The TMS320 and 56000 devices are able to continue processing from internal memory spaces while the external busses are inactive. The DSP32 has direct memory access (DMA) support on chip, as another method of accessing memory. The three DSP32 DMA channels can use 'cycle stealing' to make transfers between memory and serial or parallel ports without processor intervention.

The 56000 and the DSP32 are designed so that they can be

connected to the address and data busses of another processor without buffering hardware. The 56000 has a bootstrap program in on-chip ROM, which is able to load a program from the host computer. Thus, as a co-processor, the DSP need have no off-chip memory at all. An additional co-processor facility on the 56000 provides a vectored interrupt system which allows the host to initiate up to 32 functions.

18.11 CONCLUSION

Which is the 'best', or even the 'most typical' of these digital signal processors? These three DSPs have some features in common, such as fast, synchronous serial ports, special addressing modes and on-chip memory. Considering that they claim to target the same types of task, it is surprising how much they differ from each other. Whether the floating point arithmetic and DMA support of the DSP32 is more important than the strong co-processor support of the 56000 or the economy and memory to memory communication of the TMS320 can often only be judged by a careful study of the application.

REFERENCES

[1] TMS320C25 user's guide, Texas Instruments 1988.

[2] WE DSP32C digital signal processor, AT&T, December 1988.

[3] DSP56000/DSP56001 digital signal processor user's manual, Motorola, rev 1.

Chapter 19

Microcontrollers

J. D. M. Watson

19.1. INTRODUCTION

Over the last 14 years microcontrollers have provided the engineer with an attractive means of realising a broad span of increasingly complex and flexible instruments. Microcontrollers originated from the concept of the microprocessor system, typified by separate chip-level processor, memory and I/O devices. Such configurations had been used in systems for half a decade before technological advances and an appreciation of market requirements caused manufacturers to embark on the development of integrated solutions. The resulting devices combine many typical microprocessor system needs within a single device. Figure 19.1 shows the hardware configuration of an archetypical microcontroller. Key features include program execution from on-chip ROM rather than external RAM and a von Neumann architecture, i.e. shared data and instruction paths (even though data and program memory may be partitioned). Hardware and instruction set tend to be less 'general' in structure than those of a microprocessor. This biases devices to particular classes of application which require similar hardware resources; many industrial and commercial applications share common requirements and fall within these classes.

Developed before digital signal processors (DSPs), microcontrollers were aimed at I/O intensive, yet algorithmically simple, real time applications. However, the increasing functional complexity required of many contemporary products has motivated designers to look beyond microcontrollers and to consider using DSPs. Unfortunately DSPs are not usually provided with sophisticated I/O facilities (although on-chip memory is typically present) nor appropriate instruction sets, as they were conceived to implement the fast repetitive arithmetic sequences characteristic of discrete-time signal processing rather than I/O aspects of real time control.

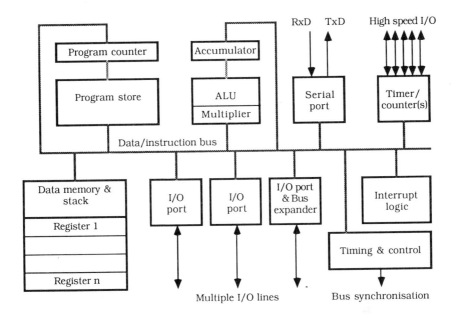

Fig. 19.1 Simplified typical microcontroller hardware structure

Contemporary applications for integrated controllers are demanding in throughput and function. The automotive industry for example, relies increasingly on high performance processors for the control of internal combustion engine firing and fuel mixture. Engine management has motivated the design and defined the architecture of several microcontrollers. The upward pressure provided by such market needs has had a two-fold effect. Recent microcontrollers have been designed with considerable hardware enhancement, particularly with respect to arithmetic capabilities and peripheral autonomy, and a new generation of DSPs, the so called digital signal controllers (DSCs) has been introduced.

19.2. HISTORICAL BACKGROUND

Intel Corporation, the company responsible for the first commercial microprocessor, the 4004 (Intel Corp (1976)), was heavily involved in the early development of microcontrollers. Since 1976 Intel has introduced several sets of devices, in particular the MCS-48, MCS-51 and MCS-96 families. Incremental development has taken place in each of these leading to greater speed, memory capacity and peripheral functionality. Now established as 'industry standards', the first two families have been replicated and enhanced by many other manufacturers. Philips, for example adopted the MCS-48 architecture in the formulation of their 8400 family, whose low cost and simple 'I squared C' on-card serial bus orientated the device towards the consumer market; Philips Components (1989). Siemens enhanced the MCS-51 to high levels of capability with the 80537/517 devices which, complete with analogue I/O and multiple ports can address sophisticated instrument requirements; Siemens (1988). The Intel MCS-96, using 16-bit internal busses, and a significant development from the 8-bit 48 and 51 families, was designed to meet the requirements of the automotive industry, including fast arithmetic and high resolution timing for injectors and ignition. In satisfying these requirements it also suited itself to a variety of demanding industrial tasks such as those associated with velocity and position control.

Many other manufacturers produce high performance microcontroller families; Hitachi, Motorola and Zylog are among important contributors. The choice of Intel devices for inspection in the following section does not imply that these are especially significant in performance; they, in the author's opinion, represent a historical strand in the evolution of the technology. Two examples of recent microcontrollers with latest-generation features will be described in a later section.

19.2.1. MCS-48

The MCS-48 family has separate data and program stores which share a common 8-bit data bus; Intel Corp (1982).The device requires fifteen clock periods per machine cycle. Principle hardware features include an ALU with accumulator, a register bank, a 1K byte program store ROM, a timer/event counter and multifunction bus and parallel I/O port controllers.

The instruction set was innovative when the device was introduced, having, for example, logical instructions capable of operating on ports directly and allowing branching on the state of selected accumulator bits. By present standards it seems restricted; the fact that all operations are based on the accumulator, and the lack of bit addressing and restricted arithmetic capability limit the MCS-48 family throughput.

19.2.2. MCS-51

As in MCS-48 devices, the MCS-51 uses an 8-bit architecture with separate program and memory stores; Intel Corp (1982), Intel Corp (1985). Design refinements permitted machine cycles of twelve clock periods duration and clock frequencies above 12MHz, yielding instruction times of between 1µs and 4µs. Throughput is significantly increased over the 8048 family. The cycle time improvement is augmented by the inclusion of significant hardware and instruction enhancements including an expanded arithmetic capability with subtraction, multiplication and (short) division. An ALU with accumulator, register bank, program store ROM, two 16-bit timer/event counters, developed interrupt controller, USART and multifunction bus and parallel port controllers are also provided. Address range capability is much increased over the 8048, with 64K byte spaces available (with external expansion) to both program and data memory. Contiguous stack address range is, however limited to 256 bytes.

The instruction set, whilst remaining inhomogeneous, is flexible and powerful in its design applications. As the accumulator is no longer required to be in the argument of move or logical operations, modifications of direct addressed locations (including I/O resources) are more efficient. Multiplication and division instructions facilitate fast scaling operations, the former taking typically 4μs to yield a 16-bit product. A sequence of multiply, add and move instructions can produce a 32-bit product from two 16-bit arguments in around 80μs. A further significant improvement over the 8048 family is the provision of Boolean variable manipulation through a single-bit processor. A section of data memory is defined to be bit-addressable and move, logical and branch operations are available on this area and on the carry bit, which serves as the single-bit accumulator. The bit-addressable data memory comprises two partitions, one of which contains registers pertaining to the on-chip hardware resources. Thus the programmer has the convenience of being able to address and inspect or modify individual bits of registers and hardware resources in an identical way.

19.2.3. MCS-96

MCS-96 represented a radical departure from the earlier Intel microcontroller concepts; Intel Corp (1985). Conceived around a 16-bit internal bus structure, the device uses a semi-Harvard architecture which permits simultaneous instruction fetches and data transfers. Instructions require a minimum of four state times to execute, each of which takes three clock cycles. At a nominal clock frequency of 12MHz, the minimum instruction time is therefore 1μs (e.g. direct addition). The maximum, for composite instructions (e.g. indirect with auto-increment multiplication), is 38 state times, corresponding to 9.5μs. Rather than being accumulator-centred, the 8096 uses a Register/Arithmetic Logic Unit (RALU) to allow direct operations on the register bank and registers associated with hardware resources. This obviates the need, in many cases, for move instructions prior to arithmetic operations. The ALU operates on 16-bit arguments and is supported by incrementers and shifters. While both program and data stores fall within a common 64K byte memory space, address ranges separate the functionally different memory areas. A three byte instruction queue implemented by an on-chip memory controller improves throughput by pipelining; post-branch flushing accounts for four state-times, however.

On-chip hardware resources are significantly extended with the inclusion of a programmable high speed I/O unit, two timers, an A-D converter with input multiplexer, a PWM D-A output channel, a serial I/O port and a watchdog timer. The high speed I/O unit (HSI/O) deserves particular attention as it has suggested an industry standard for the peripheral resources of contemporary microcontrollers and digital signal controllers. High speed input facilities allow the time of input transitions to be recorded using one of the internal timers. Four such input channels are available, and up to eight events can be recorded. For each input event a FIFO is loaded with the current time plus the state of the four input lines. The high speed output sub-system is used to autonomously trigger events at specific times. The events which may be effected include A-D conversion, resetting a timer, and switching up to six output lines. Eight or fewer events can be pending at any instant.

The instruction set is far more uniform than that of the MCS-51 or 48 families, the RALU architecture permitting efficient implementation of operations between registers, not involving an accumulator. Most classes of instructions are available with this facility, some supporting three direct register designations; two input operands and one result. The instruction set includes some operations typical of sophisticated microprocessors like normalisation of integers. Unlike the eight-bit microcontrollers previously described, the MCS-96 supports a variety of data types; bits, bytes, words, double words, short integers, integers and long integers. This gives flexibility to the assembly programmer and permits high-level language compilers to be efficiently implemented.

19.3. APPLICATIONS

Applications for microcontrollers are manifold and centred on requirements needing close-coupled, locally small-scale, real time control via Boolean and analogue I/O. Applications often lie in cost sensitive products, such as those typical of consumer or commodity electronics.

Small-scale applications, be they in cars, video recorders or industrial instruments usually share certain common characteristics;

- Some form of man-machine interface
- Deterministic operation time frames
- Multiple input / output channels
- Low functional redefinability
- Firmware based program store.

Man-machine interfaces usually lie in a complexity span which ranges between simple LED Boolean indicators and functional keys to dot-addressable liquid crystal displays with pointing devices or keypads. Program execution is typically coordinated by a simple real-time operating system, usually no more than a scheduler. Many microcontroller applications are closely coupled to transducers and actuators; e.g. limit switches, relay drivers and D-A converters. Systems using the devices typically have a low component count; bus buffering and I/O expansion largely defeat the microcontroller's object. Applications tend to be functionally specific and there is usually no need to make provision for loading new programs, other than during a prototype development cycle. Rather than executing out of RAM (loaded from disk) as in microprocessor systems, microcontroller-based products usually use an on- or off-chip ROM / EPROM firmware program store.

19.3.1. Real-time control

These tasks include basic signal processing and must therefore occur at well defined time intervals. Data acquisition, digital filtering, closed-loop compensation, range clamping and parameter adaptation are typical.

The bandwidth of closed loop control realised through microcontrollers has been until recently, low, with inner loop times of a few milliseconds being at the limit of technology. This has limited application in the field of high bandwidth servo control, particularly where plant models must be computed. Recent developments both in advanced microcontrollers based on high performance microprocessor kernels, and in the adaptation of DSPs to DSCs, have allowed this barrier to be surmounted. State-of-the-art real-time control is now practicable on time frames measured in a few hundred microseconds. The keys to this breakthrough are improved resolution (16/32-bit processors), efficient arithmetic instructions associated with dedicated on-chip arithmetic hardware resources, RISC-type architectures needing fewer clock cycles per instruction, and increased maximum clock rates.

19.3.1.1. Industrial controllers.

Since the introduction of 2nd generation 8-bit microcontrollers, it has been possible to manufacture cost- effective digital process control instruments. Market expectations are for such equipment to become progressively more compact and functionally capable.

The Eurotherm Ltd 818 provides a good example of the functions presently available in a microcontroller-based process control instrument. Measuring 96 x 96 x 219 mm it provides the following facilities:

- Input resolution of 12 bits (0.024%) on all ranges

- Linearisation for thermocouples and resistance sensors (RTDs)
- Outputs options include:
    - Relay
    - Triac
    - Logic
    - Analogue proportional

- Alarms on trapped process variable excursions

- Multi-field 4 1/2 digit vacuum fluorescent display with manual intervention through mode and raise/lower keys

- Digital communications through RS232, RS422/485 using X3.28 protocol operating at baud rates from 300 to 9600

- PID control structure with manual adjustments including:
  - Proportional band (gain)
  - Integral and differential times
  - Approach limit
  - Deadband
  - Output limits
  - Setpoint rate limit

- Auto-tune - one-shot load identification algorithm which derives settings for PID

- Adaptive tuning - continuously operates (when selected) to allow the tracking of time-varying plant parameters

This large array of functions implies considerable hardware resources. Advanced manufacturing techniques are required despite the reduced parts count to allow the instrument to fit in the prescribed enclosure.

The controller is designed for applications with cycle times in tens to hundreds of milliseconds as in control of furnaces with time constants of tens of seconds or more. Advanced microcontrollers, DSPs and DSCs make much faster loop times possible and extend process control capability to industrial motor drives and servo control systems.

19.3.1.2. Automotive systems. Many uses for microcontrollers exist within car systems. They range from the non time-critical; seat configuration, security, trip computer and fuel consumption gauge, to the highly deterministic real-time applications involved in engine control.

The 'management' of an internal combustion engine involves outputs controlling gas ignition and fuel mixture, and inputs associated with accelerator setpoint, air and engine temperature, engine revs and manifold pressure. Additionally, microphonic sensors may provide information concerning the onset of detonation or pre-ignition. Timing signals to activate for example, fuel injectors, must occur at rates of up to 80 per second per cylinder (Otto cycle engine at 10,000rpm). The required angular timing resolution is likely to be less than 1 degree, which implies internal counter increment rates of 40kHz (period of 25µs), and very low timing scatter. Firing algorithms must be highly deterministic. Early generations of microcontrollers were unable to sustain this computation rate in software, so units with on-chip support hardware, such as the MCS-96, were introduced. These are capable of handling such requirements in a semi-autonomous fashion.

The correspondence between engine timing, engine speed and manifold pressure (plus other variables) is not simple, and engine management systems often use a two or more dimensional look-up table to determine the correct timing. Efficient instructions for interrogating the table and for interpolating values are helpful. Similar table look-up activity is associated with linearisation of input transducers. Closed loop and adaptive control methods may also be employed.

Engine management, by virtue of its computational needs and large potential market value, has led to the rapid evolution of advanced microcontrollers, and may also have precipitated the development of digital signal controllers. The smaller and more diverse industrial control market might not have had the same impact on semiconductor manufacturers.

19.4. TYPICAL ALGORITHMS

Many algorithms are run on microcontroller systems. A proportion of these are trivial or 'ad-hoc', but some are significant in complexity and generality. In attempting to explore a limited sub-set of such methods it is appropriate to focus on those which highlight the relative strengths and weaknesses of microcontrollers and DSPs/DSCs.

19.4.1. Scheduling

The scheduling of tasks so that they are performed at prescribed intervals in real time is crucial to processes of physical control. In discrete-time representations of transfer functions, the sampling or loop update interval scales the effect of time constant coefficients. For predictable control performance, these intervals must be of equal length with closely defined timing scatter. Other applications of microcontrollers include the generation of firing pulses for power electronic converters and for engine ignition systems. Again, precise timing is necessary.

The concept of a scheduling system is straightforward and in simple systems, easy to implement. More complex equipment requires a real-time operating system; a flexible scheduler in which tasks can be dynamically added or removed. Basic microcontrollers are not sophisticated enough to justify the use of these in most cases. Simple microcomputer or microcontroller applications with loose timing constraints and predictable program flow (i.e. little use of interrupts) can utilise polling to schedule the invocation of tasks, but in realistic applications this is neither possible nor desirable.

Most controller applications are characterised by multiple time-frame processes. The following might apply;

- Data acquisition every 1ms
- Inner loop calculation every 50ms
- Outer loop calculation every 500ms
- Calibration cycle every 2s.

In addition, non critical keyboard scan and display refresh sequences may form a background task. Serial data arriving at random times will be received and will cause interrupts to driver routines at unpredictable intervals. Clearly a rational method of synchronised coordination is needed.

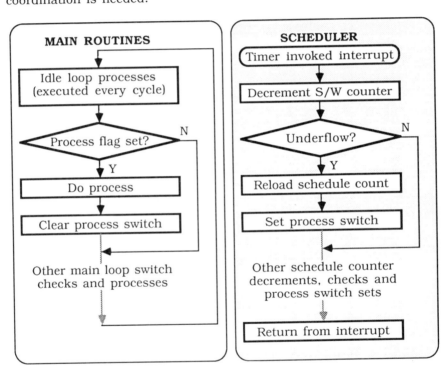

Fig. 19.2 Basic scheduler structure

A commonly used technique is shown in Figure 19.2. A hardware timer (usually provided in current microcontrollers) is configured to provide a high priority interrupt on overflow. This might happen every 256μs in an MCS-51 system. Its interrupt routine (the scheduler), comprises a number of register decrement and test operations, each associated with a scheduled task. The registers are initialised with the number of timer overflow periods corresponding to the scheduled time interval to which they relate. In the event of a register reaching zero, it is reloaded with the relevant task interval, and a process enable switch bit is set. This switch permits the scheduled task to execute at its next opportunity. Typically the decrement, test and branch operation may be efficiently performed with one 'DJNZ' type instruction. When the scheduler interrupt routine has completed its decrement and test operations for all task registers, it exits through a return from interrupt. On reentering the main routine, program flow continues through 'idle loop' processes until a task routine block is encountered. A branch instruction at its head tests the relevant switch bit which can authorise execution. If set, the task is undertaken. At its termination, the switch bit is reset, this state being maintained until the scheduler sets it again.

Obviously if the enable bit is not set, the process is skipped and program flow continues to the next task. In this way tasks can be coordinated in a non-deterministic environment which includes interrupts. Worst-case timing scatter can be calculated as a function of maximum execution times of interrupt routines, the worst case main loop time and the scheduler interrupt period. If the main loop is long, multiple tests and branches for a given process enable bit might be appropriate. Time-critical routines are thus invoked by bits set during hardware counter-triggered interrupt routines. This common function illustrates the need for hardware timer/counter resources, and the usefulness of bit addressabilty to facilitate software flags and switches.

19.4.2. PID

Proportional plus integral plus derivative compensation has long been established as a preferred technique for the majority of industrial continuous process control systems. It is robust, reliable and easily understood by the plant engineer. It has the drawback however, of only being able to fully compensate second order plant. PID controllers are readily implemented in analogue and digital hardware, and microcontrollers are now used in applications where control bandwidth requirements are modest.

In a typical single control loop, perhaps the inner loop of a cascaded system, a setpoint is fed to a difference stage where it is combined with the transduced feedback from the process variable. The difference, which is the error between desired and actual values, is fed to the controller and actuator block G1. This has three basic parameters defining gain, integral and differential times. The resulting output variable, which has been compensated and amplified to a power level, acts upon the plant. This will have one of many types of transfer function; a furnace example might approximate a second or third order response. In a continuous time representation, the compensator has transfer response:

$$G_1(s) = \cfrac{1}{1 + \cfrac{1}{sT_i} + sT_d}$$

19.1

And, if a backward difference mapping of s to z is performed to enable discrete time implementation;

$$O(t) = O(t-1) + k_p((1 + k_i + k_d)I(t) - (1 + 2k_d)I(t-1) + k_d I(t-2))$$

19.2

Where,

$$k_i = \frac{T}{T_i}$$

and

$$k_d = \frac{T_d}{T}$$

$O(t)$ and $I(t)$ are outputs and inputs at timestep t. T is the loop sample interval.

The backward difference approach is widely used, although other mappings such as the Bilinear or Tustin give a better approximation to the continuous case for lower sampling frequencies. Katz (1981) provides some useful comparisons between the discrete-time mappings for s and shows that the backward difference method or 'mapping of differentials' has good low frequency gain fidelity, but poor phase and high frequency gain performance, even when sampled at 15 times the crossover frequency. The bilinear approach on the other hand preserves both magnitude and phase performance for sampling ratios as low as 8 times the crossover frequency.

In practice, a PID controller is considerably more complex than described; the D term is normally rolled-off by at 3 - 10 times its zero corner frequency, and measures are taken to avoid 'integral wind-up' which occurs when the output variable is constrained (as it must be in a practical system). Various non-linearities are often also built in to allow optimum large and small signal responses. Astrom (1984) deals with some of these refinements.

19.4.3. Digital filtering

A comprehensive treatment of this wide ranging topic is beyond the limited space available here and other chapters will deal with it in more detail. It is however important to recognise that filtering forms part of many microcontroller applications. Low pass and notch filtering is frequently applied to the time series resulting from analogue data acquisition, and it is noted that control system compensators are nothing more than special-purpose filters.

Two general approaches for implementation are available, the infinite impulse response (IIR type) with rational pulse transfer function of the form;

$$G(z) = \frac{a_1 + a_2 z^{-1} + \dots + a_k z^{-k}}{1 + b_1 z^{-1} + \dots + b_n z^{-n}}$$

19.3

and the finite duration impulse (FIR) response form, defined by;

$$G(z) = a_1 + a_2 z^{-1} + \dots + a_k z^{-k}$$

19.4

Filters are characterised by more than their impulse response; the following are key properties:

* Amplitude response
* Phase response
* Group delay
* Magnitude response

Given a complete description, the designer can consider how the filter may be realised. A variety of discretisation methods are available, and each has strengths and weaknesses with respect to preserving particular properties of a continuous-time prototype. Katz (1981) and Rabiner (1975) explore these considerations in depth.

The FIR topology or 'all zeros' filter has the advantage of being able to offer a linear phase response (constant group delay) and guaranteed stability, but must be of a higher order than a IIR filter of equivalent characteristics.

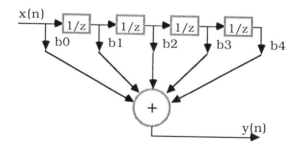

Figure 19.3 FIR filter structure

The mechanisation of the two types of digital filters is seen in Figures 19.3 and 19.4. The main operations involved are data buffer shifting (to implement unit delays) multiplication by weighting coefficients, and the formation of sums of those products. In both filter types throughput is enhanced by the availability of processor operations which both multiply and accumulate. Instructions which promote efficient indexing of data arrays, ideally incorporated into arithmetic operations, are also clearly advantageous.

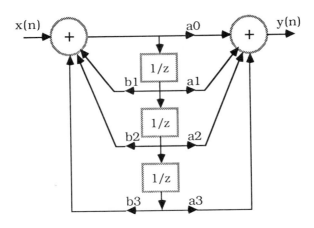

Figure 19.4   IIR filter structure

19.5. PERFORMANCE METRICS

In selecting a microcontroller or a digital signal controller for a particular task, the task specification will provide guidelines for the choice of processor. Applications which major on hardware bit manipulation and those involving signal processing require different processor resources. It is useful to identify the broad headings under which micro and signal controllers may be compared.

19.5.1. Throughput

Instruction execution speed may appear at first sight to be a clear measure of processor performance. This is not the case however, because algorithm executions per second are the significant factor in the real time control context. Benchmark programs are helpful in comparing performances, as instruction sophistication varies widely.

Processor architecture governs the feasibility of rapidly executing composite instructions. Harvard architecture with its single instruction, multiple data (SIMD), structure facilitates these, allowing simultaneous instruction fetches and data operations. DSPs have adopted this approach and typically use it in a RISC framework. The strengths of microcontrollers have been developed through the on-chip embodiment of I/O - efficient instructions and semi-autonomous hardware resources - USARTs, timer counters and A-D converters.

In assessing throughput the following should be considered;

• The activity in which the application spends most of its computational effort

• Speed of relevant instruction executions in a target processor

• The evolution of benchmark programs and their application in hand or machine simulation studies

• Execution of benchmarks in target emulator/development environment

• Assessment of associated 'housekeeping' tasks in which the selected processor may have indifferent capability, e.g.
  • Handling interrupts
  • Implementing a real-time scheduler
  • Setting and checking individual bits in registers and I/O
  • Obtaining constant table-based values.

19.5.2. I/O count

I/O channel count; analogue and digital (serial and parallel) is an important indicator of processor capability. Some microcontrollers offer external port expansion, but if this has to be used in average size systems, the advantages of the microcontroller are questionable. External I/O resources usually involve more elaborate instruction sequences for access than similar facilities in the integrated on-chip I/O space, and so are undesirable both in terms of parts count and execution speed.

19.5.3. Peripheral autonomy

All microcontrollers are equipped with hardware resources having some degree of autonomy. Serial ports are now commonplace, yet perform a function which would be very time-consuming in software. They are usually autonomous to the extent that a byte loaded will be transmitted with an interrupt generated as the last bit is emitted. Similarly, on the receive side, up to two bytes may typically be received before the processor is interrupted.

19.6. ARCHITECTURE

An impression of microcontroller architecture will have been gained from Figure 19.1 and the previous sections.

19.6.1. Separate data and program memory partitioning

In low and medium performance microcontrollers, data registers and general-purpose RAM space are often partitioned from program store ROM or EPROM either by address ranges or by separate memory enabling. In such arrangements, I/O space is frequently mapped into register space, where it can be treated on a bit or byte level. Although the architecture may appear to be of the Harvard type, program and data highways are often shared, leading to a performance bottleneck.

19.6.2. On-chip peripherals

A variety of hardware enhancement have been included in microcontrollers. Early units such as the TMS1000 family which were closely related to calculator ICs, included keypad scanning and LED display multiplex hardware. More recently developed units embody sophisticated communications handlers (C.F. Intel 8044) and other facilities such as high speed input/output units.

19.6.2.1. Timer/counters.
Timer/counters may be configured to count either a divided version of the system clock or transitions on an input pin. Current count value is often accessible as a register variable, and counter overflows or underflows can trigger system interrupts.

Such systems typically operate in several modes. These may include a frequency synthesis capability where, on over/under flow, the timer/counter is automatically reloaded from a reload register. Frequency is defined by the loaded period value. Useful for generating scheduler interrupts, this type of arrangement often sources, through internal configuration, a baud rate clock for serial communications.

19.6.2.2. Serial I/O.
One or more serial interface ports (USARTs) are often provided. These can operate in asynchronous (without clock channel) or synchronous (with clock channel) modes, covering rates from a few tens of baud to several megabaud. Full duplex operation (separate transmit and receive channels) is customary and receive errors typically set status bits to allow faults to be trapped and reported.

Enhancements may include an address recognition mode (for use in multi-drop configurations) in which the CPU is only notified by an interrupt if the USART receives a unit address rather than a data frame. Further enhancements may extend to DMA handling of serial I/O data blocks, and hardware support for various protocols.

19.6.2.3. Parallel I/O.
Typically provided through several byte-wide ports, these are often multi-purpose and used for address and data words when the microcontroller is operating with external memory or I/O. Ports may be designated bidirectional, being provided with weak pull-up resistors to define the output high state. When required to be used as inputs, '1's must be written to the appropriate bits to release them.

19.6.2.4. Arithmetic enhancements. Arithmetic hardware typically includes a multiplier in addition to the addition and subtraction circuits found in the ALU. Hardware enhancements to the program flow control systems can also be made through the inclusion of program counter and index register incrementers. The former obviates the need to tie up the ALU and system busses when performing the trivial operation of indexing the next instruction for execution.

19.6.2.5. Comprehensive interrupt systems. Interrupt systems are often prioritised to two or more levels, and interrupt source masking is supported. Usual interrupt sources are;

- Timer counter overflows/underflows

- Serial channel byte received or transmit buffer empty

- Hardware pin inputs

- A-D conversion ready

19.6.2.6. A - D converters. Despite the considerable technical difficulties of integrating an analogue function with high noise susceptibility onto a digital LSI, several manufacturers offer microcontrollers with integrated data acquisition systems. Typically these operate through the method of successive approximation and are limited, presumably by the fabrication process and substrate noise, to an accuracy of 8 bits and a resolution of 10 bits or less. A means of varying the internal reference voltage under program control is provided by some manufacturers and gives a limited facility for input scaling. Conversion time is of the order of 40µs for 8 bits accuracy. It is usual for the manufacturer to provide multiple analogue input channels which are selected for conversion by an on-chip analogue multiplexer. Anti-aliasing filters (and, if not integrated, sample-holds) must be arranged outside on a per-channel basis.

19.6.2.7. Watchdog timers. A means of restoring correct program flow in the event of a hardware or software fault (causing 'wild' execution) is essential in instrument-scale microcontroller systems. This function is often provided by an external retriggerable monostable whose output is tied to the processor reset line. The monostable is retriggered within its time-out period by a pulse train emitted by the microcontroller. This may be generated by a scheduler-driven main loop process. In the event of spurious program flow, the main loop path will be missed and the watchdog reset pulse train will cease, leading to the monostable triggering and processor reset being asserted.

This type of functionality is now often provided in digital form as an on-chip resource. A hardware timer takes the place of the monostable. Special instructions, often with a two-level interlock, are provided to reset the counter, which may have a time out period of 10 - 20 ms.

19.7. INSTRUCTION SETS

As discussed under specific example microcontroller headings, instruction sets are orientated towards simplifying the types of operations commonly encountered. Early devices were constrained by architecture and silicon real-estate and this limited opcode functionality. Current examples with complexity similar to 'top-end' microprocessors do not suffer from these restrictions, and offer a range of sophisticated and highly composite instructions.

19.7.1. Structure & efficiency

A key aspect of the structure of an instruction set is the uniformity of application of each operation type across the available repertoire of operand reference modes. These might include;

- Immediate
- Direct
- Indirect
- Indexed

It is advantageous to have the facility to perform arithmetic or logical operations with all these types. The hardware feature of a RALU also enhances efficiency by allowing direct - to - direct operations without the intermediate bottleneck of an accumulator. Efficiency is improved through the provision of composite instructions, for example those which perform an arithmetic operation on an indirectly referenced variable and which also increment the pointer to that variable. In sophisticated microcontrollers like the Motorola 68332 this type of facility has been enhanced to the point of providing an instruction which can produce a linearly interpolated value from a look-up table.

Clearly the provision of arithmetic instructions for each of a variety of data types (categorised by word length) can save a number of short operand instructions and result combine operations. Multiply and divide instructions further reduce both code lines and execution times.

It has become increasingly fashionable to write applications code in high-level programming languages (for example 'C'). Instruction sets of recent microcontrollers have been designed with at least some recognition of the needs of compiler writers.

19.8. ATTRIBUTES

Microcontroller attributes are (by design) well suited to small and medium-scale instrumentation applications. Their strengths lie in;

- Efficient Boolean variable manipulation

- Autonomous on-chip hardware resources

- Appropriate I/O structure and I/O instructions

- Fast multivector interrupt capability

- Highly integrated structure allowing low parts count implementations

Weaknesses are largely overcome in current and future generations of 'super' microcontroller. Restrictions associated with earlier families are centred on;

- Limited program and address space

- Inhomogeneous instruction sets

- Low performance stack systems

- Poor support for high level languages

- High cost of expansibility of memory and I/O spaces

- Limited arithmetic capability making devices unsuitable for applications needing intensive signal processing.

19.8.1. Differentiation from DSPs

It is interesting to highlight the differences between DSPs and microcontrollers. Both classes of device have their own areas for deployment, but these increasingly overlap in the realms of fast real time control.

DSPs were originally designed to handle the signal processing tasks typical of filtering and allied operations. In common with microcontrollers they were intended to fit into environments with minimum associated support components, and to execute programs and manipulate data from on-chip ROM and RAM, respectively. Unlike these parts however, little on no provision was made for sophisticated multi-port I/O. Signal processing needs few data channels, and these are typically connected to A-D and D-A converters. A limited I/O capability therefore sufficed. A data bus and limited address and control lines were all that was necessary to fulfil these requirements. The scheduling of processing loops was typically trivial in early applications as one one or at most a few deterministic tasks were being executed. Timed software loops could therefore be used to synchronise the activity. One early DSP was designed for executing in-line operations and was provided with a single branch instruction for returning program flow to the beginning of the sequence.

Microcontrollers on the other hand, were originally produced as a means to replace hardwired logic and so emphasis was placed more on flexible and multiple I/O facilities than on fast arithmetic. Internal architecture was ad-hoc to fulfil the needs of cost-effective implementation. Current applications show a trend towards fast real time control using substantial algorithms and with a multiple I/O requirement. The convergence of required capability between microcontrollers and DSPs has reached a point where advanced devices in each category are less easily differentiated.

19.9. RECENT MICROCONTROLLERS

Mention has been made of recent microcontrollers with sophisticated functionality, and it is appropriate to select two of these for further description. The chosen examples are from a wide range of current devices, and are pitched at different performance levels.

19.9.1. H8

The H8 microcontroller from Hitachi is a member of the set of 'H' devices which include 8, 16 and 32 bit architectures. Applications span medium scale control through to high performance information processing; Hitachi (1988).

The H8/330 has certain similarities to the MCS-51 and MCS-96 families in its support of up to 16 bit transactions in an environment using both 8 and 16 bit busses (the latter in the important RAM-CPU linkage). Using a RISC philosophy, instructions require a minimum of only two states for execution, these being comprised of two clock cycles. The maximum crystal frequency is 20MHz, so minimum instruction time is 200ns. More complex instructions such as multiply and divide require 14 cycles, implying an execution time of 1.4µs. This is a significant throughput improvement over previous microcontroller families. The device does not appear to be accumulator-centred, and register - to - register operations are available. Byte and word types are supported for most instruction categories, with bit operations also provided within their own instruction categories. Program and data stores fall within the same address space, which is partitioned by address ranges into on-chip registers and RAM, on-chip ROM and external address space. The total address range is 64K bytes. Three modes of memory configuration are available allowing all or some of this range to be mapped to external memory.

On-chip hardware resources are well selected for high performance applications. A particularly useful facility is a dual-port RAM through which the device can communicate with master processors. This comprises 15 data bytes and one control byte. Other resources include 16 bit and 8 bit timers, a dual-channel PWM timer, a serial I/O port and an 8 channel A-D converter. A number of parallel I/O lines whose configuration depends on device usage are also available. The timers are supported by three internal and one external clock source, and are connected to two independent comparators, enabling the generation of two synthesised waveforms. Input capture is also provided, enabling four channel time-stamping of events signalled by rising or falling edges.

The instruction set is RISC (Reduced Instruction Set Computer) -like with all instructions 2 or 4 bytes in length. 57 instructions are provided, with a general register-to register architecture. Bit manipulation is supported, allowing set, clear, not and logical operations. Byte and word data types are available with most operations. An 8 by 8 multiply yielding a 16 bit result, and a 16/8 divide instruction are included.

19.9.2. MC68332

This highly sophisticated microcontroller was launched by Motorola during 1989 and was apparently motivated by the automotive markets. Based on a 68020 core processor, it elevates microcontroller performance to heights previously associated with workstation-style CPUs. The MC68332 uses 32-bit data paths and embodies some very high performance hardware peripherals including a high speed autonomous timed I/O facility and a queued serial communications module; Motorola (1989).

Using a CPU similar to the 68020 and implementing the majority of its instruction set, the device is provided with 2K bytes of standby RAM. Programs are executed from external memory, as in a microprocessor, and a 24-bit address bus is supported. Memory is partitioned into separate user and supervisor spaces, each with a stack pointer. System clock rate is 16.78MHz and it is estimated that typical instruction times are in the range 400ns - 4μs depending upon instruction complexity. The device implements virtual memory (address range of 16M bytes), has a loop mode of instruction execution and fast multiply, divide and shift operations.

Hardware enhancements include an intelligent 16-bit timer (TPU) with 16 programmable channels in which any channel can perform any of the functions; input capture, output compare, PWM generation and others. A microcoded local processor handles event timing and supports advanced functions including stepper motor control. A queued serial module (QSM) is incorporated giving the 68332 two serial interfaces. One, a full duplex synchronous channel, embodies an I/O queue and is intended for communication with peripherals and other microcontrollers. The other provides a standard USART facility. Programmable chip select logic is a further valuable hardware resource. This facilitates address space decoding and selection of external devices without the need for off-chip logic.

Bearing a strong resemblance to the 68020, the 68332 is equipped with a powerful and sophisticated instruction set, well suited to the implementation of HLL compilers. Two instructions have been added to the base set of the 68020, they are 'table lookup and interpolate' and 'low power stop'. The former performs a linear interpolation between adjacent ordinate values, whilst the latter forces the device into a low power mode until a suitable interrupt occurs. Certain 68020 instructions involving bit fields, call/return module, compare and set, coprocessors and pack/unpack BCD are not implemented.

19.10. DIGITAL SIGNAL CONTROLLERS

Digital signal controllers have evolved relatively recently from DSPs. Announced by Texas Instruments in 1988, the TMS320C14 was the first device to combine the peripheral functionality of a microcontroller with the performance of a DSP; Texas Instruments Inc. (1988). A version of the device is now also available from Microchip Technology Inc. under the designation DSC320C14.

19.10.1. Origins

DSCs arose as a concept from the increasing awareness that a large market might exist for high-performance microcontroller style processors. Directed towards the automotive and servo control sectors, current DSCs were not designed using a 'clean sheet' approach by their manufacturers, who instead chose to use a combination of proven structures in their initial implementations.

19.10.1.1. Digital signal processors. The processor section of the TMS320C14 is based on the first generation Texas Instruments DSP family, the TMS320C15-25. This part (Texas Instruments Inc. (1987)) uses a Harvard architecture and offers a 160ns cycle time, 256 words of RAM and 4K words of ROM, both on-chip. The latter can be mapped off chip to allow execution from external program store. Internal data bus width is 16 bits and hardware arithmetic resources include a 32-bit ALU / accumulator, a 16 x 16 parallel multiplier yielding a 32-bit result and a 0 to 16 bit barrel shifter. The TMS320C15-25 offers only eight I/O channels which restricts its capability in control applications. Interrupt capability is unsophisticated, with only one external source provided.

Instructions supported by the TMS320C1X devices are finely tuned for DSP activity, (Kun-Shan Lin (1987)) particular strengths being the ability to form a 16 x 16 product and add it to an accumulator in one clock cycle. This facilitates code and time-efficient implementation of FIR and IIR filters. Table read/write operations are provided to circumvent constant access problems in the Harvard architecture. Direct, indirect (two index registers) and immediate addressing modes are available for data memory access. The instruction set is not orientated towards bit manipulation, and a sequence of instructions are needed to test or set a bit.

19.10.1.2. Microcontrollers. The hardware peripheral section of the TMS320C14 bears a resemblance to that of 16-bit microcontrollers. It comprises an event manager, a serial port, four independent timers and a parallel I/O port. Bank selection is used to map these functions into the limited I/O space of the TMS320C1X family.

The event manager provides capture and compare subsystems using two of the hardware timers. Time-stamping of input transitions (programmable edge sensitivity) on four channels is available in the former mode with time values buffered in FIFO stacks. Interrupt sources are associated with the FIFOs to signal their loading. Compare facilities permit the contents of compare registers to be matched against timer values. On a match, various actions can be programmed, including set, reset or toggling of an output pin, and the activation of one or two interrupts. A high precision PWM mode capable of 14-bits resolution is also provided by the event manager.

The serial port is a full-duplex USART capable of operating in asynchronous and synchronous modes at up to 400K bits/second and 6.4M bits/second, respectively. Double buffering, address detect and match, and a local baud rate generator are provided. Separate interrupts are sourced for transmit and receive events.

19.10.2. Applications

Applications for digital signal controllers lie in the field of algorithmically-intensive real-time control on an instrument scale. Such needs are frequent in today's products. Texas Instruments' sales information on the TMS320C14 enumerates a few examples;

- Automotive; Engine control, braking, active suspension

- Disk drives; Servo control of head position and spindle motor

- Robotics; Multiple axis control

- Motor control; AC and DC industrial motor drives

- Plotters; Pen positioning.

Many others can be envisaged given the published benchmarks of 16 by 16 multiplies in 160ns, (3 x 3).(1 x 3) matrix multiply in 4.3µs, PID algorithm in 2.2µs and an adaptive control algorithm on six parameter identification in 34.8µs.

19.10.3. Attributes

The TMS320C14 represents a significant watershed in DSP design. It has provided for the first time in one package, much of the functionality of both microcontroller and DSP . Being a first generation part however, there are certain restrictions which arise as a result of its direct evolution from the TMS320C1X family. These restrictions are associated with the limited I/O channel count of the TMS320C1X and with the unsophisticated (relative to current microcontrollers) interrupt system. Certain aspects of the instruction set are less than ideal for the support of on-chip peripheral bit-addressing, although provision for bit set and clear is made for the parallel I/O port.

As a means of circumventing the channel count restriction, a bank switch register is incorporated. To address a given peripheral register the bank number must first be loaded, then the given configuration word written to the appropriate port address. Adjustment of individual bits in these registers (except for parallel I/O) or in the 7 channels of external I/O space requires a sequence of load and logical masking instructions. Although the interrupt system allows individual source enabling/disabling, only one vector is supported. Hence on an interrupt, program flow vectors to a location where a handler must interrogate the individual possible originator bits in a flags register. As this interrogation is not supported by a bit test capability, a sequence of shift, mask, and conditional branch operations are required. This is less code and time - efficient than providing a separate interrupt vector for each interrupt source as in many microcontrollers.

19.11. CONCLUSIONS

It is likely that the future will see further convergence of digital signal controllers and microcontrollers. Next generation DSCs are likely to offer peripheral handling, bit manipulation and interrupt implementation comparable to state-of-the-art microcontrollers. Microcontrollers will continue to be available across a broad span of capability and cost, as many applications will not need fast signal processing capability. On the other hand, as low cost high performance processing power becomes available in DSC-class devices, new applications will be possible and families of industrial and consumer products not presently envisaged will appear.

19.12. REFERENCES

Astrom, K. J., Wittenmark, B. (1984) Computer Controlled Systems. Englewood Cliffs, N.J. 07632; Prentice-Hall International ISBN 0-13-164302-9

Hitachi (1988) H8/330 Microcontroller Overview.

Intel Corp (1976) Data Catalog 1976.

Intel Corp (1982) Microcontroller User's Manual, order number 210359-001.

Intel Corp (1985) Microcontroller Handbook, order number 210918-003.

Katz, P. (1981) Digital Control using Microprocessors. London: Prentice-Hall International ISBN 0-13-212191-3

Kun-Shan Lin (ed.) (1987) Digital Signal Processing Applications with the TMS320 Family, Volume 1. Englewood Cliffs, N.J. 07632; Prentice-Hall International ISBN 0-13-212466-1

Motorola (1989) MC68332 32-bit Microcontroller, MC68300 Embedded Control Family Overview. Rev. 0.8

Philips Components (1989) Technical Handbook Book 4 Integrated Circuits Part 14 Microcontrollers NMOS, CMOS.

Rabiner, L. R., Gold, B. (1975) Theory and Application of Digital Signal Processing. Englewood Cliffs, N.J. 07632; Prentice-Hall International ISBN 0-13-914101-4

Siemens (1988) SAB80C517/80537 High-Performance 8-bit CMOS Single-Chip Microcontroller, draft data sheet 9.88

Texas Instruments Inc. (1987) First Generation TMS320 User's Guide, document number SPRU013.

Texas Instruments Inc. (1988) TMS320C14/TMS320E14 User's Guide, document number SPRU032.

Chapter 20

Systolic arrays for high performance digital signal processing

J. V. McCanny, R. F. Woods and M. Yan

INTRODUCTION

In the past few years considerable effort has been devoted to the design of VLSI Systems for real time digital signal processing applications. In particular, a number of manufacturers have developed programmable DSP microprocessor chips which incorporate facilities such as on-board multipliers [1,2]. Whilst such devices are now enjoying widespread use, their real time capability is limited mainly to low bandwidth applications such as speech applications and sonar. In many other (eg image processing and radar signal processing) the computation rates demanded significantly exceed that offered by commercial DSP microprocessors. Considerable attention has therefore been focused on systems which utilise parallel processing techniques. An important contribution to this field has been the development of systolic array systems. These types of systems are well suited to a wide range of signal processing requirements in that they are both highly parallel and highly pipelined. They also exhibit a number of properties, such as regularity and nearest neighbour connections, which make them attractive from a VLSI design point of view.

The purpose of this paper is to present an overview of some of the developments which have occurred in this field in the past few years, with the emphasis being concentrated on the authors' main area of interest - bit level systolic array architectures and chip designs based on these. The structure of the paper is as follows. A brief overview is given in the following section of the original systolic array concept and the benefits which may be gained by extending this to the bit level. Several examples are then given of architectures and, where appropriate, chip designs which have been developed for both filtering and coding applications. Important new developments in application of bit level systolic arrays to recursive digital filters are then described and a recent chip design based on these ideas discussed.

SYSTOLIC ARRAYS - BACKGROUND

The concept of a systolic array was first proposed by H T Kung and C E Leiserson [3]. In their seminal paper they showed how a number of important matrix computations such as matrix x vector multiplication, matrix x matrix multiplication and matrix LU decomposition could be computed using arrays of inner product step processors interconnected in the form of a regular lattice. In a typical systolic array, all the cells are identical except for a number of "special" cells which are required on the boundary in some applications. On each cycle of a system clock, every cell in the array receives data from its neighbouring cells and performs a specific processing operation on it. The resulting data is stored within the cell and then passed on to a neighbouring cell on the next clock cycle. As a result, each item of data is passed from cell to cell across the array in a specific direction and the term "systolic" is used to describe this rhythmical movement of data which is analogous to the pumping action of the human heart.   Since each processor can only communicate data to its nearest neighbours, it follows that all data must be input to, or output from, the array through the boundary cells.

Figure 1 shows the original Kung/Leiserson systolic array for banded matrix x matrix multiplication, which serves to illustrate the operation of such a system. This consists of a hexagonal array of processing elements whose function is as illustrated. On each cycle of a system clock the processor takes as its inputs values a, b and c. It then computes the inner product step function c <- c + a x b before making the input values a and b, together with the new value of c, available on its output lines.

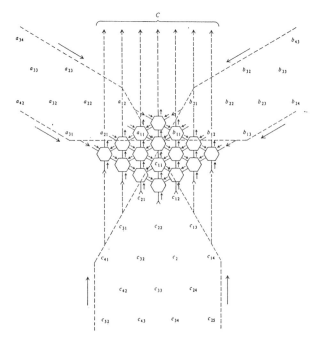

Figure 1: Systolic array for banded matrix x matrix multiplication (after Kung and Leiserson [3])

The resulting outputs are then latched into neighbouring processors on the next clock cycle as illustrated.

The circuit in Figure 1 operates by allowing each element in a given diagonal in the A and B matrices to move along a line of processors from the NW to the SE and from the NE to the SW respectively, as illustrated. As the elements in the A and B matrices meet on processors within the array, computations are performed with the results being propagated in the upward (or northward) direction to form the elements in the result matrix C.

The original work on Systolic Arrays acted as a major stimulus for research on VLSI architectures for real time matrix algebraic computations. In the past ten years or so considerable effort has been devoted to the design and optimisation of systolic systems for problems which range from convolution to Kalman filtering. An overview of these developments is given in references [4-9]. The general consensus emerging from this work seems to be that structures such as that shown in Figure 1 should be regarded as "systolic algorithms" rather than specific hardware designs. These can then be implemented in practice by mapping these algorithms onto a more general purpose system designed for such applications. A good example of this is the WARP machine developed by H T Kung and his colleagues at CMU in collaboration with General Electric [10]. This machine is implemented as a linear array of ten systolic processing nodes with each node being implemented on a single printed circuit board. The resulting system has a peak performance of 10 MFLOPs. An integrated version of the WARP processing node - the IWARP chip - is currently under development at Intel with first silicon being available later this year. The IWARP chip incorporates a 20 MFLOP floating point processor and has unidirectional byte wide ports. These run at 40 MHz giving a data bandwidth of 320 Mbytes per second.

The development of a programmable node chip also forms the basis of a major programme at STL in the UK[11]. This chip offers 24-bit floating point arithmetic at a processing rate of 60 MFLOPs. It also incorporates parallel I/O data ports allowing a total I/O data rate of 90 million floating point words per second. A major application of this chip is an adaptive digital beam forming system developed jointly by STL and RSRE [12].

BIT LEVEL SYSTOLIC ARRAYS

A major attraction of the original systolic array concept was that these architectures seemed well suited for VLSI design. However, as discussed above, the basic processing element required in the designs proposed by Kung and Leiserson [3] are of at least single chip complexity. However, subsequent research by McCanny and McWhirter [13] showed that many important front end digital signal processing (DSP) chips such as correlators and digital filters could be designed as systolic arrays in their own right if the basic processing element is defined at the single-bit level rather than at the word level. The function of the basic processing of the inner product step function. The element is then reduced to that of a gated full adder - the bit level equivalent result is that many of these processors (typically several thousand in current technology) can be integrated on a single VLSI chip, with all the inherent advantages of regularity and nearest neighbour interconnections. These features, along with bit level pipelining, can be used to ease design and produce chips capable of operating at very high sample rates. These ideas have since been applied to a wide range of important DSP functions and joint programmes established with a number of companies to see these ideas implemented as commercial chips. In this paper we give an overview of some of these developments, with attention being focused on systolic circuits for digital filtering and coding applications.

FIR FILTERING

FIR filters are probably the most common form of digital filters and have the twin advantages of being unconditionally stable and offering a linear phase characteristic. Mathematically an N point FIR filter can be written as:

$$y_j = \sum_{i=0}^{N-1} a_i \, x_{j-i} \tag{1}$$

where $a_0, a_1, \ldots, a_{N-1}$ represent a fixed set of coefficients which define the filter's frequency response and x_i ($i = 0,1,2, \ldots$) represents a sequence of input signal values. If a_i and x_i are both n-bit binary numbers then the pth bits of the output words y_j may be expressed in the form:

$$y_j^p = \sum_{i=0}^{N-1} \sum_{q=0}^{n-1} a_i^q \, x_{j-i}^{p-q} + \text{carries} \tag{2}$$

This computation is usually carried out by performing the inner summation for each value of i (ie performing the multiplication explicitly) and accumulating the results. Most conventional DSP systems based on ripple-through multipliers operate in this manner. This approach is particularly easy to understand and can be pipelined right down to the bit level [13].

Equation (2) can be computed in other ways and an alternative is, for a fixed value of q, to perform the summation:

$$y^p_j{}^q = \sum_{i=0}^{N-1} a_i^q \, x_{j-i}^{p-q} \tag{3}$$

The pth bit of the result y_j can then be formed by carrying out the final sum over q. This approach has been used in the design of a bit level systolic FIR filter with multi-bit input data and coefficients [14]. The architecture of this circuit is shown in Figure 2 and is based on a bit serial data input organisation. As illustrated, the circuit comprises an orthogonal array of simple processing bit level processing whose main logic function is an AND gate plus a full adder. The circuit has been designed with the coefficient bits remaining on fixed sites, with each cell in the rth row storing one bit of the coefficient a_{r-1}. The circuit also incorporates unidirectional data flow which is important since it permits the use of extra delays when driving signals from chip to chip or bypassing a faulty row of cells if a fault tolerant design is required.

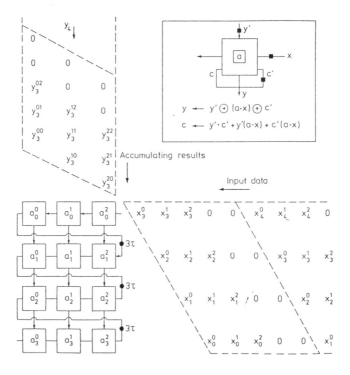

Figure 2: Architecture of the systolic multibit convolver

The operation of the circuit in Figure 2 is as follows. Data words enter the array through the top right-hand cell (least significant bit first) and are clocked from right to left. Once a given bit has traversed any row it is delayed for several cycles (in this case three) before being input to the row below. As they move through the array, these data bits interact with the coefficient bits on each cell to form partial products of the form $a_i^p x_{j-i}^{p-q}$. Each partial product is passed to the cell below on the next clock clock cycle. The net result is that the sum over i of all partial products in equation (3) is formed within a parallelogram shaped interaction region that moves down through the array. The final result is formed by summing over q (ie accumulating all partial products of the same significance using two extra rows of cells at the bottom of the main array (not shown in Figure 2).

A commercial FIR filter has been designed by Marconi Electronic Devices Ltd (MEDL)[15] which is based on the architecture shown in Figure 2. A block diagram of the chip is shown in Figure 3. This chip has been designed in 3 micron CMOS/SOS and consists of two separate 16-stage FIR filters with 8-bit wide coefficients. This allows it to be configured internally to operate either as two separate 16-stage filters with 8-bit data and coefficients (eg for use with complex signals). Alternatively these can be cascaded internally to form one 32-stage device. The chip can also be configured to form one 16-stage device with 16-bit coefficients. In each mode of operation the input data length is programmable to one of four different values and the device has been designed so that it can accept either two's complement or unsigned magnitude data. Additional circuitry has also been provided so that chips can be cascaded to form filters with up to 256 stages. This chip also allows coefficients to be updated during normal circuit operation, a feature which is useful in adaptive filtering applications. The device has been designed to operate at clock rates of more than 20 MHz with a power consumption of 0.5 W.

Since the circuit operates in a bit serial manner the output rate depends on input word length (plus word growth), a typical example being one megaword per second for 16-bit input data.

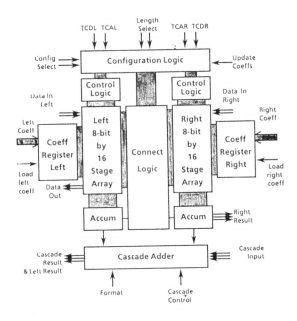

Figure 3: Block diagram of MEDL convolver chip (after Boyd [15])

The above circuit operates in a bit serial manner. It is also possible to design FIR filter systems with bit parallel data organisation. One way to do this is based on a word level convolution architecture of the type shown in Figure 4. In this system each coefficient is associated with a word level processing element with the data words and result words being propagated in a uni-directional manner through the array. The word level operation of this circuit depends on the values of the x words being delayed by one cycle more than the y words as these propagate through the array. In accordance with the cut theorem an arbitrary number of delays can be incorporated into data lines of Figure 4 provided the same number is introduced into the result lines. The processing elements in Figure 4 can therefore be replaced by multiply/accumulate structure based on systolic multiplier circuits (such as those described in references 13 and 16) which are pipelined at the bit level.

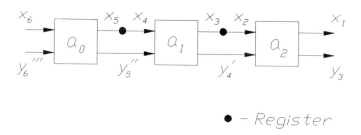

Figure 4: Word level systolic array for convolution

An alternative approach is to employ a bit sliced approach and this forms the basis of the architecture used in a commercial matched filter/correlator chip also developed by MEDL [17]. In this approach a single bit of each of the N coefficient words is stored on a single chip with bit parallel input data being propagated across these. A full correlation/convolution operation is then performed by cascading n of these bit slice chips together.

VECTOR QUANTISATION

Vector Quantisation (VQ) is an efficient coding technique which is assuming increasing importance as a means of transmitting high quality speech and image signals at medium to low bit rates [18]. Data compression is achieved by matching an input sequence (ie vector) of data samples with the entries (ie the codevectors) in a pre-classified database known as a codebook. This is done by computing a distortion value between the input vector and the appropriate entries in the codebook. The codevector which produces the best match is chosen and its index transmitted to the receiver. The signal is then reconstructed at the receiver using a simple table look-up procedure. One of the problems with vector quantisation is that the encoding process is often computationally intensive (particularly in image coding applications) and in many instances is well beyond the capability of programmable DSP chips.

An important aspect of many VQ systems is that it is only the index of the match codevector that is required for transmission and not the actual distortion value nor the codevector itself. As discussed in detail elsewhere [18], the main pattern matching computation required for most of the common distortion measures (least squares, weighted least squares and Itakura/Saito distortion measures) is an inner product computation between the input data vector and the appropriate vectors in the codebook. The index of the codevector which produces the maximum inner product value is then chosen.

A system for a full VQ codebook search is illustrated schematically in Figure 5 [19]. This consists of a total of p bit level systolic array circuits each of which is associated with a single vector y_j (j = 1, ..., p) in a p vector codebook.

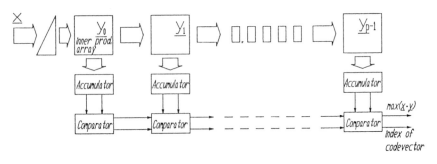

Figure 5: VQ full search system

The function of each systolic array is to compute the inner product of the input vector x with each of the codevectors y_j. These results are then passed through an accumulator circuit at the bottom of the array, to a comparator. The function of this circuit is to compare the output from a given inner product array with the result passed from the previous comparator. If it is bigger then this result, along with its associated index j, it is passed on. If not, the previous result along with its index is passed on.

A bit level systolic array circuit for computing the inner product is illustrated in Figure 6. This is similar in structure to the array in Figure 2 in that the bits of each element in a given codevector y_j remain on fixed sites. The array differs from that in Figure 2 in that all the elements in the data vector are input in parallel to the array with no links between successive rows as in Figure 2. Apart from this the basic operation of the circuit is similar. The data bits propagate across the array in a bit serial manner with partial products of the same significance being accumulated at the bottom of the array. The circuit shown in Figure 5 has been designed for one-dimensional vectors (eg speech signals) but the concepts can readily be extended to two-dimensional data blocks of the type required in image coding applications [21]. The circuit described has been designed for a VQ system in which a full codebook search is assumed. However, quite a number of other VQ techniques have been developed in which codebook size is significantly reduced. Good examples are Gain/Shape Vector Quantisation (GSVQ) and methods in which the codebook is organised in a tree structure. As is discussed in detail elsewhere[19-21], the inner product array in Figure 6 can be used to construct circuits for these applications.

Whilst the structure shown in Figure 5 is well suited to high bandwidth systems the hardware requirements may be excessive for lower bandwidth applications such as

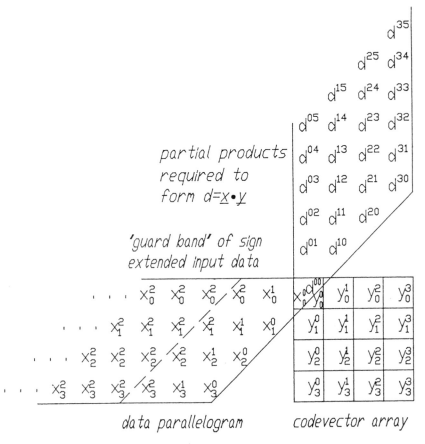

partial products
required to
form $d = \underline{x} \cdot \underline{y}$

'guard band' of sign
extended input data

data parallelogram   *codevector array*

Figure 6: Inner product Systolic Array

speech coding. An alternative circuit can be designed which requires only a single inner product array. This can be achieved by interchanging the data vector and codevector bits in Figure 6 ie the data vector bits should be held on fixed sites with successive codevector bits being propagated across these. With this system the chosen codeword is available once the entire set of codevectors have propagated across the data vector. The stored data vector can be then updated and the process repeated.

1D AND 2D DISCRETE COSINE TRANSFORM

The Discrete Cosine Transform (DCT) has widespread applications in both speech and image coding systems. Applications include mobile radio communications, high quality video telephone and high definition television. A number of highly regular bit level systolic arrays for computing both the 1D-DCT and 2D-DCT and its inverse can be derived from the Winograd algorithm[19,23,25]. This allows the 1D-DCT $\underline{y} = (y_0, y_1, \ldots, y_{N-1})$ of a 1D input sequence $\underline{x} = (x_0, x_1 \ldots, x_{N-1})$ to be written as:

$$\underline{y} = CDA\underline{x} \qquad (4)$$

where A is an (M x N) matrix and C is an (N x M) matrix (where M is usually slightly greater than N) and D is an (M x M) diagonal matrix. For almost all cases of interest, the A and C matrices contain elements that only take the values +1, -1 and 0. The M non-zero (diagonal) elements in the D matrix represent a set of fixed coefficients which multiply each of the elements in the intermediate vector $A\underline{x}$ in a pointwise manner.

A bit level systolic array system for computing the 1D-DCT is illustrated schematically in Figure 7(a). As is described in detail elsewhere [22,19] this circuit also operates in a bit serial word parallel manner, with the elements of both the A and C matrices being stored on fixed sites on the array [Figure 7(b)]. The results of the first matrix x vector multiplication are input to a bank of serial/parallel multipliers before being passed to the second matrix x vector multiplication array (the C array), as shown.

The extension of the above ideas to the 2D-DCT follows that described by Blahut[24]. This allows the 2D-DCT matrix block Y of a two dimensional input

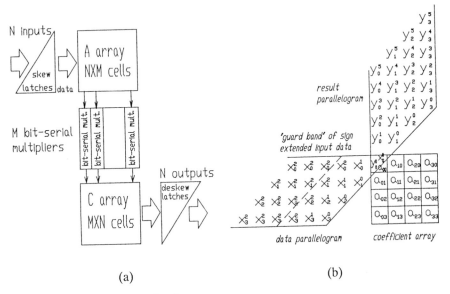

(a)   (b)

Figure 7: Systolic 1D-DCT circuit

data block X to be written in the form:

$$Y = (W1 \otimes W2)X \qquad (5)$$

where W1 and W2 represent the 1D Winograd row and column transforms respectively and W1 = W2 = CDA in the case of an (N x N) point 2D-DCT. The symbol $\otimes$ represents the Kronecker product of the matrices W1 and W2 and this can be interpreted as a column-row transposition operation. Equation (5) can be expanded in a number of ways to give a range of possible architectures[25]. One possibility for an N x N transform is to rewrite:

$$W1 \otimes W2 = (C \otimes C)(D \otimes D)(A \otimes A) \qquad (6)$$

which has the advantage of grouping together the diagonal D matrices and allows the Kronecker product D' = D$\otimes$D to be precomputed. Written in this way the 2D-DCT can be computed using a highly regular three dimensional circuit in which all the interconnects are nearest neighbour. This is illustrated schematically in Figure

8. In this circuit the A and C matrix blocks have the same function as in Figure 7, as have the serial/parallel multipliers.

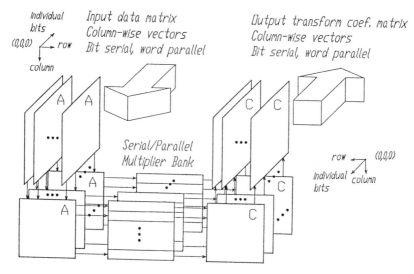

Figure 8: Three dimensional circuit for computing the 2D-DCT

The row/column transposition implicit in equation (6) is accomplished by rotating successive A and C matrix arrays through 90° as shown. This circuit operates in a word parallel, bit serial manner and produces an output with a similar format.

IIR FILTERING

Whilst the bit level systolic technique has been applied fairly widely to the design of non-recursive components such as FIR filters, correlators etc, it has, until recently, had limited application in the design of devices such as IIR filters which involve recursive computations. Bit level systolic arrays for this type of application are much more difficult to design because the effect of introducing M pipelined stages into such a system is to introduce an M cycle delay into the feedback loop. In a bit parallel system the input word rate is equivalent to the clock rate and so the pipeline latency corresponds to M word delays. The potential speed of the device is therefore reduced by a factor M.

An important contribution to the problem of pipelining recursive filters has been made by Parhi and Messerschmitt [26] in proposing "scattered look ahead" recursive filters. In this approach recursive equations are iterated so that the next result depends on the result computed M cycles previously and not the previous cycle as is normally the case. Unfortunately, this approach requires an increase in system complexity with hardware being required to compute the additional terms generated when recursive equations are iterated.

An alternative approach to this problem which allows recursive equations to be implemented directly has been proposed by Woods et al [28-30]. This can be explained by considering a simple first order IIR filter section. The computation for this function can be expressed as:

$$y_n = b_1 y_{n-1} + u_n \qquad (7)$$

where

$$u_n = a_0 x_n + a_1 x_{n-1} \qquad (8),$$

x_n is a continuously sampled data stream and a_0, a_1 and b_1 are coefficients which determine the filter frequency response. As discussed above, the non-recursive part of this computation can be implemented using conventional bit parallel pipelined multiplier/accumulators. The difficulty arises with the recursive part (ie the computation of $b_1 y_{n-1}$) and this is illustrated in Figure 9. This shows how this computation may be implemented using a pipelined shift and add multiplier /accumulator array in which the output is fed back to the input and in which it has been assumed that the input data is fractional. From this circuit it will be seen that

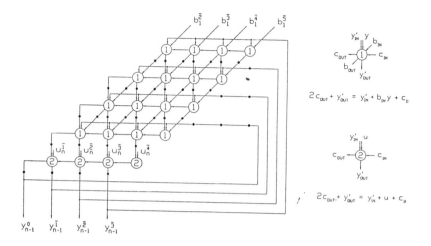

Figure 9: Pipelined shift and add array with feedback

it takes a total of five cycles (in general $p+1$ cycles, where p is the wordlength) before the bits required, which must be fed back into the array for the next iteration - the most significant bits - are available at the array output.

On the face of it this appears to be an intractable problem. However, this is only a consequence of using conventional arithmetic which dictates that computations need to be carried out most significant (msb) rather than least significant bit first (lsb). This is demonstrated in Figure 10 which shows a hypothetical multiplier array in which the computation at bit level proceeds msb first. It will be noted that with this system the timing of output bits is such that they can be fed back immediately with the result that the problem of latency and the resulting problem of throughput rate can be avoided.

In practice, the problem with implementing the circuit shown in Figure 10 is that one cannot use conventional binary arithmetic since this would necessitate the

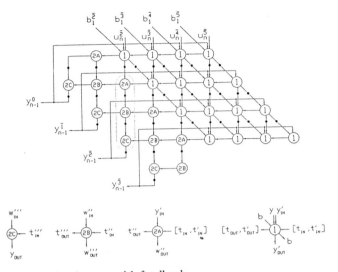

Figure 10: Conceptual msb array with feedback

propagation of carries from the lsb to the msb. However this problem can be avoided if redundant arithmetic schemes are employed which allow arithmetic to be carried out most significant bit first. A good example of such redundant number systems are the Signed Digit Number Representations (SDNRs) which were originally introduced by Avizienis [31] to reduce carry propagation in parallel numerical operations such as add, subtract, multiply and divide. A number of systolic circuits have been designed based on both radix 4 and radix 2-SDNRs which embody the principles illustrated in Figure 10 [28-30]. Space does not allow a detailed description of these arrays. However, the ability to limit carry propagation so that a computation carried out at a given level of significance only affects results a limited number of levels of significance higher (typically between one and two places) means that the latency within the pipelined loop can be reduced to a small number of cycles - typically two to three cycles. Moreover, the latency and hence the throughput rate is independent of the word length of the input data or coefficient words.

A prototype bit parallel systolic IIR filter chip based on these ideas has been designed at Queen's University Belfast. This is illustrated in Figure 11. The chip in question operates on 12-bit parallel two's complement input data with 13-bit parallel two's complement coefficients to produce a 15-bit two's complement output. The design was undertaken using VLSI Technology's Design Express Integrated Circuit Design system with the chip being fabricated in their 1.5 micron double layer metal CMOS technology.

Figure 11: Bit parallel IIR filter chip

The internal architecture of the circuit is based on radix-2 SDNR arithmetic and input/output interface circuitry is incorporated on board which allows conversion between two's complement and radix-2 SDNR. As discussed in detail in reference [32] this interface circuitry represents a small overhead in terms of chip area and does nothing to reduce data throughput rate. The chip in question can operate up to 15 Megasamples per second.

An important conclusion which can be drawn from this design is that typical cell sizes required in a system employing redundant arithmetic are only about 60% larger than a conventional bit level cell. Moreover, Knowles and McWhirter[33] have recently shown that conventional carry save arithmetic can be used to implement the main multiplier array with signed digit number representations only being required along the periphery. This means that systolic IIR filters with bit parallel input data can be implemented in the same amount of hardware as structures built from conventional pipelined multiplier arrays.

DISCUSSION

It has only been possible in this paper to describe a few of the DSP operations which may be implemented as bit level systolic arrays. The list of applications is numerous and includes for example the computation of Fourier transforms, Walsh Hadamard transforms and front-end image processing operations such as rank order filtering, two dimensional convolution and correlation.

The devices fabricated to date provide hard evidence that the bit level systolic array architecture has a number of important advantages for high-performance VLSI chip design. Since the arrays are completely regular the task of designing an entire VLSI chip is essentially reduced to optimizing the design and layout of one or two very simple cells.

It is also very easy to carry out a functional validation of an array using a hardware description language such as ELLA or VHDL, less than 100 lines of ELLA code being sufficient, for example, to describe the convolution circuit described earlier. As a result, the typical design time for a bit level systolic array is very much less than that required for a random logic circuit with the same number of transistors. It is important to realise that bit level systolic arrays are not just regular in the geometric sense; they also exhibit complete electrical regularity by virtue of the fact that each cell is connected only to its nearest neighbours. The corresponding absence of long interconnects also enhances the circuit performance by reducing parasitic capacitance (and hence RC time delays) and leads to extremely high circuit packing densities.

Since the circuits described above have been pipelined down to the bit level, their throughput rates depend only on the propagation delay through a single cell (typically of order three to four gate delays). As a result, we have been able to achieve high performance data rates at relatively low levels of power consumption. In fact it was possible with some of the earlier systolic arrays (eg the MEDL correlator and convolver chips) to obtain functional throughput rates of 5×10^{11} to 10^{12} gate Hz/cm^2 with a 3 micron process, a figure which exceeded the then targets specified for phase 1 VHSIC 1.25 micron devices. Although most of the emphasis in this and previous sections has been on pipelining circuits right down at the single bit level, it should be noted that for some circuit technologies where the propagation delay and physical size of a latch is comparable with that of the processing element, it may be better to pipeline the circuits at a higher level. In such cases the basic cell can be designed as the logical equivalent of a 2 x 1, 2 x 2, or larger array of single bit cells but without any internal latches. It is possible in this way to retain the VLSI design advantages of bit level systolic arrays without compromising the cell design in terms of the optimum area, speed, and power consumption for a given technology.

Although the systolic array approach to chip design eliminates most of the global connections within a circuit, it has been argued that many of the problems associated with the interconnect lines have simply been transferred to the clock. This may be true to some extent. However, since the problem has been centralized, the circuit designer can concentrate his attention on the problems of clock distribution, confident that the function of the array is otherwise determined by that of a single cell. With a more conventional architecture it is essential to minimize clock skew over the entire chip. In a fully systolic design, where processors communicate only with their neighbours, it is only the incremental skew between processors that must be minimized, and this can be kept small through careful design and by running all the clock lines in metal. In most of the chips designed to date, the master clock is bussed from an input pad along the bottom or top of the chip, the clock signal to each row or column being driven by its own clock buffer. Using this type of scheme, clock skews of less than 1 ns have been achieved across entire rows or columns of the chips described above.

In this paper attention has been focused mainly on research undertaken by the author and his colleagues. Considerable effort has also been devoted in many other laboratories world-wide on research on VLSI array processor architectures which are pipelined at the bit level and suitable for high performance DSP chip design. Quite a number of these architectures have been used as the basis of chip designs. Further information on these designs are available from a number of sources including those listed in references [4-7].

Recent research on systematic design techniques such as those based on the use of dependence graphs[4,34-38] have shown that many specific bit level array processor designs can be derived by applying various mappings, transformations and "cuts" to dependence graphs. In the simple case of binary multiplication, for example, a whole range of well known ripple-through, systolic and serial/parallel multipliers can be derived from a single graph which illustrates the dependencies between the various bit level computations required for this operation[34]. Similar insights are obtained using methods such as those based on functional programming languages[39,40]. These tools provide a simple and intuitive mechanism which enables the VLSI chip designer to examine the relative trade-offs between levels of pipelining, chip area, power consumption, latency and throughput for a given application. They also provide a means whereby a non-specialist can rapidly map his desired function onto an efficient bit level processor array.

REFERENCES

1. *AT and T WE DSP32 and DSP32C Reference Manual*, 1988.

2. *Digital Signal Processing Applications with the TMS 320 Family*, 1987, ed. Kun-Shan Lin, Prentice Hall International.

3. Kung, H.T. and Leiserson, C.E., 1980, "Algorithms for VLSI Processor Arrays" in *Introduction to VLSI Systems*, C. Mead and L. Conway, Addison-Wesley, Reading, Mass.

4. Kung, S.Y., 1988, *VLSI Array Processors*, Prentice Hall International, Englewood Cliffs, New Jersey.

5. Moore, W., McCabe, A., Urqhart, R., 1986, *Systolic arrays*, Adam Hilger, Bristol.

6. Bromley, K., Kung, S.Y. and Swartzlander, E.S., 1988, *Proc. of the Int. Conf. on Systolic Arrays*, San Diego, May 1989, IEEE Computer Society Press.

7. McCanny, J.V., McWhirter, J.G. and Swartzlander, E.S., 1989, *Systolic Array Processors*, Prentice Hall International.

8. "Systolic Arrays", 1987 Special Issue, Computer July 1987, Computer Society of the IEEE.

9. McWhirter, J.G. and McCanny, J.V., 1987, "Systolic and Wavefront Arrays", Chapter 8, *VLSI Technology and Design*, Academic Press, eds. McCanny, J.V. and White J.G., pp. 253-299.

10. Menzilcioglu, O., Kung, H.T. and Song, S.W., 1989, "A Highly Configurable Architecture for Systolic Arrays of Powerful Processors" in *Systolic Array Processors*, eds. McCanny, J.V., McWhirter, J.G. and Swartzlander, E.S., Prentice Hall International, pp. 156-165.

11. Ward, C.R., Hazon, S.C. Massey D.R., Urqhart, A.J., Woodward, 1989, "Practical Realisations of Parallel Adaptive Beamforming Systems" in *Systolic Array Processors*, eds. McCanny, J.V., McWhirter, J.G. and Swartzlander, E.S., Prentice Hall International, pp. 3-12.

12. McCanny, J.V. and McWhirter, J.G., 1987, "Some Systolic Array Developments in the United Kingdom", IEEE Computer, July, pp. 51-63.

13. McCanny, J.V. and McWhirter, J.G., 1982, "Implementation of Signal Processing functions using 1-Bit Systolic Arrays", Electron. Letts., Vol. 18, no. 6, pp. 241-243.

14. McCanny, J.V., Evans, R.A. and McWhirter, J.G., 1986, "Use of Unidirectional Dataflow in Bit Level Systolic Arrays", Electron. Letts., Vol. 22, no. 10, pp. 540-541.

15. Boyd, K.J., 1989, "A Bit Level CMOS/SOS Convolver", in *Systolic Array Processors*, eds. McCanny, J.V., McWhirter, J.G., Swartzlander, E.S., Prentice Hall International, pp. 431-438.

16. Hoekstra, J., 1985, "Systolic Multiplier", Electron. Letts., Vol. 20, pp. 995-996.

17. White, J.G., McCanny, J.V., McCabe, A.P.H., McWhirter, J.G. and Evans, R.A., 1986,"A High Speed CMOS/SOS Implementation of a Bit Level Systolic Correlator" Proc. IEEE Int. Conf, on Acoustics Speech and Signal Processing, Tokyo, pp. 1161-1164.

18. Gray, R.M., 1984, "Vector Quantisation", IEEE ASSP Magazine, Vol. 1, pp. 4-29.

19. Yan, M., McCanny, J.V. and Kaouri, H.A., 1988, "Systolic Array System for Vector Quantisation using transformed Sub-band Coding", IEEE Proc. of Int. Conf. on Systolic Arrays, San Diego, May 1988, pp. 675-685.

20. Yan, M. and McCanny, J.V., 1989, "A Bit Level Systolic Architecture for implementing a VQ Tree Search", Journal of VLSI Signal Processing, to be published.

21. Yan, M., 1989, "VLSI Architectures for Speech and Image Coding Application", Ph.D. Thesis, The Queen's University of Belfast.

22. Ward, J.S., McCanny, J.V. and McWhirter, J.G., 1985, "Bit Level Systolic Array Implementation of the Winograd Fourier transform Algorithm", IEE Proc., Pt. F, Vol. 132, pp. 473-479.

23. Ward, J.S. and Stannier, B.S., 1983, "Fast Discrete Cosine transform for Systolic Arrays", Electron. Letts., Vol. 19, pp. 58-60.

24. Blahut, R.E., 1985, *Fast Algorithms for Digital Signal Processing*, Addison Wesley.

25. Yan, M. and McCanny, J.V., 1989, "VLSI Architectures for Computing the 2D-DCT" in *Systolic Array Processors*, eds. McCanny, J.V., McWhirter, J.G. and Swartzlander, E.S. Prentice Hall International, pp. 411-420.

26. Parhi, K.K. and Messerschmitt, D.G., 1987, "Concurrent Cellular VLSI Adaptive Filter Architectures", IEEE Trans. on Circuits and Systems, Vol. CAS-34, no. 10, pp. 1141-1151.

27. Parhi, K.K. and Messerschmitt, D.G., 1988, "Pipelined VLSI Recursive Filter Architectures using Scattered Look-Ahead and Decomposition", Proc. IEEE Int. Conf. on Acoustics, Speech and Signal Processing, New York, pp. 2120-2123.

28. Woods, R.F., Knowles, S.C., McCanny, J.V. and McWhirter, J.G., 1988, "Systolic IIR filters with Bit Level Pipelining", Proc. Int. Conf. on Acoustics, Speech and Signal Processing, New York, pp. 2120-2123.

29. Knowles, S.C., Woods, R.F., McWhirter, J.G. and McCanny, J.V., 1989, "Bit Level Systolic Architectures for High Performance IIR filtering", Journal of VLSI Signal Processing, Vol. 1, no. 1, pp. 1-16.

30. Woods, R.F., McCanny, J.V., Knowles, S.C. and McWhirter, J.G., "Systolic Building Blocks for High Performance Recursive Filtering", Proc. IEEE Int. Conf. on Circuits and Systems, Helsinki, June 1988, pp. 2761-2764.

31. Avizienis, A., 1961, "Signed Digit Number Representations for Fast Parallel Arithmetic", IRE Trans. on Electronic Computers, Vol. EC-10, pp. 389-400.

32. McCanny, J.V., Woods, R.F. and Knowles, S.C., 1989, "The Design of a High Performance IIR filter chip", *Systolic Array Processors*, eds. McCanny, J.V., McWhirter, J.G. and Swartzlander, E.S., Prentice Hall International, pp. 535-544.

33. Knowles, S.C. and McWhirter, J.G., 1989, "An Improved Parallel Bit Level Systolic Architecture for IIR filtering", *Systolic Array Processors*, eds. McCanny, J.V., McWhirter, J.G. and Swartzlander, E.S., Prentice Hall International, pp. 205-214.

34. McCanny, J.V., McWhirter, J.G. and Kung, S.Y., 1989,"The use of Data Dependence Graphs in the Design of Bit Level Systolic Arrays", IEEE Trans. on Acoustics, Speech and Signal Processing, to be published.

35. Quinton, P., 1984, "Automatic Synthesis of Systolic Arrays from Uniform Recurrent Equations", IEEE 11th Int. Symp. on Computer Architectures, pp. 208-214.

36. Moldovan, D. and Forbes, J.A.B., 1986, "Partitioning and Mapping Algorithms into Fixed Size Systolic Arrays", IEEE Trans. on Computers C-35(1), pp. 1-12.

37.   Rao, S.K., 1985, "Regular Iterative Algorithms and their implementation on Processor Arrays", Ph.D. Thesis, Stanford University, USA.

38.   Van Dongen, V., 1989, "Quasi-regular Arrays: Definition and Design Methodology", *Systolic Array Processors*, eds. McCanny, J.V., McWhirter, J.G. and Swartzlander, E.S., Prentice Hall International, pp. 126-135.

39.   Sheeran, M., 1985, "Designing Regular Array Architectures using Higher Order Functions" in *Functional Programming Languages and its Applications*, ed. Jovannaud, J.P., Springer-Verlag.

40.   Luk, W., Jones, G. and Sheeran, M., 1989, "Computer-based tools for Regular Array Design", *Systolic Array Processors*, eds. McCanny, J.V., McWhirter, J.G. and Swartzlander, E.S., Prentice Hall International.

Chapter 21

Throughput, speed and cost considerations

M. Carey

21.1 BACKGROUND

We are all familiar with the rapid advances in digital component technology. These advances have followed in rapid procession over the last 25 years.

The first major drive was to compress the maximum functionality into a single chip device. However, early VLSI design techniques increased the power density resulting in devices that ran hot and, as a result, had low reliability.

With the introduction of CMOS and other low power technologies the power density problem was reduced and high reliability devices could be produced.

However, most manufacturers concentrated on microprocessors, memories and attendant support devices. For those requiring DSP related functions, especially at high throughputs, it was necessary to produce special-to-purpose ASICS. Although many improvements have been made in the ASIC design and fabrication route, it still represents a relatively expensive and time consuming option. This can rule it out for many commercial application areas.

Over recent years we have seen the emergence of devices aimed specifically at DSP application areas. Such devices are optimised to carry out DSP functions, at high data rates, that previously would have required ASICS.

21.2 PERFORMANCE/COST TRADE-OFF

One could argue that the potential application areas for such devices are virtually unlimited. However, in practical terms there are a number of possible variations in the architecture of DSP's. This has led manufacturers to target application areas appropriate to specific DSP architectures. This inevitably means that there are variations in the performance/cost trade-off dependant upon the DSP architecture chosen.

In some cases an inappropriate choice of device can lead

to little improvement being obtained.

Some application areas have benefitted more than others as a result of the improvements in DSP technology. In particular, many computationally intense digital algorithms have been known for some time. Work on these had usually been carried out through non-real-time simulation on general purpose computers. For many applications these calculations required such processes as frequency spectral analysis, convolution and correlation. For such operations fast, accurate digital multiplication is required. As little as a decade ago economic processors capable of carrying out these functions were not available.

During the 1980's things have changed dramatically with the development of an increasing range of high speed digital processors. Algorithms which had previously merely been the subject of laboratory experiment could now be economically implemented to run in real-time. This opened up a wide potential market for DSP products, this in turn encouraged DSP manufacturers to devote more time and resource into the development of better, faster more flexible devices. We are now on the threshold of a rapidly expanding DSP revolution which will see a great many, previously just theoretical, ideas turned into practical, marketable products.

21.3   IMPACT ON SYSTEM DESIGN

The impact of DSP technical advances on system design, from the point of view of throughput, speed and cost has been very significant. The impact has been more marked in some areas than others and the following sections consider the effects on on some of the application areas.

21.3.1 DSP Application Examples

Such systems might be regarded as the more obvious application areas, such as:

* Telecommunications. This would include cellular, cordless and secure phones, high speed modems and transcoders.

* Computer Peripherals. Such items as tape drives, printer/plotter control and disc drive servo controllers.

* Automotive. Sophisticated engine monitoring and control systems, also suspension steering and braking.

* Military. This is a very wide ranging field covering, among others, image and signal processing for radar, navigation, sonar and guidance.

Clearly many of these application areas have widely differing criteria for optimising the performance/cost trade-off. For instance, the automotive industry would favour lower cost options, possibly at the expense of some performance aspects. For some military applications the performance is vitally important and higher costs will often be accepted.

One particular area in which there have been significant advances in the use of DSP is in the field of instrumentation.

Most of the signal processing functions required for instrumentation, measuring and analysis have traditionally been the stronghold of analogue techniques. Processes such as filtering and spectral analysis have, for some time, been far more economically implemented using analogue components.

This situation has been gradually changing as the higher accuracy and long term stability and reliability of digital techniques has become appreciated.

A typical DSP based system is shown in figure 21.1.

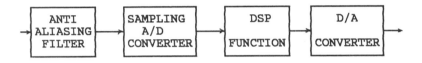

Figure 21.1   DSP Based System

The DSP function can be implemented in a number of ways. It could be a dedicated ASIC, designed specifically for the purpose. This approach would produce a high speed, high throughput solution but at a relatively high cost since it would be necessary to fund the fabrication of a special to type IC. An early technique was to use a software programme running on a general purpose processor. This solution was only really suitable for low speed applications which severely limited its usefulness. Hence the trade-off was traditionally one of speed and cost.

The recently available single chip digital signal processors have the potential to provide the speed of an ASIC but at a significantly lower cost. This lower cost occurs not just as a result of reduced component cost, but also as a result of reduced design cycle times.

It is difficult to provide categoric figures for the likely throughput as it depends on a number of criteria. In particular, the extent to which parallelism can be exploited. Many DSP operations can be readily decomposed into multiple parallel paths and as such will acheive the highest throughput gains.

Instruction cycle times of 100-200nsec are common and many
DSP devices improve coding efficiency by optimising the
instruction format to execute typical digital processing
algorithms with the minimum of instructions. In particular,
many have been optimised to perform a high speed FFT. There
are many ways of implementing an FFT but most rely on the
Butterfly operation and DSP chips have been designed with
architectures particularly suited to this operation. In
addition some employ dual data bus architectures to speed up
the arithmetic operations involving the real and imaginary
parts of complex data formats.

21.4   IMPACT ON SYSTEM DEVELOPMENT

There are many areas that, as a result of the advances in
DSP technology, are now moving out of the laboratory into
commercial applications.

The successful development of such products have their own
unique problems. A particular example of this is the area
of speech processing and the following sections describe
a practical example in the development of a commercial
speech processing system, with specific regard to the
performance/cost trade-off.

21.4.1   Introduction to the Example

This example describes the advantages and disadvantages of
developing DSP applications on floating point processors
using high level languages. A speech recognition system is
used as the exemplar in which it was required to transfer
the algorithm from a VAX computer to an AT&T DSP32
processor. The VAX version was written in Fortran and the
DSP32 version consisted of a `C' language programme calling
assembler routines. The bench mark is an already existing
version of the algorithm implemented in assembler for
real-time operation on a Texas Instruments TMS320 digital
signal processor.

21.4.2   Development Background

During 1987 Ensigma developed the SRS-1 speech recognition
system which was based on a Texas Instruments TMS320C17
digital signal processor. The aims of this development
were:

* Accurate recognition

* Speaker independance

* 16 word vocabulary

* Robust operation

* Low cost

* Single device realisation

* Isolated word operation

At that time it was believed that the implementation of a speech recogniser on a low cost DSP would lead to a variety of applications in the consumer and toy markets. During 1987 and 1988, the recogniser was marketed by Ensigma and Texas Instruments within Europe. The market feedback received led to a revision of the design aims, to the following:

* Telephone network operation

* Accurate recognition

* Speaker independance

* 16 word vocabulary

* Robust operation

* Moderate cost

* Single board realisation

* Isolated word operation

The difference between the two can be summarised in that the market required units which could work over the telephone network, albeit at a rather greater size and cost than the original design. Ensigma then undertook a programme of research and produced an algorithm which could cope with the addition of noise and the impairments to frequency response which are characteristic of telephone line operation. This was completed in mid 1988. At that time the algorithm was running on a Dec VAX under VMS and was written in Fortran. It was then decided to develop the algorithm to work in real-time on a digital signal processor.

21.4.3   Development Options

It was immediately clear that the increased complexity of the algorithm would not permit it to be implemented on the TMS320C17 with the required vocabulary. Therefore, it was necessary to consider how to port the algorithm to a more powerful processor. While a wide range of digital signal processing devices could have been used for the real-time implementation, two devices were regarded as being particularly suitable, given the background of the project.

The first device was the TMS320C25, the second generation Texas Instrument part. Since this device has an instruction cycle time which is half that of the TMS320C17, and a more advanced instruction set, It clearly had the extra processing power required for the enhanced algorithm. It also had the advantages of moderate cost and a substantial code compatability between the two TI processors. However, previous experience with the design of speech recognition systems had demonstrated that a series of experiments would have to be carried out to optimise the operation of the algorithm once it was working in real time. It was felt that the use of an assembly language programme on a fixed point processor like the TMS320C25 would lead to the optimisation being carried out somewhat slowly, as had been the case with implementing the SRS-1 on the TMS320C17.

During 1988 low cost PC compatable boards, containing the AT&T DSP32 processor, had become available. The DSP32 has the advantage of floating point operation and the support of a `C' compiler. Together these allow the code to be developed much more quickly by the use of high level constructs and by relieving the programmer of the scaling and rounding problems associated with fixed point devices. The disadvantage of using the DSP32 relate to the much higher price, compared with the TMS320C25, coupled with the very high power consumption and the slower cycle time.

It was felt that the power consumption problem would be solved when the CMOS DSP32 became available, while the price difference would be offset by the reduction in engineering costs associated with the time saved by using this device. The cycle time increase could be offset by the saving in the required number of cycles given by the use of floating point arithmetic.

It was therefore decided to continue the development using the DSP32. This required the following steps:

* Transcode VAX Fortran to VAX `C'

* Port VAX `C' to DSP32

* Optimise `C' for real-time operation

* Assembler code key routines

The first two steps are straightforward and need no further discussion. However, in order to be able to understand the optimisation and the assembly coding of key routines, it is necessary to have an appreciation of the operation of the speech recognition algorithm employed.

21.4.4   Speech Recognition Process.

The main elements of the speech recogniser are shown in figure 21.2.

Speech from the input undergoes transformation by a feature extractor. This produces a representation of the spectrum of the speech.

In this particular case, the spectrum is represented as the output of a set of filters in a filter bank.

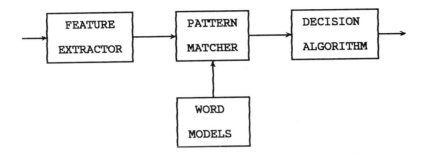

Figure 21.2 Speech Recognition System

The filters in the filter bank are implemented as a cascade of two recursive sections, as shown below:

$$w(n) = gx(n) + a_1 w(n - 1) + a_2 w(n - 2) \quad \ldots\ldots(1)$$

$$y(n) = w(n) + b_1 y(n - 1) + b_2 y(n - 2) \quad \ldots\ldots(2)$$

Each of these filters forms a fourth order Bessel filter. The running estimate of the power of the output of the filters is given by the following expression:

$$p(n) = y^2(n) + \alpha p(n - 1) + \beta p(n - 2) \quad \ldots\ldots(3)$$

The filtering operation takes place every 125msec and there are typically 10 to 20 filters in the bank. The DSP must therefore carry out approximately 100,000 filtering operations per second.

The output of the feature extractor is passed to a pattern matcher which compares the recent spectrum of the input with a set of spectral models of the words to be recognised.

The output of the pattern matcher is then fed to a decision algorithm which determines when a valid word has been recognised.

The pattern matching operation is carried out in two stages. In the first stage an estimate is made of the probability that a given spectrum, b, comes from a given state in the word model.

This is given by equation (4).

$$p(b|\theta) = \prod_{i=1}^{N} \exp \left[- \frac{(b_i - \theta_i)^2}{2\sigma_i^2} \right] \quad \ldots (4)$$

Although it is more usually computed in the logarithmic domain as shown in equation (5).

$$\ln p(b|\theta) = - \sum_{i=1}^{N} \frac{(b_i - \theta_i)^2}{2\sigma_i^2} \quad \ldots (5)$$

Where θ is the mean and σ is the standard deviation of the state.

The second stage is to calculate the probability $\alpha_{t+1}(j)$.

This is the probability of being in a given state j, in the word model at time t+1. This is given by the Viterbi algorithm. This is a breadth-first dynamic programming algorithm which is computed as shown in equation (6).

$$\alpha_{t+1}(j) = \min [\alpha_t(i) + \alpha_{ij}] + \beta_j(\theta_{t+1}) \quad \ldots (6)$$

Where

$$\beta_j(\theta_{t+1}) = \ln p(b|\theta) \quad \ldots (7)$$

And a_{ij} is the transition probability, state i to state j.

Each word model contains approximately 10 states. Therefore with 16 words it is necessary to perform 160 state computations in each 20msec frame period. Since the number of filters is approximately equal to 10, each weighted difference summation in the probability calculation (equation (5)) must be carried out approximately 1600 times per frame, or 80,000 times per second. From this it is clear that the probability calculation and the filtering operation are the computationally intensive parts of the algorithm.

21.4.5   Optimisation

Table 21.1 summarises the percentage real-time usage at each stage of the optimisation process.

PROCESS STAGE	% REAL-TIME USAGE
MICROVAX II	1200
DSP32	625
POINTERS IN FILTER BANK	413
POINTERS IN PATTERN MATCHING	288
FBA IN ASSEMBLER	113
PATTERN MATCHING IN ASSEMBLER	52

Table 21.1   Percentage Real-time Usage

Running on the Microvax, the algorithm executed in 12 times real-time. The initial transfer to the DSP32 approximately halved the amount of time required. The algorithm was then modified to replace the array computation with pointerised code using register variables. As can be seen from Table 21.1, this produced further useful improvements in the performance of the algorithm. However, it was still nearly 3 times slower than was required for real-time operation. At this point it was decided to re-code the feature extractor in assembler. This halved the time taken, allowing performance just outside real-time operation. Finally, the pattern matching algorithm was re-coded in assembler, which halved again the execution time and enabled better than real-time operation.

The original TMS320C17 had taken more than 5 man months of effort to optimise and correct. The `C' version on the DSP32 reduced this time to just over 1 man month. This demonstrates the benefit to be gained by using a high level language compiler on a floating point processor.

21.5   CONCLUSIONS

An example has been described of the coding of a speech recognition algorithm onto a floating point digital signal processor by means of a high level language compiler. It has been shown that this can considerably reduce the engineering effort required, compared with that for a fixed point DSP using assembly language. It must be concluded however, that it is necessary to use assembly coding for computationally intensive routines if real-time performance is to be acheived.

When devices are used in production volumes, the extra cost of floating point over fixed point devices will not be justified by the savings in engineering costs. However, if the cost of engineering effort involved in programming the DSP is significant, when compared with the total cost of the DSP devices, then the extra cost for a floating point device can be justified by the saving of incremental engineering costs. This is particularly the case if further modifications of the algorithm may be required.

Chapter 22

A case study of a DSP chip for on-line monitoring and control in anaesthesia

S. Kabay and N. B. Jones

22.1 INTRODUCTION

High Frequency Jet Ventilation (HFJV) is a recognised form of mechanical ventilatory support that is used in both anaesthesia and critical care medicine. The technique differs from conventional modes of ventilatory support in both its relative tidal volume and respiratory rate. Several studies have shown that HFJV is capable of maintaining adequate gas exchange in cases where conventional methods have either failed or proved to be impractical. The main advantages of HFJV include lower peak and mean airway pressures (Heijman et al., 1972), a reduction in pulmonary barotrauma (Keszler et al., 1982) and less disturbance to cardiovascular function (Sjostrand, 1980). A more detailed consideration of the subject may be found in several recent reviews (Smith, 1982; Drazen et al., 1984; Kolton, 1984).

General acceptance of HFJV has been inhibited by lack of understanding of the mechanism by which HFJV maintains gas exchange, and also by a lack of practical guidelines for clinicians on when to apply HFJV and what ventilator parameters to set for optimum gas exchange. One approach to the better understanding of gas exchange during HFJV treats the patient's respiratory system as an acoustic resonator whose characteristics vary markedly over the range of HFJV frequencies (Lin and Smith, 1987). Preliminary results based on an animal model have supported the hypothesis that the respiratory system behaves as an acoustic resonator (Smith and Lin, 1989).

This chapter presents a case study of computer-aided instrumentation for on-line measurement and control in the area of critical care medicine. The system has been initially developed to test the hypothesis that acoustic resonance of the respiratory airways represents an optimal state for alveolar gas exchange. Real-time signal processing algorithms have been implemented around a user-friendly shell to aid in the identification of transfer function relationships between respiratory data. This analysis is a first step towards the development of a self-tuning algorithm for automatic control and management of ventilator parameters to optimise gas exchange in patients. The instrumentation can be used for extremely high resolution identification of respiratory dynamics in a fraction of the time taken by previous workers with only minimal changes to existing jet ventilation procedures.

22.2 TRANSFER FUNCTION ANALYSIS OF THE RESPIRATORY SYSTEM

The purpose of the instrumentation described in this chapter is to characterise the respiratory system dynamics in a systematic manner. The aim is to provide the framework on which stimulus-response experiments can be carried out. This should reveal information about the functional characteristics of the system which will enable a good hypotheses to be made on its structure.

The instrumentation should take into account inherent characteristics of the system:

- non-linearities, which often are essential for optimal functioning
- limited durations of experiments because of great variability in the data
- low signal-to-noise ratio in the measurements

The identification task is to be carried out through a series of stimulus-response experiments, it is important to design these experiments so as to maximise the information obtained from the system. The following sections will describe the jet ventilator circuit and its application as a tool for identifying respiratory system dynamics.

22.2.1 High Frequency Jet Ventilator

The Penlon high frequency jet ventilator used in this study is an electrically and pneumatically operated time-cycled jet ventilator device of the type described by Smith (1985). The ventilator circuit shown in Fig. 22.1 is a modified version of the commercial design, which includes a proportional controller valve to support continuous pressure excitation of patients' airways. A brief description of the ventilator is given below.

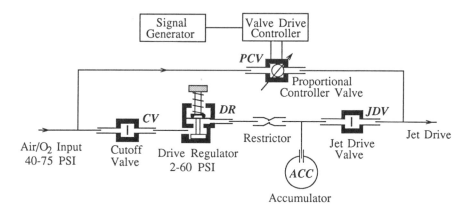

Fig. 22.1 The modified high frequency jet ventilator circuit.

Gas from a high pressure (40-75 PSI) source is delivered into the ventilator via the inlet cutoff valve (CV) and drive pressure regulator (DR) to an accumulator (ACC). An electronic time-base controller energises the jet drive valve (JDV) so that gas flows from the ACC to the jet line and patient periodically depending on the ventilation rate (variable over the range 40-200 BPM) and inspiration time control. The inspiratory:expiratory (I:E) ratio can be varied over the range 10-60 % of the respiratory cycle. The inspiratory flow is determined by the drive pressure regulator and the respiratory airways impedance of the ventilator/patient system.

A proportional controller valve (PCV), connected in parallel to the jet drive valve (JDV), introduces small low amplitude pressure oscillations which is hydraulically summed with the binary output of the jet ventilator. The PCV controller accepts an analogue signal over the range 0-10 V and produces a continuous valve output which tracks the reference input. A white-noise circuit (Horowitz and Hill, 1983) is used to drive the PCV with a pseudo-random analogue noise signal with bandwidth selectable over a range of frequencies. This allows for white-noise testing of respiratory airways.

The jet ventilator can operate in either a) binary, b) proportional, or c) binary plus proportional mode. The various ventilator modes provide maximum flexibility, enabling respiratory dynamics testing for different stimuli.

22.2.2 Swept Sine-Wave Analysis

The standard Penlon high frequency jet ventilator (*mode a*) provides the patient with binary pressure pulses. Such stimuli can, at best, be approximated to sine waves, and swept sine-wave analysis performed to determine the system's response. This method is tedious, time-consuming, and particularly cumbersome in the clinical environment. In general, the technique suffers from many problems and may be impractical for applications where:

- the system adapts to the applied signal or shows fatigue
- the amplitude of each *sinusoid* must be adjusted so that it remains within the linear range of the system at a given frequency
- the output variables have long settling times before the system is in steady-state.

Since physiological systems exhibit many of these characteristics, sine-wave testing can lead to erroneous results. For example, Smith and Lin (1989) used this approach to respiratory dynamics testing in animal studies with some success. However, the measured output variables had relatively short time-constants and did not require the delay associated with blood gas stabilisation.

With the swept sine-wave method of identification we must make many arbitrary assumptions about the system in order to postulate a mathematical description of its structure. The identification task is then reduced to estimating the parameters of the model based on data collected from the stimulus-response experiments. The success of this approach depends solely on the skill and imagination of the modeller. Apart from the time-consumed in making the measurements, the task of designing differential equations to fit a given set of data becomes a laborious one. Finally, the model is usually valid for the chosen set of stimuli and not satisfactory for other types of input. Thus, adding new information about the system often means that the modelling process must be repeated from the beginning.

22.2.3 White-Noise Identification

The system identification objectives set out above imply that the black-box approach to determining the transfer characteristics of the system may be more suitable. This method of identification does not rely on any assumption about the internal topological structure of the system. This approach becomes a search for the function $H[x(t)]$ where

$$y(t) = H[x(t)] \tag{22.1}$$

and x(t) is the stimulus and y(t) is the response, the identification task consists of estimating the system functional H.

A white-noise stimulus with a flat power spectrum over the frequency range of interest is equivalent to applying a range of sinusoids simultaneously. This technique inherently provides a method for rapidly determining respiratory dynamics over an expanded frequency range with a frequency resolution greater than is practicable with other techniques. This is far superior to using swept sine-wave stimuli, where representation is over a limited and pre-determined region of the function space.

To use the white-noise approach the ventilator must be operated in either *mode b* or *c*. Mode b is extremely useful for:

- determining the frequency response characteristics of respiratory pressure, flow and displacement transducers
- performing respiratory impedance measurements on patients; significantly simplifying existing methods of measurement.

Mode c is intended for routine clinical use since it enables on-line identification of respiratory dynamics during normal high frequency jet ventilation. Once preliminary decisions are made, the white-noise method can be used for characterising the respiratory airways over a short period of time.

22.3 INSTRUMENT SPECIFICATION

The method of analysing real data involves sampling of the stimuli and response signals and performing spectral analysis using the fast Fourier Transform (FFT). From these spectra the frequency response function and measures of the linearity of the system can be determined (Bendat and Piersol, 1971).

The sampling interval determines the maximum frequency f_{max} which can be analysed correctly before aliasing becomes a significant problem and is defined according to the Nyquist sampling theorem. In practice, analogue low-pass filters with sharp cut-off close to f_{max} are used to minimise the undesirable energy content of signals prior to digitisation.

The maximum bandwidth of HFJV measurements varies between 30-300 breaths-per-minute (BPM) or 0.5-5 Hz. Assuming that all significant spectral energy is contained within 10 times the fundamental frequency, this sets the maximum bandwidth per channel to be 50 Hz. Thus, to satisfy the Nyquist sampling criterion, the data acquisition system must have a minimum bandwidth of 400 Hz.

A maximum of four channels of data must be logged, including upper airways pressure and flow signals, abdominal and thoracic wall displacements. Several methods of analysis must be implemented to assess the relationships between stimuli-response data. These include the ability to compute and display power spectral density functions, two sets of transfer function gain, phase and coherency spectra from the raw time-domain signals in real-time.

Finally, the instrumentation is intended for use by nonexperts. Hence the man-machine interface should be an environment which presents the functional power of the system in a clear and concise manner. A friendly menu-driven system based on the type of system architecture described by Hailstone et al. (1986) is desirable.

22.4 REAL-TIME MEASUREMENT SYSTEM

22.4.1 System Hardware

Since its introduction in 1982 the single component digital signal processor (DSP) has become a significant device for tackling high-speed signal processing tasks. The DSP is more than just a peripheral device for performing front-end real-time signal conditioning under the control of a host microcomputer. The DSP device features increased functionality and computational power which enables it take on the role of the central processing function of a system.

The hardware implementation adopted is that of a host IBM-AT compatible microcomputer and a DSP front-end. This dual processor configuration was selected since it separates a task into two functional sub-units and then allocates the work amongst the two processors according to their relative strengths.

A high speed DSP comprising a TMS320C25 running at 40 MHz with 64 kWords of dual-ported memory is used as the front-end processor. The host can read and write to the DSP memory at any time, even whilst the DSP is working on other tasks. An on-board timer can be used to generate vectored-interrupts, allowing complex applications to be developed which make full use of the DSP bandwidth.

The DSP device offers distinctive features in terms of mathematical capability/speed and incorporates on-chip the functions needed to achieve maximum throughput and standalone operation. The main advantages of the device include:

- a wide dynamic range of data representation with sufficient resolution to achieve the required degree of performance and minimise the effects of quantisation and truncation errors.
- fast on-chip memory organised to support a high degree of parallelism and maximise throughput. The on-chip RAM is sufficient to download execution of a complete 256-point complex FFT.
- a large number of hardware registers provides the flexibility required for complex calculations.
- a wide variety of software support tools are commercially available.

An analogue interface card comprising 4-input and 2-output channels connects directly to the DSP. Each input channel contains an anti-aliasing filter and sample-and-hold amplifier. This system permits the simultaneous sampling of 4 channels of data which are multiplexed out to a single 3 μs conversion 12-bit analogue-to-digital converter. This permits 4-channel data acquisition of up to 60 kSamples per second. Two 12-bit digital-to-analogue converters with a typical settling time of 3 μs are also available for transmitting analogue signals to external devices.

22.4.2 Host Software

The host software is responsible for supervising the man-machine interface and redirection of data depending on the state of the system. A generic core implements the kernel, window management, softkey processors, display, and a variety of library functions accessible by the instrument. All DSP-specific code and data are downloaded to the DSP board via the host.

The man-machine interface is probably one of the most significant factors determining whether an instrument will be accepted as a valuable tool for everyday use in the clinical environment. Complex multi-mode instruments that use traditional front-panels have to choose between a) implementing a large number of controls and displays, many of which are not relevant to an individual measurement and only contribute to front-panel clutter, or b) using the same controls and displays for multiple functions, each of which requires its own graphics screened in different colours on the panel. Both of these options are potentially confusing to the user.

The man-machine interface running on the host provides a mechanism whereby only those components that are relevant to the current measurements are displayed. If a function requires further information, the window manager will activate a dialogue box instructing the user of the required actions. Once the correct parameter has been set, the dialogue box vanishes, restoring the screen to its original state. This scheme avoids major re-definition of the screen and thereby avoids a complex appearance to the system. A more concise description of the system is given by Kabay et al. (1989).

22.5 ALGORITHMS FOR REAL-TIME SPECTRAL ANALYSIS

The spectral analysis of data, based on the FFT, is usually the first operation to be performed and is often a preliminary to further processing. However, the FFT is a very computation intensive operation and is the major bottleneck hindering the widespread development of real-time spectrum analysers. The DSP offers several features which make it ideal for computation intensive digital signal processing applications such as the FFT. This section describes the design considerations necessary for implementing a non-trivial real-time DSP system.

22.5.1 Overlapping Real-Time FFT

An overlapping sample buffer can be used as part of a spectral averaging scheme to improve the signal-to-noise ratio of measurements. For example, a 75 % overlap section-add facility will significantly minimise errors introduced by the spectral estimation process (Rabiner and Gold, 1975). Hence, a complete N-point FFT must be performed for each new (N/4) input samples. A circular buffer memory algorithm was implemented on the DSP for handling this overlap scheme.

For a 1024 data points with a 75 % window overlap, the total memory requirement for a single channel will be 1280 words. This consists of five equal blocks of 256 words B_0 - B_4. Interrupt-driven DSP software is used to implement overlapped data acquisition. All signal processing tasks are carried out between interrupts. For example, whilst the interrupt service routine (ISR) is filling memory B_0 with data, the DSP computation loop is processing data currently stored in blocks B_1-B_2-B_3-B_4 (Fig. 22.3a). In this case, the age of the data decreases from B_1 to B_4. During the next data acquisition cycle (Fig. 22.3b), samples will be stored into B_1 whilst blocks B_2-B_3-B_4-B_0 are processed.

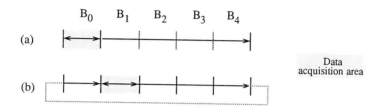

Fig. 22.3 Overlapping sample buffer.

The DSP timing for these operations are critical. For example, the DSP must have performed all processing on the current set of data and transferred the results to the host before the next frame of 256 samples is transferred to the main calculation buffer. With this configuration, data acquisition memory is optimised to be 256 samples per channel.

The overall bandwidth of this data acquisition scheme can be calculated from the following timing relationship:

$$t_{DSP} + t_{XFR} + t_{PC} < t_{256} \tag{22.2}$$

where:

- t_{DSP} = time taken to perform signal processing routines ≈ 60 ms,
- t_{XFR} = time taken to transfer data over to the PC ≈ 3 ms,
- t_{PC} = time taken to plot the signal onto the screen ≈ 250 ms,
- t_{256} = time taken to fill 256-point acquisition buffer $= 256/f_s$.

The timing for t_{DSP} includes computation of two sets of:

- cross-spectra calculations
- sets of transfer function magnitude squared
- transfer function phase
- transfer function coherency
- the power spectrum of four input channels.

The spectral data transferred to the host includes 4-channels of raw data and power spectra and two sets of transfer function gain, phase and coherency spectra.

Thus, $\qquad t_{DSP} + t_{XFR} + t_{PC} = 60 \text{ ms} + 3 \text{ ms} + 250 \text{ ms} < t_{256}$

where, $\qquad\qquad t_{256} = \dfrac{256}{f_s} = \dfrac{256}{2.56 f_{max}} = \dfrac{100}{f_{max}} \tag{22.3}$

$\Rightarrow$ $\qquad\qquad$ $f_{max} < 320$ Hz.

The maximum signal bandwidth that can be captured is approximately 320 Hz. The main limitation is due to t_{PC} which is imposed by the graphics system. For example, if a graphics coprocessor were used for handling display information (t_{PC} would be negligible) then the overall bandwidth would be $f_{max} \approx 5$ kHz per channel.

The requirement was initially specified for real-time operation up to a spectral bandwidth of 50 Hz. This can be easily achieved with the existing system. The bandwidth is user selectable over the range 1-50 Hz, providing spectral resolutions of the order 0.001-0.050 Hz. This includes plotting a signal onto the screen at intervals determined by the sampling rate.

22.5.2 FFT Implementation

The fast Fourier Transform (FFT) is a method of determining the spectral properties of a signal. The maximum frequency which can be analysed is determined by the sampling rate, and this together with the data record length determines the size of Fourier transform required. Since a high spectral resolution was major requirement, the FFT record-length was selected to be 1024-points. For the range of HFJV frequencies (10 Hz) this will give a minimum frequency resolution of 0.025 Hz.

The speed at which the FFT is executed plays a crucial part in determining the overall bandwidth of signals that can be analysed in real-time. The processing time of the DSP is summarised in the definition of t_{DSP} in Sec. 22.5.1. To obtain the required spectral resolution over the desired bandwidth meant that a high-performance FFT algorithm had to be implemented using in-line assembly code. The algorithm uses an 1024-point complex transform and a mixture of radix-2 and radix-4 butterflies.

The first two stages of a N-point FFT involves multiplications with known constants. Hence, stages 1 and 2 are implemented as a special radix-4 butterfly. Stages 3-8 are implemented by repeated execution of a 256-complex point FFT which is executed from on-chip DSP RAM for maximum efficiency.

The algorithm is implemented using a DIT, radix-2, in-place FFT routine. Assuming N = 2^n, where n = total number of stages of decimation then N = 1024 and n = 10.

Thus, the butterfly operation at any given stage m of the FFT can be defined as:

$$\left. \begin{array}{l} X_{m,i}(k) = X_{m-1,2i}(k) + W_\alpha^k \, X_{m-1,2i+1}(k) \\[2ex] X_{m,i}(k) = X_{m-1,2i}(k) - W_\alpha^{k'} \, X_{m-1,2i+1}(k) \end{array} \right\} \quad \begin{array}{l} 0 \le k \le \dfrac{\alpha}{2} - 1 \\[2ex] \dfrac{\alpha}{2} \le k \le \alpha - 1 \end{array} \qquad (22.4)$$

where $W_\alpha^k = e^{-j2\pi k/\alpha}$, $\alpha = N/2^{n-m}$, k' = k $-\dfrac{\alpha}{2}$, $1 \le m \le n$ and $X_{m,i}$ represents the decimated samples after stage m of the FFT algorithm $0 \le i \le (N/2^m)-1$. The DIT redundancy reduction is achieved by dividing the input sequence into odd and even sample sequences. Thus, at stage-1, an N-point sequence is transformed by combining the DFTs of these two N/2-point sequences. The index i represents the number of samples at the input to each sequence.

After stage-1 of the FFT algorithm we have:

$$\left.\begin{array}{l} X_{1,i}(k) = X_{0,2i}(k) + W_{\alpha}^{k} X_{0,2i+1}(k) \\ X_{1,i}(k) = X_{0,2i}(k') - W_{a}^{k'} X_{0,2i+1}(k') \end{array}\right\} \quad \begin{array}{l} k = 0 \\ k = 1,\ k'= 0. \end{array} \qquad (22.5)$$

for $0 \le i \le (N/2^1)$-1, $\alpha = N/2^{n-1} = 2$, where m = 1, n= 10 and N = 1024.

After stage-2 of the FFT algorithm we have:

$$\left.\begin{array}{l} X_{2,i}(k) = X_{1,2i}(k) + W_{\alpha}^{k} X_{1,2i+1}(k) \\ X_{2,i}(k) = X_{1,2i}(k') - W_{\alpha}^{k'} X_{1,2i+1}(k') \end{array}\right\} \quad \begin{array}{l} 0 \le k \le 1 \\ 2 \le k \le 3,\ k'= k\text{-}2. \end{array} \qquad (22.6)$$

for $0 \le i \le (N/2^2)$-1, $\alpha = N/2^{n-2} = 4$, where m = 2, n = 10 and N = 1024.

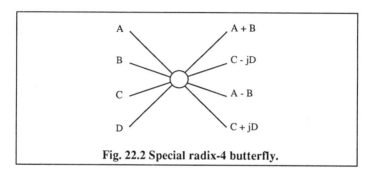

Fig. 22.2 Special radix-4 butterfly.

Thus the first two stages of an N-point FFT can be performed simultaneously with a special radix-4 butterfly which avoids multiplication operations to enhance execution speed (Fig. 22.2). For N=1024 the first radix-4 butterfly is given by substituting i=0 in equations (22.5) and (22.6):

$$\left.\begin{array}{l} X_{2,0}(0) = [X_{0,0}(0) + X_{0,1}(0)] + [X_{0,2}(0) + X_{0,3}(0)] \\ X_{2,0}(1) = [X_{0,0}(0) - X_{0,1}(0)] - j[X_{0,2}(0) - X_{0,3}(0)] \\ X_{2,0}(2) = [X_{0,0}(0) + X_{0,1}(0)] - [X_{0,2}(0) + X_{0,3}(0)] \\ X_{2,0}(3) = [X_{0,0}(0) - X_{0,1}(0)] + j[X_{0,2}(0) - X_{0,3}(0)] \end{array}\right\} \qquad (22.7)$$

Later stages of the FFT require multiply-and-accumulate operations on raw-data with non-integer W_{α}^{k} twiddle factor terms. To maximise the execution speed, the twiddle factors are pre-computed and stored in a look-up table. The butterfly operations are implemented as macro-calls that the main computation loop comprises in-line code. This provides a further speed advantage by eliminating the overhead associated with conditional branch instructions. Wherever possible, special butterfly operations are used for integer-valued W_{α}^{k} terms. Automatic scaling is also incorporated into butterfly operations to avoid any possibility of overflow.

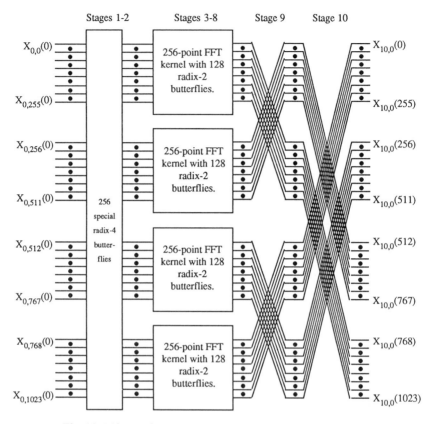

Fig. 22.4 1024-point decimation-in-time radix-2 FFT kernel.

Fig. 22.4 shows the flow diagram for a 1024-point FFT based on an in-place decimation-in-time algorithm. The computation loop between stages 3-8 are processed using a smaller 256-point complex FFT which is executed utilising the on-chip memory of the DSP. This scheme provides marked increase in speed of performance since on-chip memory data accesses are single cycle instructions. Data is transferred between external and internal memory at high speed via the pipeline capability of the TMS320 device.

The time required to compute a 1024-real point FFT and its squared magnitude terms using the developed algorithm is 8 ms. This is a four-fold increase in performance compared with the in-line FFT algorithm reported by Papamichalis and So (1986), and an eight-fold increase over that of Burrus and Parks (1985). The latter workers have implemented a similar algorithm in Fortran on a PDP 11 and achieved a timing performance of 1400 ms. A more general discussion of the FFT may be found in Chapter 6 of this book.

22.5.3 Transfer Function Analysis

This section will derive a method of computing a transfer function where both the magnitude and phase information can be easily obtained over a wide range of frequencies. This method of computing a transfer function makes efficient use of computer resources and does not make assumptions on the character of the input signal, i.e. the input signal does not have to be Gaussian white-noise.

The Fourier transform S_x of a time-varying signal x(t) can be expressed as a complex number of the form:

$$S_x = X_r + jX_i \tag{22.8}$$

where j = -1, X_r and X_i are the real and imaginary parts of S_x at frequency $\omega = 2\pi f$, and contain magnitude and phase information of this frequency component of x(t).

Consider a single-input x(t), single-output y(t) linear system h(t), the transfer function $H(j\omega)$ is:

$$H(j\omega) = S_y / S_x \tag{22.9}$$

If x(t) is random white noise, $H(j\omega)$ can be found at all frequencies in x(t). However, due to synchronisation problems and statistical considerations related to the randomness of the input x(t), the computation of $H(j\omega)$ by this method is not practical. These problems can be overcome by using the concept of power spectra.

The input power spectrum is:

$$G_{xx} = S_x S_x^* = X_r^2 + X_i^2 \tag{22.10}$$

where S_x^* is the complex conjugate of S_x. The term G_{xx} no longer contains phase information. Similarly,

$$G_{yy} = S_y S_y^* = Y_r^2 + Y_i^2 \tag{22.11}$$

The cross-power spectrum is defined as:

$$G_{yx} = S_y S_x^* = Z_r + jZ_i \tag{22.12}$$

where $Z_r = (X_r Y_r + X_i Y_i)$ and $Z_i = (X_r Y_i - X_i Y_r)$ which preserves the phase relationship between input and output and can be used to compute the transfer function from:

$$H(j\omega) = S_y/S_x = G_{yx}/G_{xx} \tag{22.13}$$

and modulus, $\qquad |H| = |G_{yx}|/G_{xx} = (Z_r^2 + Z_i^2)/(X_r^2 + X_i^2) \tag{22.14}$

and phase, $\qquad \arg(H) = \tan^{-1}(Z_i/Z_r) \tag{22.15}$

The coherence function can also be derived from these calculations to provide a measure of the extent to which the response is due to the stimulus and not to extraneous sources of noise. The coherence function γ^2 is given by:

$$\gamma^2(j\omega) = |G_{yx}| / (G_{xx}G_{yy}) \qquad 0 \le \gamma^2 \le 1 \qquad (22.16)$$

The coherence γ^2 is a real term which, for any frequency ω give the fraction of power at the response that is due to the input. Whenever γ^2 is less than unity, either the system is non-linear or the signals do not have a causal relationship, or both. The deviation of γ^2 from unity is a quantitative measure of these conditions. Since physiological systems are both nonlinear and have high noise content, the coherence function should provide a valuable measure of confidence in spectral estimates.

22.5.4 Phase Angle Calculation

The DSP is a fixed-point device with no support for trigonometric functions. This section describes an efficient algorithm for computing the transfer function phase spectrum. The phase is calculated using an inverse-tangent look-up table so as not to compromise the DSP performance.

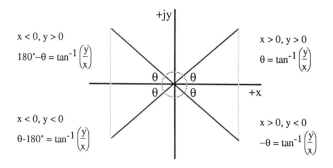

Fig. 22.5 A method for computing the argument of a vector.

The phase angle θ of a complex number of the form $z = (x + jy)$ can be calculated using $\tan(\theta) = \left(\dfrac{y}{x}\right)$. Using the Q_{15} number notation defined in Rabiner and Gold (1975), we can represent $-180^\circ \le \theta \le +180^\circ$ as an integer in the range $-32767 \le \theta \le 32767$.

The Argand diagram of Fig. 22.5 shows a method for computing the phase θ of the vector $x + jy$ depending on the quadrant it occupies. Hence, by defining $0^\circ \le \theta \le +89^\circ$ we can calculate θ for any value in the $-180^\circ \le \theta \le +180^\circ$. Using a look-up table of 2048 θ-values over the range $0^\circ \le \theta \le +89^\circ$, we can achieve a maximum phase resolution of 0.044°.

A system had to be devised which could be used to compute $\theta = \tan^{-1}\left(\dfrac{y}{x}\right)$ from values of the ratio $\left(\dfrac{y}{x}\right)$:

$$\frac{y}{x} = \left(\frac{57.9n}{2048}\right) \qquad (22.17)$$

Since $0 \le \tan\theta \le 57.29$ over the range $0^\circ \le \theta \le +89^\circ$ and $0 \le n \le 2047$.

The phase look-up table values are calculated using:

$$\theta_n = \left(\frac{16383}{90°}\right) \tan^{-1}\left(\frac{y}{x}\right) = 182.04 \tan^{-1}\left(\frac{n}{35.75}\right) \qquad (22.18)$$

Hence, the correct phase angle for a given ratio of $\frac{y}{x}$ is found by calculating the index value n, where $n = \left(\frac{36y}{2048}\right)$.

22.6 A CONTROL SYSTEM FOR AUTOMATIC VENTILATION

The prime task of artificial ventilation is to oxygenate patients incapable of performing spontaneous breathing, by transfer of oxygen from the inspired air to the blood stream via the alveoli, and at the same time remove carbon dioxide from venous blood and exhaust it out to the environment. The aim is to produce a ventilator which automatically adjust its parameters in response to changes in the patient's respiratory state (i.e., lung and chest wall compliance, airways impedance and partial pressure of alveolar gases) during HFJV.

The objectives of this study are to verify the hypothesis that a patient's respiratory airways behaves like an acoustic circuit with a resonant frequency determined by the geometry and mechanical properties of its branches (Smith and Lin, 1989). The approach taken to assess the clinical significance of high frequency ventilation in the light of this hypothesis can be summarised:

- perform an identification study on a large group of patients, with varying respiratory physiology, to determine resonances over the range of HFJV frequencies
- investigate the correlation between ventilation frequency and gas exchange in human subjects
- implement a self-tuning ventilator for optimal ventilation and monitoring of the patient's respiratory state.

The preceding sections have discussed the detailed implementation of a real-time system for characterising the respiratory dynamics of patients receiving high frequency jet ventilation. The next design phase of this study is to implement a self-tuning algorithm for automatic control and management of ventilator parameters to optimise gas exchange in patients.

One approach to automatic control of the jet ventilator is to use a model-reference adaptive control (MRAC) scheme. The basic principle of an MRAC system (Fig. 22.6) is to express the desired performance of the patient in terms of a *reference* model for a given command stimulus. The *patient* model, in this case the respiratory system of the subject, is a variable parameter system which is outside our control. The error e(t) represents the difference between the outputs of the patient and reference model. The adaptive control algorithm has two parameters (i.e. gain terms K_1 and K_2) that are changed based on the error and other inputs. The adaptive controller gains are set according to Lyapunov's stability theorem to ensure that the system will remain stable and convergent provided certain criterion are satisfied (Astrom and Wittenmark, 1989).

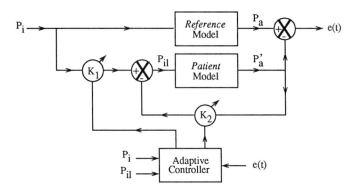

Fig. 22.6 Model-reference adaptive control scheme.

The mathematical description of each model is in terms of the input pressure P_i and P_{il} (derived from the ventilator) and the output alveolar pressure P_a and P'_a. The state variables must be indicative of respiratory dynamics since they form the basis of the control law which drives the ventilator.

The computation power of the DSP can be conveniently exploited to implement a model-reference adaptive control scheme in real-time. The mathematical routines required for this purpose are well-suited to DSP architecture which is optimised to implement digital filters. Where state estimation is necessary, subroutines may be written to perform matrix manipulations.

22.7 CONCLUSIONS

The use of HFJV as a form of respiratory support is known to have certain advantages over conventional methods of ventilation. However, a general acceptance of the technique has been inhibited by lack of understanding of the underlying fluid dynamics and a lack of practical guidelines for clinicians on when to apply HFJV and the ventilator settings that should be used for optimal gas exchange.

A computer orientated approach offers an attractive solution to these problems, as it can cope with the volume of information that needs to be considered in such an application, and with the level of complexity that the HFJV technique entails. A real-time instrumentation system has been developed based around a PC and DSP environment. The white-noise method has been incorporated into the system to allow for a systematic procedure of characterising a patient's respiratory system. The method is simple to apply and gives good results over a short period of time.

A ventilator control system based around a model-reference adaptive scheme is currently undergoing simulation tests. Based on the results of the simulation study and the identification data collected from patients receiving jet ventilation as part of their therapeutic regimen, a control scheme will be embedded into the existing instrumentation to produce a self-tuning ventilator.

22.8 REFERENCES

Astrom, K.J. and Wittenmark, B. (1989): *Adaptive Control*, Addison-Wesley Publishing Company, Reading, MA.

Bendat, J.S. & Piersol, A.G. (1971): *Random data: analysis and measurement procedures*, John Wiley & Sons, NY.

Burrus, C.S. and Parks, T.W. (1985): *DFT/FFT and convolution algorithms*, John Wiley & Sons, NY.

Drazen, J.M., Kamm, R.D., and Slutsky, A.S. (1984): "High frequency ventilation". *Physiol. Rev.*, v64, 505-.

Hailstone, J.G., Jones, N.B., Parekh, A., Sehmi, A.S., Watson, J.D. and Kabay, S. (1986): "Smart instrument for flexible digital signal processing". *Med. & Biol. Eng. & Comput.*, v24, 301-304.

Heijman, K., Heijman, L., Jonzon, A., Sedin, G., Sjostrand, U., and Widman, B. (1972): "High frequency positive pressure ventilation during anaesthesia and routine surgery in man". *Acta Anaesthesiologica Scandinavica*, v16, 176-187.

Horowitz, P. and Hill, W. (1983): *The art of electronics*, Cambridge University Press, 444-445.

Kabay, S., Jones, N.B. and Smith, G. (1989): "A system for real-time measurement and control in high frequency jet ventilation". *IFAC-BME Decision Support for Patient Management: Measurement, Modelling and Control*, England, 151-160.

Kolton, M. (1984): "A review of high frequency oscillation". *Can. Anaesth. Soc. J.*, v31, 416-.

Lin, E.S. and Smith, B.E. (1987): "An acoustic model of the patient undergoing ventilation". *Br. J. Anaesth.*, v59, 256-264.

Papamichalis, P. and So, J. (1986): "Implementation of fast Fourier transform algorithms with the TMS32020". *Digital Signal Processing Applications with the TMS320 Family*, Texas Instruments, 92-.

Rabiner, L.R., and Gold, B. (1975): *Theory and application of digital signal processing*, Prentice-Hall, Englewood Cliffs, 386-388.

Sjostrand, U.H. (1980): "High frequency positive pressure ventilation (HFPPV)". *Critical Care Medicine*, v8, 345-364.

Smith, B.E. (1985): "The Penlon Bromsgrove high frequency jet ventilator for adult and paediatric use". *Anaesthesia*, v40, 700-796.

Smith, B.E. and Lin, E.S. (1989): "Resonance in the mechanical response of the respiratory system to HFJV". *Acta Anaesthesiologica Scandinavica*, v33, 65-69.

Smith, R.B. (1982): "Ventilation at high frequencies". *Anaesthesia*, v37, 1011-.

Chapter 23

A case study on digital communication systems

E. M. Warrington, E. C. Thomas and T. B. Jones

23.1   INTRODUCTION

The methods by which digital data may be encoded and transmitted over communication links have been discussed in Chapter 15. Communication may, however, be over many different media, including copper wires, fibre-optic cables and radio links at frequencies ranging from a few hertz to a several hundreds of gigahertz. Various problems are associated with all of these systems, and some of the special problems encountered in radio systems employing high frequency (ie. in the frequency range 3–30 MHz) ionospherically propagated signals are discussed in this chapter.

High frequency (HF) radio signals are, in general, propagated between transmitter and receiver after reflection from the earth's ionosphere. The propagation characteristics result in various distortions being imposed upon the signal. Furthermore, the HF spectrum is shared by many users and although the various national regulatory bodies attempt to reduce sharing of frequency allocations, signals are almost invariably subject to some degree of interference from co-channel transmissions.

Several techniques have been devised in order to overcome some of the problems associated with both the propagation effects and with the presence of other signals. These techniques can be divided into several categories which include:

a) Measurements of the data transfer characteristics and interference background of several radio channels may be employed to select the best channel to use for communication.
b) Measurements of the channel characteristics may be employed to enable the system to adapt to the prevailing propagation conditions and interference environment. The adaption may include alteration of the signalling rate, the addition of error correcting codes or changes in the modulation.
c) Application of spread spectrum techniques in which the signal is transmitted using a bandwidth greatly in excess of that required to accomodate the data.

In the following sections of this chapter, the application of some of these techniques are discussed. The treatment given to all of the topics, particularly those areas concerned with the propagation aspects, is necessarily brief, however more detailed information may be obtained from the various references.

23.2   HF RADIO SIGNALS

23.2.1   An introduction to HF radio propagation.

An extensive description of HF propagation is given by Davies [1], and several summary chapters are contained in Hall and Barclay (editors) [2]. Some aspects are, however, discussed in outline below.

The earth's atmosphere above about 60 km is weakly ionised and as such is able to reflect radio signals up to a frequency determined by the peak electron density. For vertically incident signals, this maximum frequency is known as the critical frequency, f_0. At oblique incidence, the maximum frequency which may be reflected is greater than f_0 and increases as the angle of incidence to the ionosphere increases. The maximum frequency which can be propagated over any given path is known as as the maximum usable frequency (MUF). Frequencies above the MUF will not be received. As the signal passes through the lower regions of the ionosphere, it is subject to absorption processes which are greatest at the lower frequencies. This results in a lower frequency limit for which the signal has an acceptable signal to noise ratio (SNR). This is known as the lowest usable frequency (LUF).

Since the electron density in the ionosphere exhibits a diurnal variation, the LUF and MUF for a particular path vary throughout the day in a manner illustrated in Figure 1. The ionosphere also exhibits changes on an hourly, day-to-day, seasonal and yearly basis, and although average values of LUF and MUF may be determined for a particular circuit, the values at a particular time may be considerably different from the average values.

The signals travelling from transmitter to receiver may undergo one or more reflections from the earth's ionosphere (eg. see Figure 2). Several such paths (referred to as modes) may occur simultaneously and the signals corresponding to each mode combine at the receiving antenna. Since the ionosphere is in constant motion, the phase of each of the constituent modes varies with time resulting in periodic variations in the overall signal amplitude. This effect is known as interference fading and when the modes have similar amplitudes may result in periods of destructive interference when no signal is detectable at the receiving antenna. Furthermore, some regions of the ionosphere cannot be considered as a plane 'mirror in the sky', but rather as a rough reflector.

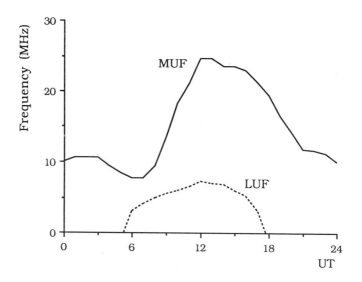

Figure 1. *A typical example of the variation of MUF and LUF for the path from the Canary Islands to London for one day.*

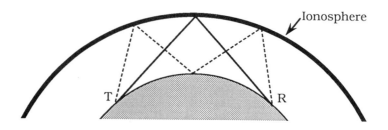

Figure 2. *Illustration of two possible paths which a signal may take between transmitter (T) and receiver (R).*

Under these conditions, many reflection points may exist for each mode (see Figure 3).

Since the motion of the ionosphere is not necessarily well correlated between the various reflection points, the spectrum of the received signal will usually contain several components. Typically, over mid-latitude paths, the frequency spreads are of the order of 0.1 Hz, but over high-latitude paths may be of the order of 10 Hz or more (eg. see Figure 4).

The effect of ionospheric propagation on a radio signal may be summarised in terms of a channel transfer function (eg. see Figure 5) in which each mode has its own attenuation due to transmission loss, its own time and frequency offsets and dispersions. For some paths, in particular trans-equatorial and trans-auroral paths, the spreads associated with each mode are sufficiently large that they merge.

23.2.2   Propagation effects on digital data signals

23.2.2.1 Intersymbol interference. The signals corresponding to each of the propagating modes travel via paths of different length and consequently arrive at the receiver at slightly different times. This results in interference between adjacent bits in a data signal and becomes significant when the difference in the times of flight of the various propagating modes become comparable with, or greater than, the duration of each bit. This effect is known as intersymbol interference and is illustrated in Figure 6.

For example, over a 1000 km path, the difference in times of arrival between 1 hop and 2 hop F–region signal is typically 1 ms. As such, intersymbol interference would be expected to become significant for bit lengths of less than about 10 ms (ie. for a 10% or greater bit overlap). This corresponds to a data rate of 100 baud.

23.2.2.2 Frequency selective fading. A further problem arises as a result of multi-moded propagation. The relative phases of each of the constituent modes measured at the receiving antenna is frequency dependent. As the frequency is changed, the various signal components combine with different relative phases resulting in changes to the overall signal amplitude. This effect is known as frequency selective fading and may result in the loss of the reception, or a distortion of the relative magnitudes, of the tones employed in FSK data transmissions. This effect is illustrated in Figure 7 for both quiet and disturbed conditions.

23.2.2.3 Phase stability. Each of the propagating modes is subject to Doppler shifts of between 0 Hz and (under extreme circumstances) approximately 20 Hz. The magnitude of the frequency shifts usually vary with time in an unpredictable manner. Consequently, the phase of the signal at the receiver will also vary in an unpredictable manner. This imposes limitations on the types of phase modulation which may be employed.

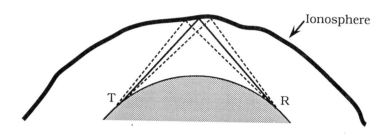

Figure 3. Some regions of the ionosphere cannot be considered as a smooth reflector.   Under these circumstances, many reflection points may exist for each mode.

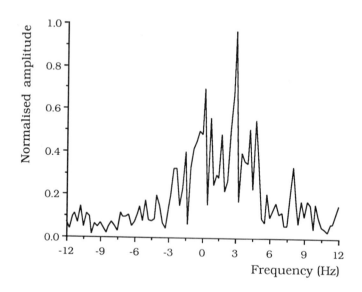

Figure 4. Example of Doppler spreading introduced on high latitude paths. (Data are for a 17.515 MHz signal transmitted from Clyde River, Canadian NWT to Leicester, UK.   15 UT, 2 Oct 1989.)

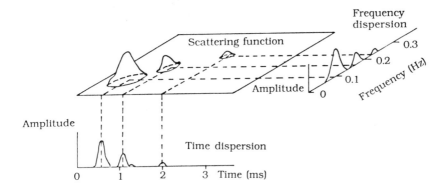

Figure 5. Channel transfer function for a three moded signal (Bradley, in [2]).

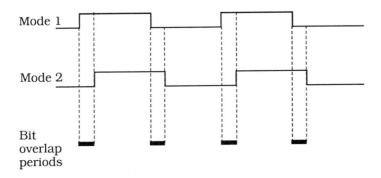

Figure 6. Illustration of intersymbol interference. The signals arrive at the receiver via modes 1 and 2 at different times due to the difference in path length. During the periods indicated, signals from adjacent bits are received.

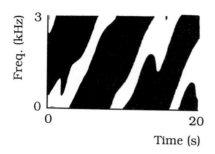
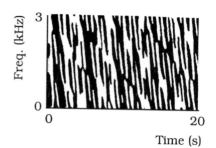

Quiet conditions   Disturbed conditions

Figure 7. Examples of channel transfer functions for cases of quiet and disturbed conditions. The shaded areas indicate low amplitude due to frequency selective fading. (Gouteland, Caratori and Nehme [3])

23.3   REAL TIME CHANNEL EVALUATION

Many communications systems include techniques which enable the quality and propagation characteristics of a radio channel to be evaluated in real time. Real-time channel evaluation (RTCE) may be based upon the measurements of a variety of parameters which may be used either singly or in combination.

A definition of RTCE (as adopted by the CCIR) is as follows: RTCE is the term which describes the process of measuring appropriate parameters of a set of communication channels in real-time and of employing the data thus obtained to describe quantitatively the states of those channels and hence their relative capabilities for passing a given class, or classes, of communication traffic.

The simplest parameter which can be determined is the amplitude of the signal detected by the receiver. For cases when the signal can be discerned from the noise and interference background, further parameters may be measured which are indicative of the signal quality. Examples of these parameters include [4]:

a) Signal frequency.
b) Signal phase.
c) Ease of synchronisation.
d) Noise and/or interference level.
e) Channel impulse response.
f) Received signal-to-noise or signal-to-interference ratio.

g) Baseband spectrum.

h) Received data error rate.

i) Telegraph distortion.

j) Rate of repeat requests in an ARQ system.

The RTCE techniques fall into three categories which are considered separately in the following sections.

23.3.1   Passive monitoring

The simplest form of evaluation is to monitor the radio channel before the message is sent to determine the presence or absence of co-channel signals within the signal baseband. This passive monitoring (no cooperative transmission is required) may be extended to determine the amplitudes of any co-channel signals which may be compared with the amplitude to be expected from the link transmitter.

Often, the spectral occupancy of the channel is measured (usually with the aid of an FFT) to determine whether or not the data transmission would be subject to significant interference. For example, an FSK signal is sent by alternating the signal between two frequencies within the radio channel. If any co-channel signals do not occupy parts of the spectrum required for the FSK signal, the interference will not be significant. The spectrum of a typical 50 kHz segment of the radio spectrum is presented as Figure 8.

23.3.2   Active probing

A special data sequence may be transmitted to evaluate the channel performance. This special signal may take the form of a known data sequence for which the transmission errors may be evaluated at the receiver. One alternative to this is described by Jones and Hayhurst [5] in which a series of short (0.7 ms) pulses are inserted at periodic intervals into the transmitted data sequence. This allows the modal structure of the signal to be measured (each mode has a different time of flight due to the difference in path length). Typical examples of this type of probing signal are presented in Figure 9. High bit error rates are expected at times when the pulsed sounding indicates the presence of several propagation modes of similar amplitude.

23.3.3   Use of the normal data signal

Both of the above techniques have the disadvantage that some interruption in the normal data transmission must occur. Various schemes can be devised whereby the normal data signal is monitored to determine the signal quality. In some cases the signal includes known data sequences, which are often included for synchronisation purposes, which can provide an evaluation of the channel. If error recognition or

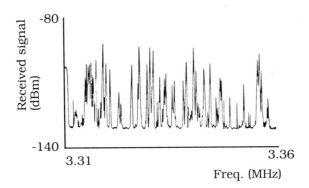

Figure 8.   A typical example of a 50 kHz segment of the HF spectrum. (Doany, Wong and Gott [11]).

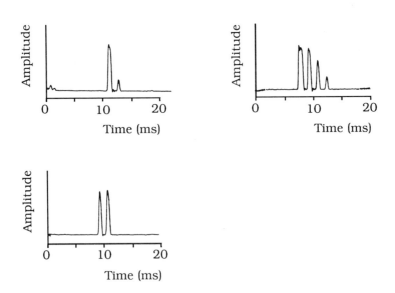

Figure 9.   Typical examples of the probing signal received on the system described by Jones and Hayhurst [5].   Data are for a 4.9 MHz signal transmitted over a 122 km path in the UK.

correcting codes are transmitted over the link, then a measure of signal quality could be based on the rate at which errors are detected.

In addition to direct bit error measurements, other parameters may be determined which can be related to the bit error rate. For example, in a series of experiments undertaken by the University of Leicester, an FSK signal was transmitted over a link from Baffin Island to the UK. The transmitted data consisted of a continuously repeated (known) pseudo-random data sequence combined with a Barker coded sequence to enable synchronisation between transmitter and receiver. Data rates of 50, 75 and 150 baud were employed. Since the transmitted data signal was known, the bit error rate could be directly measured. As a further indication of the received signal quality, a quality factor, QF, was defined in terms of the signal amplitudes detected in both the mark and space tone detectors (A_m and A_s respectively) as indicated in Equation 23.1. This quality factor is a measure of the difference in amplitude of the signals measured at the two FSK frequencies, normalised to the sum of the amplitudes.

$$QF = \frac{|A_m - A_s|}{A_m + A_s} \qquad (23.1)$$

Under ideal conditions, all of the received power should be received at either the mark or space frequency giving a quality factor of 1, and when the measured amplitudes are equal (unable to distinguish a mark tone from a space tone), the quality factor is zero (for this case, no information is received). This parameter, measured for each received bit, is then averaged for each transmission period and an overall quality indication determined. The average quality factor is found to be well correlated with the directly measured bit error rate (see Figure 10) except when the nature of the interference is such that the received power is in excess of that due to the wanted signal in one or both of the FSK tones. In this latter case, high quality indication is also associated with a bit error rate of 50% (the maximum likely to occur since random or uncorrelated data would be expected to agree on 50% of occasions). This apparent discrepancy in this quality indicator emphasises the need to assess the presence and characteristics of any co-channel signals which may cause significant interference.

23.4 ADAPTIVE SYSTEMS

The real-time channel evaluation (RTCE) described in the previous section could enable a communication system to adapt automatically to the prevailing propagation and interference conditions. Such systems usually require communication in both directions to be established.

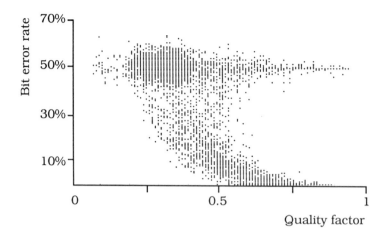

Figure 10. Measured FSK quality factor and bit error rate for a transmission from Clyde River, Canadian NWT to Leicester, UK.

However, communication in one direction may only be for the control of the main data flow and therefore only a low data rate is necessary for that part of the link. Usually this channel can be established when data transfer is possible in the main direction.

Various systems have been developed which assess the quality of a range of available frequencies and select the channel with the best characteristics for operational use. Such a system, employing in excess of 60 possible channels distributed over the HF spectrum, is described by Williams and Clarke [6]. In their system, test messages are sent on predetermined channels spread across the entire allocated band of frequencies by a main control station. The remote station(s) measure the bit error rate and the signal strength of the received signal and transmit these back to the originating station. From these data and a knowledge of the path length, various propagation parameters are calculated (certain propagation modes may support high data rates). This information is then employed in various algorithms to select the optimum channel for communication.

Presented in Figure 11 are plots of the channels which the system described in [6] found to be available on two days with significantly different ionospheric conditions. For comparison, estimated values of the maximum usable frequency for the path obtained with the aid of a computer based prediction program are also shown. One such prediction program, developed by the CCIR, is described in [7]. It is clear from examination of these figures that although the prediction may indicate a general trend, this does not closely match the measured value obtained from the channel assessment process.

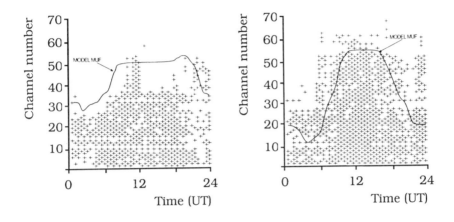

Figure 11. Channel availability on two days with different ionospheric conditions as determined by the channel evaluation system described by Williams and Clarke [6]. For comparison, the predicted MUF is also shown.

Further possibilities exist for the adaption of a communications system to the available channels and propagation conditions. Although several channels may be found which are capable of supporting acceptable communication, some of these are preferred since they may be able to support higher bit rates or yield lower error rates. The signal may, therefore, be adjusted by means of variations in the bit rate and/or the inclusion of various levels of error correcting codes to maximise the message throughput on the link. Under good conditions, simple codes may be employed with a high data rate, whereas under poor conditions, more complex (error correcting) codes are required with a lower data transmission rate. The data throughput may be continuously monitored and adjustments made as appropriate or operation may be switched to one of the other available channels. Scanning of all of the available channels is usually repeated periodically in order to accommodate changes in the propagation conditions or interference background.

In some systems, the channel assessment includes not only the signal strength and bit error rate (as described above), but the spectral content of the noise and interference on the channel. This information can allow the system to adapt the modulation to make best use of the channel. One such example is described by Hague, Jowett and Darnell [8] in which the frequencies of the two FSK tones are adjusted to positions in the spectrum where there is least noise and interfering signal.

23.5   SPREAD SPECTRUM

In many communication systems, the principal concern is to provide efficient utilisation of bandwidth and power.   However, there are situations where these constraints are relaxed in order to meet certain other design objectives (for example,   the rejection of interference from other signals or to provide secure communication such that the transmitted signal is not easily detected or recognised by unwanted listeners).   This type of requirement may be catered for by modulation types which occupy bandwidths greatly in excess of that necessary to convey the data.   These are collectively known as spread-spectrum modulation techniques.

The definition of spread spectrum may be stated [9] as:

a) Spread spectrum is a means of transmission in which the data of interest occupies a bandwidth in excess of the minimum bandwidth necessary to send the data
b) The spectrum spreading is accomplished before transmission through the use of a code that is independent of the data sequence. The same code is used in the receiver (operating in synchronism with the transmitter) to despread the received signal so that the original data may be recovered.

Although many of the standard modulation techniques, such as frequency and pulse-code modulation, occupy bandwidths in excess of the minimum requirement ((a) above), they are not considered as true spread-spectrum techniques since they do not satisfy both parts of the definition. For the purposes of this chapter, however, the constraint of the second condition will be relaxed to include diversity techniques.

23.5.1   Frequency diversity

One of the simplest forms of modulation which occupies a bandwidth in excess of the minimum requirement (and may be considered as spread spectrum) is one in which the data are sent using simple FSK modulation techniques but several mark and space tones are employed.   The separation between the tones pairs is sufficiently large as to significantly reduce the correlation of interference between the various tone pairs. A popular dual diversity signal format for low speed (50 - 100 baud) data is illustrated in Figure 12. In this case, the separation of the tones is such that they occupy frequencies at either side of a voice band frequency allocation.   This decorrelates the interference since statistically independent interference is obtained for separations in excess of 1 kHz [10].

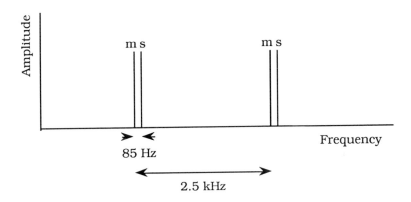

Figure 12. A popular dual diversity signal format for low speed data transmission. The mark and space tones are indicated as m and s respectively.

This concept has been extended to include higher order frequency diversity formats, an example of which is given by Doany, Wong and Gott [11]. These authors employ a sixth order diversity FSK signal (see Figure 13) for low speed communication, and show that the probability of not being able to receive at least two of the FSK tone pairs is 90% when each pair has a probability of being unusable of 50%. The probability of at least one pair being available is considerably higher.

23.5.2   Frequency hopping

An alternative approach in avoiding interfering signals is to periodically change the carrier from one frequency to another according to a schedule agreed between transmitter and intended recipients of the data. Although interference may occur on some of the frequencies, it is unlikely that this will occur on all the allocated frequencies. This method has an additional advantage when data security is of importance, since if a pseudo-random frequency hopping algorithm known only to intended recipients of the data is adopted, then the signal is difficult to intercept and is more resilient to deliberate interference. Furthermore, in a frequency hopping system, the receiver is able (if required) to monitor a channel in advance of transmission (passive RTCE) to take account of, and possibly adapt to, the interference background.

Systems have been built which employ both slow hopping in which the period for which the signal remains on a particular frequency is in excess of one data bit and usually longer. An alternative approach, which is more resilient to interference and interception, is fast hopping in which several frequency changes are performed within the space of a single data bit.

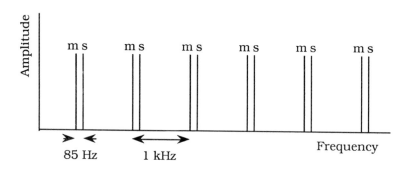

Figure 13.   Illustration of the sixth order diversity signal employed by Doany, Wong and Gott [10].

23.5.3   Suppression of the effects of multi-path propagation

In addition to providing resilience against interference and security against unwanted reception of the signal, frequency hopping techniques may be employed to overcome some of the effects of multi-path propagation. In general, the signals from each of propagation modes arrive at the receiving antenna at different times due to the differences in path length. The effect of multi-path propagation is diminished provided that the carrier frequency of the transmitted signal hops sufficiently fast such that only the signal from the path with the shortest time delay is received.

For example, for a path length of 1000 km, the differences in arrival times between the signals for each mode are typically 1 ms. To overcome the effects of multi-path propagation in this case, the carrier signal must, therefore, change frequency at least every 1 ms.

This objective may also be achieved by sweeping the transmitted carrier frequency across a range of frequencies (known as chirp) in synchronism with the receiver. The signals from each mode will appear to the receiver to be offset in frequency by an amount proportional to the propagation delay. Providing the modulation is such that the frequency separated components don't overlap, the components may be separated by filtering.

For the example given above, if the frequency is swept at 100 kHz s^{-1}, the signals corresponding to each mode will be separated in frequency by 100 Hz.

23.6   CONCLUDING REMARKS

The special problems associated with data communication over high frequency radio links have been discussed and several techniques to overcome some of the limitations imposed by the transmission medium considered. Systems have been designed which incorporate combinations of the principles discussed. For example, Doany, Wong and Gott [11] describe a system which combines frequency diversity and slow frequency hopping. Although such systems considerably improve the reliability and data throughput, they are not intended to preclude the use of forward error correcting codes.

23.7   REFERENCES

1.   K. Davies. *Ionospheric Radio Propagation.* Dover Publications Inc., New York, 1966.

2.   M.P.M. Hall and L.W. Barclay (editors). *Radiowave Propagation.* Peter Peregrinus Ltd., London, 1989.

3.   C. Goutelard, J. Caratori and A. Nehme. Optimisation of H.F. digital radio systems at high latitudes. AGARD conference proceedings number 382, 1985, 3.5.1-3.5.18.

4.   M. Darnell. An overview of real-time channel evaluation techniques. *In:* HF Frequency Management. IEE Colloquium digest 1986/122, 1986, 7/1-7/4.

5.   T.B. Jones and P.L. Hayhurst. An ionospheric mode detection system for HF communication applications. AGARD conference proceedings number 363, 1984, 57-61.

6.   D.J.S. Williams and E.T. Clarke. Practical real time frequency management for automated HF networks. *In:* Proceedings of the Fourth International Conference on HF Radio Systems and Techniques. IEE conference publication 284, 1988, 61-65.

7.   CCIR. LIL252, HF Skywave Field Strength Prediction Program, Version 84.11. Geneva, 1984.

8. J. Hague, A.P. Jowett and M. Darnell. Adaptive control and channel encoding in an automatic HF communication system. *In:* Proceedings of the Fourth International Conference on HF Radio Systems and Techniques. IEE conference publication 284, 1988, 17-23.

9. S. Haykin. *Digital Communications.* John Wiley and Sons, New York, 1988.

10. G.F. Gott, S. Dutta and P. Doany. Characteristics of HF interference with application to digital communications. *Proc. IEE,* part F, August 1983.

11. P. Doany, S.W. Wong and G.F. Gott. A frequency hopping modem for reliable slow rate data transmission. *In:* Proceedings of the Third International Conference on HF Communication Systems and Techniques. IEE conference publication 245, 1985, 141-145.

Chapter 24

Digital filters — a case study

T. J. Terrell and E. T. Powner

24.1 INTRODUCTION

The purpose of this case study is to present a development of the step-invariant approach to highpass digital filter design [1] - a topic only briefly reported in the literature [2-5]. It will be shown that the transfer function, $G(z)_{sif}$, obtained via the step invariant design method, may be made to approximate closely to the transfer function, $G(z)_{bl}$, obtained via the bilinear z-transform method (Chapter 9). The analysis presented justifies this step-invariant design method, and shows that it is possible to convert a time domain (step-invariant) filter, $G(z)_{si}$, to one that satisfies a frequency domain specification, $G(z)_{sif}$. This is achieved by observing certain conditions and by employing a suitable gain term. The validity of the method is demonstrated using a practical example of a simple highpass filter and a digital phase-advance network.

It was shown in Chapter 9 that the impulse-invariant design method for bandlimited filters, is based on the application of standard z-transforms, whereby an analogue filter transfer function, $G(s)$, is transformed to an equivalent digital filter transfer function, $G(z)$, that is

$$G(s) = \sum_{i=1}^{m} \frac{k_i}{(s + p_i)} \quad \rightarrow \quad G(z) = T \times \sum_{i=1}^{m} \frac{k_i}{1 - \exp(-p_i T) z^{-1}}$$

For the analogue filter, the impulse response, $g(t)$, is defined as. $L^{-1}[G(s)]$ Similarly, for a digital filter, the impulse response, g_k, is defined as $z^{-1}[G(z)]$. To be impulse-invariant $g_k = g(t)$ for $t = 0, T, 2T, ...$, where T is the sampling period. Furthermore, the frequency response of the digital filter, $G(\exp(j\omega T))$, will approximate to the frequncy response of the analogue filter, $G(j\omega)$, if aliasing errors have been minimised by bandlimiting $[G(s)]$ and by correct choice of sampling period T.

A common approach to the design of non-bandlimited digital filters (highpass or bandstop) is to use the well-known bilinear z-transform (Chapter 9), that is, by directly substituting $2/T[(z-1)/(z+1)]$ for s in $[G(s)]$ a corresponding transfer function $G(z)$, is obtained. However, non-linear distortion (warping) may be introduced into the filter representation because of the non-linear relationship between the analogue filter frequency scale and the digital filter frequency scale. This problem is resolved by using prewarping techniques (Chapter 9).

An alternative approach to the design of non-bandlimited filters is to use the step-invariant method described in this case study.

24.2 HIGHPASS FILTER DESIGN

A simple analogue highpass filter has a transfer function: $G(s) = s/(s + \alpha)$, where $\omega = \alpha$ rad/s is the cut-off frequency of the filter. This is non-bandlimited and consequently the impulse-invariant design method is excluded, however, employing the suitable bilinear z-transform we obtain

$$G(z)_{bl} = \frac{1}{1 + \frac{\alpha T}{2}} \left[\frac{(z-1)}{z + \frac{\left(\frac{\alpha T}{2} - 1\right)}{\left(\frac{\alpha T}{2} + 1\right)}} \right] \qquad (24.1)$$

Referring to equation (24.1) it is seen that $G(z)_{bl}$ has a gain term equal to $1/\left(1 + \frac{\alpha T}{2}\right)$, and the corresponding z-plane representation has a zero at $z = 1$ and a pole at $z = \left(1 - \frac{\alpha T}{2}\right) / \left(1 + \frac{\alpha T}{2}\right)$, that is, the pole is at $z = \left(1 - \alpha T + \frac{\alpha^2 T^2}{2} - \frac{\alpha^3 T^3}{4} + \cdots\right)$.

Now consider $G(s) = \frac{s}{(s + \alpha)} = \frac{Y(s)}{X(s)}$, where $Y(s)$ is the Laplace transform of the filter response and $X(s)$ is the Laplace transform of the filter input signal. For the step-invariant design method the step input signal is assumed to have an amplitude of A for $t = \geq 0$, i.e.

$X(s) = A / s$, therefore

$$Y(s) = \frac{A}{s} \times \frac{s}{s + \alpha} = \frac{A}{s + \alpha} \tag{24.2}$$

Taking the inverse Laplace transform of equation (24.2) we obtain the corresponding step response of the analogue filter, thus

$$y(t) = L^{-1} [Y(s)] = A \exp(-\alpha t) \tag{24.3}$$

Transforming equation (24.2) into the z-domain (standard z-transform) yeilds

$$Y(z) = \frac{A z}{(z - \exp(-\alpha T))} \tag{24.4}$$

The standard z-transform of $X(s) = A / s$ is $X(z) = A z / (z - 1)$, therefore the corrsponding transfer function of the digital filter is

$$G(z)_{si} = \frac{Y(z)}{X(z)} = \frac{z - 1}{z - \exp(-\alpha T)} \tag{24.5}$$

Taking the inverse z-transform of equation (24.4) we obtain the corresponding step response of the digital filter, thus

$$y_k = z^{-1} [Y(z)] = A \exp(-\alpha k T) \tag{24.6}$$

Comparing equations (24.3) and (24.6) we see that y_k is equal to $y(t)$ at the sampling instants, and therefore the digital filter defined by equation (24.5) is step-invariant.

Now returning to the bilinear z-transform of the analogue filter (equation (24.1)) we allow for prewarping by making the substitution

$$\omega_a = \frac{2}{T} \times \tan\left(\frac{\omega_d T}{2}\right) \tag{24.7}$$

For example, if the specified parameters of a digital filter are:

Cut-off frequency, f_{cd} = 100 Hz, and

sampling period, $T = 10^{-3}$ s

then equation (24.7) yields the corresponding value of ω_a, which is $\omega_a \cong 650$ rad/s, which is the value of α in equation (24.1). Therefore $\alpha T / 2 = 650 \times 10^{-3} / 2 = 0.325$, and thus

$$G(z)_{bl} = \frac{1}{1 + 0.325}\left[\frac{(z-1)}{z + \frac{(0.325 - 1)}{0.325 + 1}}\right] \cong 0.755\left[\frac{z-1}{z - 0.51}\right]$$

and the corresponding frequency response is

$$G(\exp(j\,\omega T))_{bl} = 0.755\left[\frac{\exp(j\,\omega T) - 1}{\exp(j\,\omega T) - 0.51}\right] \qquad (24.8)$$

Similarly, for equation (24.5) $\alpha T = 2\pi \times 100 \times 10^{-3} \cong 0.628$ and the corresponding frequency response is

$$G(\exp(j\,\omega T))_{si} = \frac{\exp(j\,\omega T) - 1}{\exp(j\,\omega T) - 0.534} \qquad (24.9)$$

The frequency response of the analogue highpass filter is

$$G(j\omega) = \frac{j\omega}{\alpha + j\omega} = \frac{j\omega}{\left(2\pi \times 100\right) + j\omega} = \frac{j\omega}{628.32 + j\omega} \qquad (24.10)$$

A useful comparison between $|G(j\omega)|$, $\left|G(\exp(j\,\omega T))_{bl}\right|$ and $\left|G(\exp(j\,\omega T))_{si}\right|$ can be made for a frequency value of 100Hz (the cut-off frequncy value). The required calculations are summarised below:

$$G(j\omega) = \frac{j\,628.23}{628.23 + j\,628.23} = \frac{j1}{1 + j1}$$

$$\therefore |G(j\omega)| = \frac{1}{\sqrt{2}} \quad \left[20\log_{10}|G(j\omega)| = -3 \text{ dB}\right]$$

$$\omega T = \omega_c T = 2\pi \times 100 \times 10^{-3} \cong 0.628$$

$$\exp\left(\omega_c T\right) = \cos\omega_c T + j\,\sin\omega_c T = 0.809 + j\,0.588$$

$$G(\exp(j\omega T))_{bl} = 0.755\left[\frac{0.809 + j\,0.588 - 1}{0.809 + j\,0.588}\right]$$

$$\left|G(\exp(j\omega T))_{bl}\right| = \frac{0.755 \times 0.618}{0.66} \cong 0.707 \quad (-3\,\mathrm{dB})$$

$$G(\exp(j\omega T))_{si} = \frac{0.809 + j\,0.588 - 1}{0.809 + j\,0.588 - 0.534}$$

$$\left|G(\exp(j\omega T))_{si}\right| = \frac{0.618}{0.649} = 0.952 \quad (-0.43\,\mathrm{dB})$$

Clearly the bilinear design method produces a good approximation to the analogue prototype highpass filter, however, the step-invariant design method does not produce an adequate approximation. Consequently if the step-invariant method is to be used successfully, then a suitable gain term must be introduced to minimise the error of approximation.

Referring to equation (24.1) we see that the pole of $G(z)_{bl}$ is at

$$z = \left(1 - \alpha T + \frac{\alpha^2 T^2}{2} - \frac{\alpha^3 T^3}{4} + \cdots\right),$$ whereas referring to equation

(24.5) we see that the pole of $G(z)_{si}$ is at

$$z = \exp(-\alpha T) = \left(1 - \alpha T + \frac{\alpha^2 T^2}{2} - \frac{\alpha^3 T^3}{6} + \cdots\right).$$ If $\alpha^3 T^3 \ll 1$, then

the location of the pole of $G(z)_{si}$ closely approaches the location of the pole of $G(z)_{bl}$. Since $T = 2\pi / \omega_s$ the approximation holds true for $\omega_s \geq 2\pi\,\alpha$, and it should be noted that the multiplying constant, 2π, is the minimum value required. Under this condition the only significant difference between $G(z)_{bl}$ and $G(z)_{si}$ is the gain term $1/\left(1 + \frac{\alpha T}{2}\right)$ associated with $G(z)_{bl}$. Consequently, to convert the time-domain (step-invariant) filter, $G(z)_{si}$, to one that satisfies a frequency-domain specification, $G(z)_{sif}$, we use

$$G(z)_{sif} = 1 / \left(1 + \frac{\alpha T}{2}\right) \times G(z)_{si}$$

$$= 1 / \left(1 + \frac{\alpha T}{2}\right) \left[\frac{z - 1}{z - \exp(-\alpha T)}\right]_{\omega_s \geq 2\pi \alpha} \tag{24.11}$$

The pole/zero and gain matching technique used in arriving at equation (24.11) can be applied to other forms of prototype analogue filter. Another practical example is presented in the next section of this chapter.

24.3 DIGITAL PHASE ADVANCE NETWORK

The transfer function of the digital phase advance network is of the form $G(s) = (s + \alpha) / (s + \beta)$ *where* $\alpha < \beta$. Using the bilinear z-transform design method we obtain

$$G(z)_{bl} = \frac{\left(1 + \frac{\alpha T}{2}\right) \left[z - \frac{\left(1 - \frac{\alpha T}{2}\right)}{\left(1 + \frac{\alpha T}{2}\right)}\right]}{\left(1 + \frac{\beta T}{2}\right) \left[z - \frac{\left(1 - \frac{\beta T}{2}\right)}{\left(1 + \frac{\beta T}{2}\right)}\right]} \tag{24.12}$$

Referring to equation (24.12) we see that $G(z)_{bl}$ has:

(a) a gain term equal to $\left(1 + \frac{\alpha T}{2}\right) / \left(1 + \frac{\beta T}{2}\right)$ \tag{24.13}

(b) a zero at $z = \left(1 - \frac{\alpha T}{2}\right) / \left(1 + \frac{\alpha T}{2}\right)$

$$= \left(1 - \alpha T + \frac{\alpha^2 T^2}{2} - \frac{\alpha^3 T^3}{4} + \cdots\right) \tag{24.14}$$

(c) a pole at $z = \left(1 - \frac{\beta T}{2}\right) / \left(1 + \frac{\beta T}{2}\right)$

$$= \left(1 - \beta T + \frac{\beta^2 T^2}{2} - \frac{\beta^3 T^3}{4} + \cdots\right) \tag{24.15}$$

Now consider the step-invariant design method. Firstly we have

$$Y(s) = \frac{A}{s} \times \frac{s + \alpha}{s + \beta} = \frac{\frac{A\alpha}{\beta}}{s} + \frac{\frac{A}{\beta}(\beta - \alpha)}{s + \beta}$$

which then z-transforms to

$$Y(z) = \frac{\frac{A\alpha}{\beta}z}{(z - 1)} + \frac{\frac{A}{\beta}(\beta - \alpha)z}{z - \exp(-\beta T)}$$

Now dividing $Y(z)$ by $X(z) = Az / (z - 1)$ gives

$$G(z)_{si} = \frac{z - 1 + \frac{\alpha}{\beta}(1 - \exp(-\beta T))}{z - \exp(-\beta T)} \qquad (24.16)$$

Referring to equation (24.16) we see that $G(z)_{si}$ has:

(a) a zero at $z = 1 - \frac{\alpha}{\beta}(1 - \exp(-\beta T))$

$$= \left(1 - \alpha T + \frac{\alpha \beta T^2}{2} - \frac{\alpha \beta^2 T^3}{6} + \cdots\right) \qquad (24.17)$$

(c) a pole at $z = \exp(-\beta T)$

$$= \left(1 - \beta T + \frac{\beta^2 T^2}{2} - \frac{\beta^3 T^3}{6} + \cdots\right) \qquad (24.18)$$

Comparing equations (24.12) and (24.16) we see that the zero of $G(z)_{si}$ will closely approximate to the zero of $G(z)_{bl}$ if $\alpha \beta T^2 \ll 1$ and if $\alpha^2 T^2 \ll 1$. Similarly, the pole of $G(z)_{si}$ will closely approximate to the pole of $G(z)_{bl}$ if $\beta^2 T^2 \ll 1$. If these conditions are true the digital equivalent of the phase advance network is

$$G(z)_{sif} = \frac{\left(1+\dfrac{\alpha T}{2}\right)}{\left(1+\dfrac{\beta T}{2}\right)}\left[\frac{z-1+\dfrac{\alpha}{\beta}(1-\exp(-\beta T))}{z-\exp(-\beta T)}\right] \qquad (24.19)$$

A practical phase-advance network is shown in Figure 24.1. In this case $\alpha = \dfrac{1}{C_1 R_1} \cong 26.9$ and $\beta = \dfrac{\alpha(R_1+R_2)}{R_2} \cong 94.9$. Assuming that the sampling period is $T = 1$ ms, the derived conditions for valid approximation should be checked as follows:

$$\alpha \beta T^2 = 26.9 \times 94.9 \times 10^{-6} \cong 0.0026 \quad \text{(which is} \ll 1)$$

$$\alpha^2 T^2 = 26.9^2 \times 10^{-6} \cong 0.00072 \quad \text{(which is} \ll 1)$$

$$\beta^2 T^2 = 94.9^2 \times 10^{-6} \cong 0.009 \quad \text{(which is} \ll 1)$$

The conditions for approximation are satisfied.

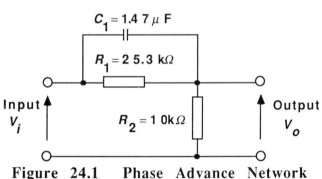

$C_1 = 1.47 \mu F$

$R_1 = 25.3 \text{ k}\Omega$

Input V_i

$R_2 = 10 \text{k}\Omega$

Output V_o

Figure 24.1   Phase Advance Network

The transfer function of the network is

$$G(z)_{sif} = \frac{\left(1+\dfrac{26.9\times10^{-3}}{2}\right)}{\left(1+\dfrac{94.9\times10^{-3}}{2}\right)}\left[\frac{z-1+\dfrac{26.9}{94.9}\left(1-\exp\left(-94.9\times10^{-3}\right)\right)}{z-\exp\left(-94.9\times10^{-3}\right)}\right]$$

$$= 0.9675\left[\frac{z-0.9743}{z-0.9095}\right]$$

The corresponding linear difference equation is

$$y_k = 0.9675x_k - 0.9426x_{(k-1)} + 0.9095y_{(k-1)}$$

This digital representation of the phase advance network is suitable for implementation using a microprocessor-based DSP system, and has been used sucessfully in a servo control system.

24.4 REFERENCES

[1]   T. J. Terrell and R. J. Simpson, *A Step-Invariant Design Method for Highpass Digital Filters*, Int. J. Elect. Engineering Educ., Vol
1   7, pp 335-342, Manchester U. P.

[2]   D.K. Frederick and A. B. Carlson, *Linear Systems in Communications and Control*, Chapter 13, pp 526, (J. Wiley, 1971).

[3]   A.V. Oppenheim and R. W. Schafer, *Digital Signal Processing*, Chapter 5, pp 203, (Prentice-Hall, 1975).

[4]   W. D. Stanley, *Digital Signal Processing*, Chapter 7, pp 190, (Reston Publishing 1975)

[5]   S. D. Sterns, *Digital Signal Analysis*, pp 157-162, (Hayden Book Co., 1975).

Chapter 25

Speech processing using the TS32010—a case study

R. E. Stone

25.1 INTRODUCTION

This case study illustrates the now classical Linear Predictive Coding (LPC) method of speech compression used to reduce the bit rate in digital speech transmission systems.

The chapter begins with a non-rigerous introduction to the theory of linear predictive coding. This is followed by an explanation of the real-time algorithms used to calculate the parameters required to synthesise individual pitches of voiced speech. The study ends with a description of how these algorithms are implemented on the TMS32010, the problems incurred and results obtained.

Toll quality telephonic speech covers the frequency range 300Hz to 3.4kHz which for an 8-bit pulse code modulation system sampled at 8kHz produces a bit rate of 64-kbits/second. To economise on bandwidth and hence increase channel capacity it is desirable to reduce this bit rate to a lower figure, this is possible because of the significant redundancy present in the English Language.

Linear predictive coding is one method used to exploit the redundancy in speech by assuming that the speech waveform can be modelled as the response to a linear filter as shown in figure 25.1. This system works by separating the speech waveform into two broad categories, these are 'voiced' and 'unvoiced' sounds. An 'unvoiced' sound is produced by the fast passage of air through teeth and lips such as the "ss" in speech, to reproduce these sounds the model uses a noise source as the input to the filter. A 'voiced' sound is produced when periodic pulses from the glottis excite resonances in the vocal tract and head cavities to produce vowel sounds such as the 'ee' in speech. A typical section of voiced speech is shown in figure 25.2,

in the model a series of impulses representing the glottal
pulses are used as input to the filter to reproduce these
sounds.

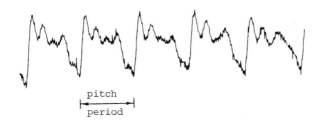

Fig 25.2 A section of voiced speech

Even though the model of fig 25.1 is simplistic it does
produce good quality speech with a significant reduction in
bit rate, typical values being around 5 kbit/second. The
complete system will consist of a transmitter where a
limited section of speech is analysed to deliver a number of
parameters which completely describe that speech section.
These parameters which are significantly fewer than the
number derived by direct sampling are then transmitted by
PCM to be used in the receiver. It is the receiver which is
shown in fig 25.1 and if the original section of speech
analysed was voiced the switch position will be as
indicated.

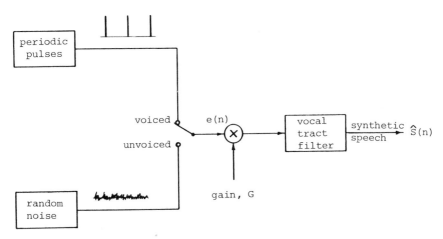

Fig 25.1 Speech synthesiser block diagram

For a voiced section of speech the extraction of the parameters can be made pitch synchronously, ie a new set of parameters every pitch. For an unvoiced section of speech the parameters are extracted after regular predetermined time intervals, eg every 20 milliseconds.

For voiced speech the following parameters are required for resynthesis:-

(i)   The pitch period PL, ie the number of samples in the pitch giving its duration.

(ii)   A figure which indicates the amount of energy in the pitch, G.

(iii)   A number of coefficients which characterise the shape of the pitch.

The coefficients of (iii) are the linear predictor coefficients, or a-parameters, and are used directly in the digital filter of the receiver which when excited by an impulse whose magnitude is governed by the value of G will reproduce the original pitch analysed at the transmitter in shape, amplitude and duration.

25.1.1 What is Linear Prediction?

The concept of linear prediction is that the present sample value of a waveform $S_{(n)}$ may be predicted from a weighted sum of its previous values, ie

$$\hat{S}_{(n)} \quad = \quad a_1 \cdot S_{(n-1)} + a_2 \cdot S_{(n-2)} + a_3 \cdot S_{(n-3)} + \ldots \ldots a_p \cdot S_{(n-p)}$$

This is more conveniently written as:-

$$\hat{S}_{(n)} \quad = \quad \sum_{k=1}^{p} a_k \cdot S_{(n-k)}$$

Thus a pth order predictor will require p 'a'-parameters and the latest p sample values to predict the next sample value. A schematic diagram of such a system is given in fig 25.3 and it can be seen that if the predicted value $\hat{S}_{(n)}$ is compared with the actual value $S_{(n)}$ by subtracting them an error $e_{(n)}$ will occur which in an ideal predictor will always be zero.

The effectiveness of the predictor will depend upon the accuracy of the predictor coefficients a_1, a_2 , ..., a_p and how many of them there are - the greater the number the better the predictor.

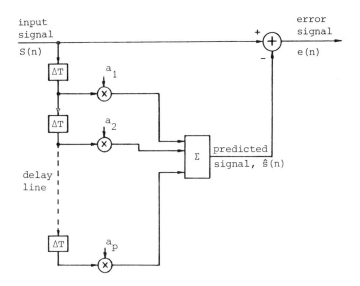

Fig 25.3   Direct form analyser used to calculate the
a-parameters from original speech

25.1.2 The Inverse Filter

The a-parameters are calculated by minimising the mean
squared error E_n over the number of samples M in the pitch,
ie

$$E_n = \sum_{n=0}^{M} e_{(n)}^{2}$$

When the a-parameters are calculated using this method
they are up-dated after every sample value until after M
samples they reach their final form which are the
coefficients of an inverse filter. What has been
manufactured is a whitening filter such that if the original
pitch were passed through it a small error signal with a
flat frequency spectrum appears at its output.

25.1.3 Resynthesis Using the Direct Filter Structure

When the a-parameters are arranged in the resynthesis
structure of fig 25.4, ie the direct filter, then the
application of a signal with a flat spectrum will reproduce
the original waveform. For unvoiced speech this will be

random noise whereas for voiced speech it will be an impulse. The size of this impulse must reflect the size of the original pitch and so is multiplied by the gain factor G. Thus a series of impulses applied at the input of the filter once every PL samples represents the glottal excitation applied to the vocal tract. The resonances in the vocal tract which produce the pitch are modelled by the all-pole filter, one-pole requiring two a-parameters. A sixth order filter is sufficient to provide a smoothed spectrum of most pitches and so twelve a-parameters are required for voiced sections of speech. For unvoiced speech the number of a-parameters can be reduced to four.

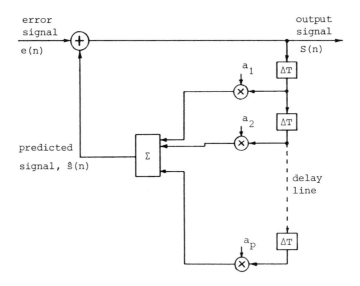

Fig 25.4   Direct form synthesiser used to reproduce
           original speech

25.2 DERIVATION OF THE 'A'-PARAMETERS

Calculation of the p predictor coefficients a_1 to a_p requires the error signal E_n to be minimised over the complete pitch under analysis, thus the function to be minimised is

$$E_n = \sum_{n=0}^{M} [S_{(n)} - \sum_{i=1}^{P} a_i \cdot S_{(n-i)}]^2$$

This is done by setting the partial derivatives of E_n with respect to a_i simultaneously equal to zero, ie

$$\frac{dE_n}{da_k} = 0, \qquad k = 1, 2, \ldots\ldots\ldots p$$

This gives p simultaneous equations with p unknowns which after expansion becomes

$$\sum_{n=0}^{M} 2 \, [S_{(n)} - \sum_{i=1}^{P} a_i \cdot S_{(n-i)}] \, [-S_{(n-k)}] = 0, \qquad k = 1, 2, \ldots p$$

Changing the order of summation this expression becomes

$$\sum_{n=0}^{M} S_{(n)} \cdot S_{(n-k)} = \sum_{i=1}^{P} a_i \sum_{n=0}^{M} S_{(n-k)} \cdot S_{(n-i)}, \qquad k = 1, 2, \ldots p$$

For a 6-pole filter this results in a 12th order matrix equation to be solved for a_i, i = 1 to 12.

Closer examination of the sample multiplications reveals that they are in fact short term autocorrelation values, thus

$$\sum_{n=0}^{M} S_{(n)} \cdot S_{(n-k)} = R_{(k)}$$

and

$$\sum_{n=0}^{M} S_{(n-k)} \cdot S_{(n-i)} = R_{(i-k)}$$

and therefore the equation to be solved can be expressed more concisely as

$$R_{(k)} = \sum_{i=1}^{P} a_i \cdot R_{(i-k)} \qquad , \qquad k = 1, 2 \ldots\ldots\ldots p$$

or in matrix form

$$[R_{(i-k)}] \cdot [a_i] = [R_{(k)}]$$

Since $R_{(i-k)} = R_{(k-i)}$ it can be seen that the matrix $[R_{(i-k)}]$ takes a special form known as Toeplitz which can be solved efficiently by an iterative method known as the Levinson-Durbin algorithm to yield the a-parameters.

Once the a-parameters for a single pitch have been found the resynthesis all-pole filter can be set up in the receiver. This filter has the transfer function:-

$$H(z) = \frac{1}{1 - \sum\limits_{k=1}^{p} a_k z^{-k}}$$

For voiced speech this function is excited by the weighted impulse function G.u(n) once every pitch period. Using only this single pulse as input the complete pitch is reconstituted sample by sample from the transfer function shown above.

The analysis and synthesis procedure just described can and has been done on the TMS32010, however, there are problems associated with this direct form implementation the most serious being computation of the a-parameters using integer arithmetic. The a-parameters in theory can take any real value and hence scaling to the maximum can easily cause round-off errors in the smaller values and these errors are magnified as the iteration progresses.

25.3 THE LATTICE FILTER

The filter described in the previous section can be implemented in a much more efficient lattice structure. The filter coefficients used in the lattice structure are obviously related to the a-parameters but have the enormous advantage of being bounded by the limits of -1 to +1. These k-parameters, as they are known, are much less sensitive than their a-parameter counterparts and much easier to calculate using the integer arithmetic of the TMS32010.

As with the direct form there is an analyser at the transmitter which is shown in fig 25.5(a), this is the lattice form of the inverse filter. The k-parameters can be derived using the original speech as the input or the first $p + 1$ autocorrelation values. Using the normalised autocorrelation values as input each k parameter is evaluated at a stage by making the error signal $e^+(n)$ on the top rail at the output of that stage equal to zero. Once k_1 is evaluated using $R_{(0)}$ and $R_{(1)}$ as inputs it does not alter and can be used to calculate k_2 using $R_{(0)}$, $R_{(1)}$ and $R_{(2)}$ which is used to calculate k_3 etc until finally k_p is found.

This problem is solved very efficiently in the Leroux-Gueguen algorithm, a flow diagram for which is given in fig 25.7.

The procedure simplifies down into four basic steps which can be implemented as subroutines.

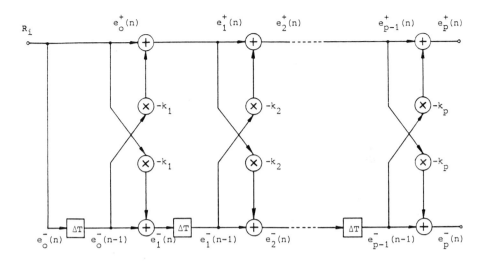

(a) Inverse lattice filter used to extract the k-parameters from a
     section of speech.

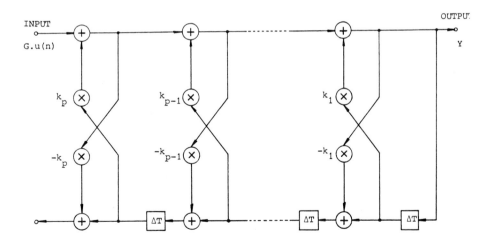

(b) Lattice filter used to resynthesise the section of speech using
     the k-parameters found in (a).

Fig 25.5

1   Add on the next section and input next
     autocorrelation value giving e_o^+ and e_o^-.

2   Evaluate the new outputs from each summing
     junction on top and bottom rails using lower
     rail outputs previously stored.

3   Store outputs of bottom rail summing junctions
     behind each time delay ready for next section.

4   Evaluate new k-parameter by dividing the input
     to the last summing junction on the top rail
     just calculated by the input to the last summing
     junction on the bottom rail found and stored in
     the previous iteration.

ie

$$k_{i+1} = \frac{e_i^+(n)}{e_i^-(n-1)}$$

A further advantage of using this method is that
continual monitoring of the error signal gives a check on
how good the filter is performing.

Resynthesis of the original pitch, as with the direct
form, uses a filter structure which is the inverse of the
lattice filter used for analysis. This resynthesis lattice
filter shown in fig 25.5(b) when excited by the weighted
impulse reproduces the original pitch. This structure is in
fact a very good model of the human speech system whereby
energy from the glottis undergoes a series of transmissions
and reflections as it passes through the vocal tract. The
magnitude of each k-parameter indicates how much energy from
the impulse is reflected back and for this reason they are
referred to as reflection coefficients.

25.4 THE TMS32010 VOCODER SYSTEM

Speech processing using the LPC vocoder model on the
TMS32010 for voiced speech requires the following components
to be calculated at the transmitter in real-time.

(i)   13 normalised autocorrelation values $R_{(o)}$ to $R_{(12)}$.

(ii)   12 k-parameters using the inverse lattice filter.

(iii) A value for gain G to set the magnitude of the
       impulse.

(iv)   The pitch period, PL.

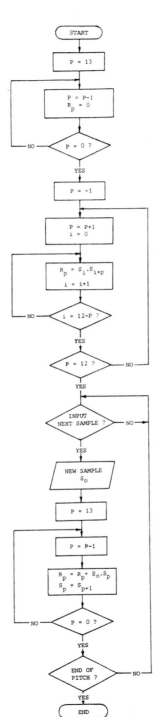

Flow diagram for real
time autocorrelation.

Fig 25.6

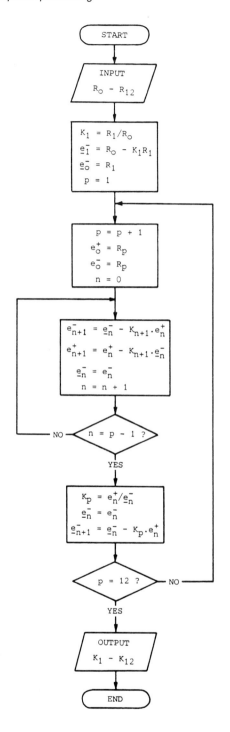

Flow diagram for extraction
of k-parameters from the
autocorrelation values.

Fig 25.7

Estimating the pitch period is a subject in its own right and will not be covered here, for the purposes of this study the pitch period will be assumed.

25.4.1 Autocorrelation on the TMS32010

Once a single pitch has been identified the first 13 normalised autocorrelation values must be calculated in real-time. An algorithm which relies on storing every sample in the pitch before autocorrelating is wasteful of both memory and computational time which are at a premium in the TMS32010. The algorithm used consists of performing 13 intermediate autocorrelations each time a new sample is received thus when the final sample in the pitch is received the final autocorrelation values are returned after 13 more computations. A flow diagram for this program is given in fig 25.6.

Using this approach only 13 data memory locations are required to hold the latest 13 samples held. A further 26 data memory locations are needed for the current autocorrelation values which are stored as 32-bit numbers and updated whenever a new sample is input.

Using a 12-bit ADC the longest pitch possible (16 milliseconds) when sampled at 8kHz will give 128 sample values. Even if every sample value were the maximum of 2047 (an impossible condition) a result of less than 2^{29} would result and so an overflow condition cannot occur. Once the absolute values for $R_{(0)}$ to $R_{(12)}$ have been found they are left-shifted until $R_{(0)}$ lies between 16384 and 32767 in the highest 16 bits of the accumulator. Dividing 16384 by $R_{(0)}$ gives the fraction F by which $R_{(1)}$ to $R_{(12)}$ must be multiplied to normalise them to 32767. In this way only one division has to be made followed by 12 multiplications, a much easier task for the TMS32010.

This sub-program requires 75 program memory locations, 46 data memory locations and is executed in 20 microseconds at each new sample input.

25.4.2 Lattice Analysis on the TMS32010

The Leroux-Gueguen program shown in fig 25.7 which implements the lattice analyser of fig 25.5(a) requires the first thirteen autocorrelation values from the previous section. $R_{(0)}$ to $R_{(12)}$, all temporary intermediate calculations and the 12 k-parameters are stored in data memory as 16-bit words.

When these k-parameters are compared to those obtained from the IBM using the Levison-Durbin algorithm and floating point arithmetic the results are very close. In every case the first 3 k-parameters are identical and higher order k-parameters show a maximum discrepancy of 0.1% showing the resistance of the Leroux-Gueguen algorithm to the inevitable round-off errors of the TMS32010.

This sub-program requires 84 program memory locations, 59 data memory locations and is executed in less than half a millisecond. This program will operate in the background while the next pitch autocorrelation values are also being computed.

25.4.3 Calculating the Gain Factor, G

For voiced speech the input to the filter is an impulse and so the total energy in each pitch must come from its magnitude, G. This magnitude can be calculated from the p normalised autocorrelation values and the a-parameters from the equation:-

$$G^2 = R_{(0)} - \sum_{k=1}^{p} a_k \cdot R_{(k)}$$

When this formula is expressed in terms of the k-parameters it becomes:-

$$G = \sqrt{\prod_{n=1}^{P} (1-k_n)^2} = \sqrt{N}$$

From this it can be seen that the gain G can be calculated in the receiver where more time is available.

As a general rule the higher the order of k value the smaller it is, thus to keep as much accuracy as possible the order of evaluation is k_p to k_1 .

The main problem for the TMS32010 was writing a fast accurate program to perform the square root in integer arithmetic. Newtons method gives an accuracy of better than 0.1% after only two passes provided the initial guess from a look-up table is judicious.

Using integer arithmetic bounded by ±1 the square root of a number N will always give an increase in value as indicated by the look-up table below:

INTEGER VALUES OF N			INITIAL GUESS FOR G	
0			0	
1	-	4	256	(= $\sqrt{2}$)
5	-	16	512	(= $\sqrt{8}$)
17	-	64	1024	(= $\sqrt{32}$)
65	-	256	2048	(= $\sqrt{128}$)
257	-	1024	4096	(= $\sqrt{512}$)
1025	-	4096	8192	(= $\sqrt{2048}$)
4097	-	16384	16384	(= $\sqrt{8192}$)

An example will illustrate the process.

N = 0.00042725 = 14 as an integer

Initial guess for G = 512. This guess is passed through Newtons equation twice, ie

$$G' = \frac{1}{2} \left[\frac{14 \times 2^{15}}{512} + 512 \right] = 704$$

$$G'' = \frac{1}{2} \left[\frac{14 \times 2^{15}}{704} + 704 \right] = 678$$

As a fraction 678 = 0.02069 whereas the correct answer is 0.02067 an error of less than 0.1%.

This program written as in-line code takes up 61 program memory address locations, 4 data memory address locations and on average will take 8 microseconds to perform.

25.4.4 Pitch Synthesis on the TMS32010 Using the Lattice Filter

The resynthesis lattice of fig 25.5(b) is implemented on the TMS32010 by the flow diagram of fig 25.8. This part of the vocoder is located in the receiver section and each of the 12 k-parameters are stored in data memory address as single 16-bit words. As with the analyser in the transmitter all intermediate values which need to be temporarily stored are also stored as single 16-bit words in data memory address locations.

For the purposes of single pitch evaluation the pitch is made periodic by repeated application of the weighted impulse.

This program requires 47 program memory locations and 34 data memory locations. Each output is evaluated in 30 µs but must wait another 95 µs before it can be delivered to the DAC for a sampling rate of 8kHz.

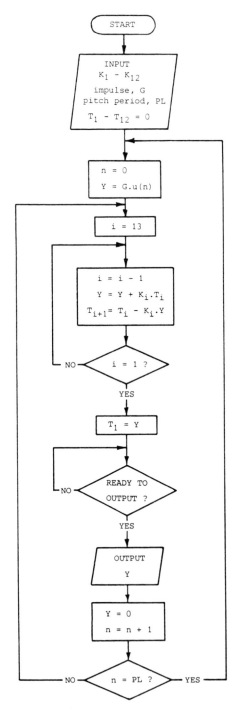

Flow diagram for
lattice synthesiser.

Fig 25.8

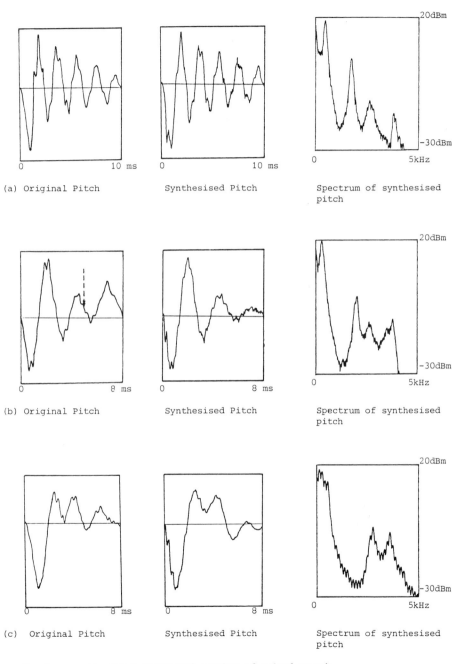

(a) Original Pitch   Synthesised Pitch   Spectrum of synthesised pitch

(b) Original Pitch   Synthesised Pitch   Spectrum of synthesised pitch

(c) Original Pitch   Synthesised Pitch   Spectrum of synthesised pitch

Vocoder results for 3 individual pitches of voiced speech.

Fig 25.9

25.5 RESULTS AND CONCLUSIONS

The result for synthesis of three separate pitches are given in fig 25.9. In all these pitches the TMS32010 evaluation module was used first as an analyser to extract the k-parameters and then as a synthesiser to reproduce the pitch. From the time domain representations it can be seen that the lattice filters formulated for these pitches are all stable giving decreasing energy with time. In fact all the pitches synthesised by this method have always proven stable.

The frequency domain plots show the smoothed spectrum obtained from this 6 pole (12th order) filter. Where the original pitch contains 6 formant frequencies these peaks are clearly represented showing also that pitches which have fewer major resonances can be represented by a lower order filter.

Although the process of linear predictive coding can seem initially very mathematical it is in essence an uncomplicated concept. The idea of modelling the human vocal tract by a linear filter needing only one impulse is an oversimplification which can easily be exposed and is illustrated in fig 25.9(b) where a partial glottal impulse increases energy mid-pitch which cannot be modelled using this simple approach.

Despite its shortcomings LPC is one of the most successful methods of producing high quality synthesised speech and modifications to this basic model, eg multipulse, have and continue to improve the process.

REFERENCES

Atal, B.S. & Hanauer, S.L.; "Speech Analysis and Synthesis by Linear Prediction of the Speech Wave"; JASA Vol. 50 No.2; 1971.

Markel, J.D. & Gray, A.H.; "Linear Prediction of Speech"; Springer Verlag; ISBN 0-378-07563-1; 1976.

Makhoul, J.; "Linear Prediction - A Tutorial Review"; Proc. IEEE, 63, 561-80, 1975.

"TMS32010 Users Guide 1983"; Texas Instruments; ISBN 0-904047-38-5.

Rabiner, L.R. & Schafer, R.W.; "Digital Processing of Speech Signals"; Prentice-Hall; ISBN 0-13-213603-1: 1978.

Chapter 26

A case study in digital control

E. Swindenbank

1 INTRODUCTION

The continuing boom in semiconductor technology, and associated improvements in computing power have been widely acclaimed in control literature. In this field however there seems to be an ever widening void between the expectations of new technology and reality. Advances in control theory have far outstripped technological capability. The advantages of modern digital control techniques however promise great improvements in system performance, and implementation of such schemes must be achieved. The problems imposed by endeavouring to apply complex algorithms in the real world however are many.

In recent years there has been a continual growth in power system networks and interconnections, with a corresponding economic demand to install larger generating sets. Improvements in machine design have led to generators with lower inertias, and higher reactances. Load centres which are placed large distances from generation plant add to problems of instability. Over the past years, work at The Queen's University has concentrated on power system control. From early work involving on-line modelling of full size generator units [1], has developed a programme to design and implement self-tuning and adaptive control strategies. The results of this research have been extremely encouraging, and significant improvements in controller performance have already been shown on a physical generator model [2]. Many implementation problems have already been solved using heuristic programming techniques which are essential for supervision of the control algorithms. As a consequence the number of instructions to be executed every sample interval has risen dramatically, forcing a reduction in sample speed. The lack of computing power offered by conventional microprocessors has severely restricted the advancement of this work.

Improvements in microprocessor architecture and speed have temporarily eased implementation problems. Concurrent processing however provides an open-ended solution for the continuation of this research. In order to perform the overall control function optimally in terms of execution speed, task partitions must be carefully considered. Three distinct areas can be identified in this application, and are sufficiently independent to be implemented on separate microprocessors.

1) Data acquisition and measurement estimation.

2) Control algorithm.

3) Control algorithm supervision.

Each of these functions execute in similar times. As communication overheads can be high, the interfaces between these sub-systems must be

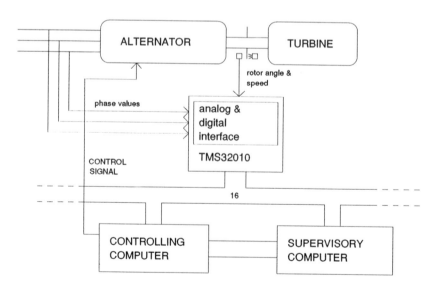

Figure 1 CONTROL SYSTEM STRUCTURE

standardised to ensure modularity and ease of expansion (Figure 1). The asynchronous nature of real-time multi-processing necessitated the use of interface buffering. This allows different microprocessor architectures with varying response times to be interconnected.

In this chapter, the design, implementation, and performance of the measurement system will be considered. The function of this sub-system is to provide measurement information to the second line of processors. On-line tests on a laboratory alternator are used to compare an existing single computer system with the new configuration.

2 THE LABORATORY ALTERNATOR

This is a 3KVA physical simulation of a full size generator. A simple analog simulation of boiler-turbine characteristics provides the torque representation to drive a dc motor. Instrumentation on the system provides analog information in the form of phase voltages and currents at the machine terminals. An optically sourced representation of machine speed and rotor position is also available. The machine is connected to the grid by a pi line simulator which can be used to generate various fault conditions.

For several years, a PDP 11/73 was used to carry out all computing tasks. These not only involved measurement and control, but machine compensation for simulation inadequacies, and transmission line fault control. This arrangement has been used with some success, and self-tuning controllers have been implemented. The minimum sample interval which can be achieved however is four times the standard set by industry.

3 MEASUREMENT SYSTEM FUNCTIONS

Many of the requirements of the measurement system are imposed by the control techniques to be used.

The existing system has exposed limitations in the accuracy of derived signals, under unbalanced loads. These can cause severe measurement errors during fault conditions. Alternative methods for deriving RMS terminal quantities are required. Conventional rectified summation of voltage and current signals is unacceptable due to the relatively long time constants involved in filtering the resulting noisy signals. Transient performance of this type of measurement is poor. Several alternative software techniques were considered. These are based on synchronous sampling and Fourier analysis techniques, and do not suffer from problems introduced by imbalance of the power system. The algorithms operate on windows of phase voltage and current, typically twelve points wide. They are computationally intensive however, some involving several square root calculations.

Rotor angle and machine speed are essential quantities for the assessment of Automatic Voltage Regulator (AVR) performance. These signals are directly measurable from the optical transducer.

The robustness of adaptive control schemes relies completely on the identification of current machine status. During long periods of steady operation, measured data contains little useful information for updating the model parameters from which the controller is derived. Fault conditions also have a detrimental effect on the parameter estimator. If these situations can be identified quickly, the controller can be protected more effectively.

From discussion with manufacturers of digital AVRs, a control interval of 10ms is the accepted value. Obviously the measurement system must be capable of producing information at this rate at minimum.

These requirements suggest a dedicated measurement system capable of providing the required data at a fast rate. It became obvious that conventional microprocessors were unsuitable for implementation of these tasks, resulting in the choice of a Digital Signal Processor.

4 THE TMS DIGITAL SIGNAL PROCESSOR FAMILY

The TMS Digital Signal Processor family has been in production for more than seven years. The first devices were 16/32 bit fixed point with speeds of 5 MIPS. The high speed was due to the use of a Harvard architecture and the inclusion of specialised silicon. Programming was in assembly language, again to attain the best performance, and the cost was around £300 per device. The TMS family has advanced to second generation floating point chips working at 12.5 MIPS. Many options are now available including on-board EPROM, data acquisition interfaces, and other support chips. Great advances have also been made in software development facilities, with C compilers, simulators and assemblers being available. The cost of these devices has reduced dramatically.

In this application, a TMS32010 was used. This has 144 words of on-board data memory, and 4K words of program memory residing off-chip. This is enough capacity for embedded applications. Due to the high access speeds required, the RAM program memory was loaded from ROM before program execution. Unlike strict Harvard devices, the TMS has the ability to transfer information between program and data memory, enabling storage of coefficients within the program address space.

The device has a 16 bit multiplier on-board, giving a 32 bit result within 200ns. Special registers are included to greatly improve the performance of the device.

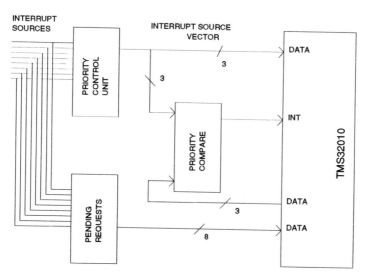

Figure 2   INTERRUPT CONTROL LOGIC

Interface to the TMS is through the data bus. Three multiplexed address lines are used to select appropriate devices through a 3 to 8 line selector. Dedicated mnemonics are used to read and write data. A maximum of 128 input and 128 output lines are available. As only 16 lines are available at any instant, handshaking control requires that data be stored in intermediate latches.

5 HARDWARE IMPLEMENTATION

The measurement system consists of two sub-systems. Acquisition of signals used to derive the terminal quantities is primarily an analog function, while rotor angle and speed are measured digitally. These two functions work asynchronously, and are interrupt driven. As the TMS has only one interrupt line, device identification and priority are handled externally (Figure 2). An Intel 8214 Priority Interrupt Control Unit was used to arbitrate between devices. The priority of the requesting device is compared to that currently executing. Should the interrupt be successful, a device identification vector is read in the first statements of the service routine. The memory required to store machine status restricts the system to a maximum of four levels of interrupt. External logic ensures that stack overflow does not occur. Valid pending interrupt addresses are stored in memory and executed on completion of the current ISR.

Device polling is a useful method of implementing real-time systems. A special instruction BIOZ is associated with a dedicated line to the microprocessor (BIO). This is connected to the A/D status pin. When the BIO line is activated, the current sample is deemed ready, and the processor may read it. This facility prevents the need for a specific input line via the data bus, and speeds up hardware polling.

5.1 Data transfer

Due to the differing speeds between the TMS and conventional microprocessors, transferring information is performed using First-In-First-Out (FIFO) memories. This successfully decoupled the two systems which were mutually excluded from the FIFO by suitable logic. The buffer size is 16 values each 16 bits wide.

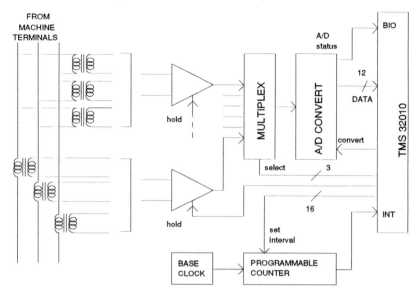

Figure 3   ANALOGUE INPUT HARDWARE

5.2 Measurement of rotor angle and slip

The optical transducer produces TTL signals proportional to rotor angle and machine speed. These are used to gate a high frequency source which is connected to the clock input of 16 bit counters. The tri-state output registers of the counters are connected directly to the data bus.

5.3 Analog input

Six phase quantities are taken from current and voltage transformers via differential amplifiers. Several features are included specifically for implementation of the chosen measurement algorithms (Figure 3). Instantaneous sampling of all phase quantities is a pre-requisite for accurate measurement. Sample-and-hold devices perform this task. The signals are connected sequentially to the A/D converter through an analog switch. As some of the measurement techniques require a precise number of samples per machine cycle, a sample interval control clock is included so that the sample interval can be adjusted on-line by writing a value to a counter preset register. This clock rate is adjusted in response to the measured machine speed.

5.4 Hardware development

All hardware was designed on a CAD system with schematic capture and PCB layout facilities. Due to the high frequencies on the system, copper ground planes were inserted on all non-connection areas. Decoupling capacitors were used on all IC packages, and power supply lines decoupled at various points on each board. Interconnection of the boards is through a 64 way backplane. The system is housed in a standard double eurocard rack.

6 SOFTWARE

6.1 Software Implementation

The TMS has a rich instruction set which operates in three addressing modes. The majority of these execute in one machine cycle. Arithmetic, logical and program control instructions are augmented by special operations for multiply, input/output, and memory manipulation. Data shifting can be implemented as an integral part of an instruction, allowing scaling of integer data to reduce quantisation errors. In this application, the ability to efficiently update data within the sample window is particularly useful. The TMS was

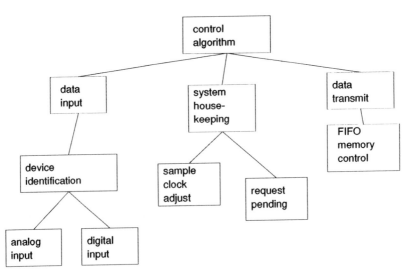

Figure 4   SOFTWARE STRUCTURE

specifically designed for the implementation of digital filters which have a multiply/accumulate structure. The cpu architecture allows simultaneous operations for data load, multiply, add, and move. These can be performed by a single instruction in one machine cycle. The measurement algorithms were arranged to use these facilities where possible.

One of the most common software implementation problems is that of overflow and underflow. Several solutions are possible with the TMS. The best of these relies on the estimation of the data range which may be encountered during processing. Critical stages of the algorithm are checked for error conditions by a 'branch on overflow' instruction. Remedial action is then taken to saturate the offending variable at the appropriate maximum value.

6.2 Software Structure

This is shown in the data flow diagram of Figure 4. The main function of the software is to produce the terminal quantities of voltage, current, real and reactive power. This loop is performed on receipt of an interrupt from the analog hardware. The servicing ISR communicates with the main loop by setting a flag on completion of data input. The gathered information is used to update the data 'window' on which the measurement algorithm works. As shown, data inputs are through a single ISR which identifies the requesting device using the hardware described in section 5. Various housekeeping functions are performed on completion of the loop such as checking pending interrupt requests, and adjustment of the sample interval in relation to the current machine speed. The calculated data is transmitted to the FIFO memories to be read asynchronously. Access exclusion delays are kept to a minimum by programming the receiver to read data in the fastest time possible. Interrupt overrun is therefore avoided.

6.3 Software Development

The software was written entirely in assembly language. Whereas this has a disadvantage in prolonged development time, performance in terms of execution speed and memory usage is improved. The modular nature of the software enabled coding in a top-down manner. The various control algorithms were written on an IBM PC under DOS. These routines were then tested by a simulation package allowing the simultaneous development of hardware and software. The use of a familiar environment for program development was a significant advantage as editors and high level language compilers could be used to check results from TMS code, and produce simulation data. All non-target dependent code was developed using this stub concept. On completion of the hardware, a XDS320 emulator was used to debug embedded code.

7 PERFORMANCE

The accuracy of the measurement system was difficult to assess as the only benchmark was the existing PDP computer. The TMS based system was capable of generating and transfering data (6 values) at rates in excess of 1KHz. Figure 5 shows a comparison of responses generated by the two measurement systems under three phase fault conditions. This shows the range of values the measurement system must be capable of dealing with. Tests such as these were performed over the operating range of the generator, proving the integrity of the new system. Noise was artificially introduced to the measured phase values. It was found that the the new tecniques produced better results even under conditions where the PDP system failed. The measurement system was also used to interrupt the controlling computer when a major fault was detected. This allowed time for preparatory code within the controlling computer to compensate for imminent transient conditions.

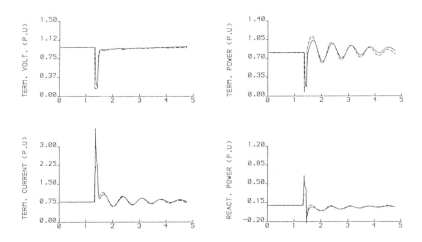

Figure 5 RESPONSE TO 3 PHASE FAULT

Due to the performance of the DSP, it is possible to offload some of the control tasks to the measurement system in order to give an overall increase in performance. On-line analysis of measured data can be used to detect long periods of steady operation as well as transient behaviour.

8 CONCLUSIONS

The system described has shown that the use of a DSP in the role of measurement for control gives advantages over conventional systems. Increased speed, and the capability to use more complex algorithms leads to improved controller performance. Although designed specifically for a laboratory system, the necessity for improved measurement systems in industry is growing. Manufacturers are now starting to implement digital controllers, the performance of which will soon be limited by a lack of accurate information at high transfer rates. Parameters which were once deemed unmeasurable can now be constructed on-line and in real-time. This opens up greater possibilities for control system designers.

Specialist hardware which was originally designed for communications processing will have an increasing role in control systems implementation due to the similarities in algorithm structure. Manufacturers have already seen this trend, and have begun to produce high performance devices specifically for these applications. It is now up to the systems engineer to recognise the advantages of the DSP for particular applications, and to use these to the full.

9 REFERENCES

1 "APPLICATION OF SYSTEM IDENTIFICATION TECHNIQUES TO MODELLING A TURBOGENERATOR"
Swidenbank E, Hogg B W

IEE Proceedings Vol. 136 Pt. D No. 3 May 1989.

2 "A LONG TERM OPERATION OF SELF TUNING EXCIATAION CONTROL ON A MICROMACHINE"
Wu Q H, Swidenbank E, Hogg B W

Chapter 27

Implementation and performance of digital controllers

D. Rees

27.1 INTRODUCTION

Over the past 30 years the application of digital control to industrial processes has grown dramatically. This has been brought about in the main by the rapid advances in VLSI technology, with the provision of increased processor complexity at a reduced cost. The application of microprocessors to control requires real time operation, and normally only fixed point two's complement arithmetic is used. In addition to this, the designer has to put in a substantial amount of effort to produce a practical implementation of the control algorithms taking account of such issues as finite word length, selection of sampling rate, efficient programming of the algorithms, scaling of all the variables and cofficients and the provision of a suitable input/output interface. The advent of digital signal processors with their specially tailored architectures, provide an ideal solution to overcome some of the limitations inherent in the use of microprocessors for control.

It has recently been shown (1,2) that such processors can be readily used to implement discrete algorithms for control system applications. The use of digital signal processors (DSP) for controller realisation overcomes some limitations associated with their design using microprocessors, as they have architectures and dedicated arithmetic circuits which provide high resolution and high speed arithmetic, making them ideally suited for use as controllers. Initial work was done using the Intel 2920, and material has already been published by the author on its use for controller implementations (2,3). The 2920 is one of the earliest of DSP's and consequently is not the most powerful in terms of speed and instruction set. It was initially decided however, to use the 2920 as a controller as this was the only processor available which had A/D and D/A converters on chip. For the particular applications considered the limited speed and size of instruction set proved no real restriction, whilst having a single chip controller requiring no external circuitry apart from a clock and power supply, was a decided advantage. The applications covered implementations of PID, Smith Predictor and Dahlin algorithms which were successfully applied to control a heating process.

The need for powerful digital signal processing devices in an expanding range of application areas has provided the catalyst for major advances in DSP architecture which has completely overtaken the 2920, and rendered it obselete. There are now available to the design engineer a wide range of devices with considerably more powerful architectures of which the TMS320 family of DSP's are representative. The need for such powerful processors in control depends very much on the level of sophistication included within the control scheme - whether real-time adaption is required, and how fast the dynamics of the process being controlled. The initial impetus for the work described in this chapter was to assess the ease with which discrete controllers could be implemented on one of the more recent DSP's. The particular processor chosen was the TMS 32010, the first member of the TMS family of VLSI digital signal processors.

27.2 TMS320 SIGNAL PROCESSOR

Digital signal processing applications present, in general, such restrictive requirements concerning speed and types of operations, that the corresponding algorithms are difficult to implement in real-time using general purpose computers. To solve this problem two different approaches have traditionally been followed: to design specific hardware architectures (4) or to use powerful general computers, like data-flow machines, systolic arrays etc (5) with a high degree of parallelism and pipeline structured modes. However, in recent years alternatives have come on the market which can be defined as problem orientated processors, of which the TMS320 is an example. The TMS family of 16/32 bit single-chip DSP's combines the flexibility of a high-speed controller with the numerical capability of an array processor offering an inexpensive alternative to custom VLSI and multi-chip bit-slice processors.

TMS320 is designed to support both numeric-intensive operations, such as required in signal processing, and also general purpose computation as would be required in high-speed control. It uses a modified Harvard architecture, which gives it speed and flexibility. In this structure the program and data memory is allocated separate sections on the chip permitting a full overlap of the instruction fetch and execution cycle (Fig. 27.1).

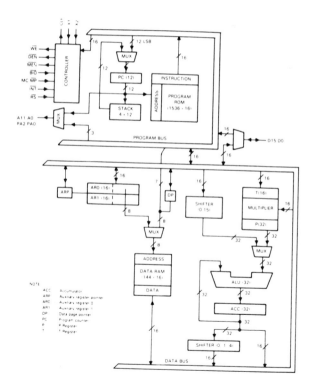

Fig. 27.1 Block Diagram of TMS320M10

The processor also uses hardware to implement functions which previously had been achieved using software. As a result multiplication takes 200ns, ie. one instruction cycle, to execute. It also has a hardware barrel shifter for left-shifting data to 15 places before it is loaded into the ALU. Extra hardware has also been included so that two auxiliary registers which provide indirect data RAM addresses, can be configured in an auto increment/decrement mode for single cycle manipulation of data values. This gives the design engineer the type of power previously unavailable on a single chip (6,7).

Data memory consists of the 144x16 words of RAM present on chip. All non-immediate operands reside within this RAM. If more data is required, then it is possible to use external RAM and then read the data into the on-chip RAM as they are required. This facility was not required in this project since the on-chip RAM was found adequate for all the programs implemented.

27.3 THERMAL PROCESS

The TMS32010 was applied to control the temperature of a process where air drawn through a variable orifice by a centrifugal blower, is driven past a heater grid and through a length of tubing to the atmosphere again. The process consists of heating the air flowing in the tube to some desired temperature. The detecting element consists of a bead thermister fitted to the end of a probe inserted into the airstream 0.28m from the heater (Fig. 27.2).

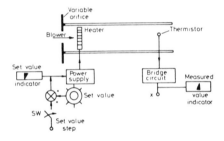

Fig 27.2 Schematic of the temperature process

On this process, a step response test was made resulting in a second order model of the form:

$$G(s) = 1.24 \frac{\exp(-0.3s)}{(0.377s + 1)(0.132s + 1)} \tag{27.1}$$

The corresponding pulse transfer functions for the continuous model for different sampling periods is shown in Table 27.1. Recall that this is the z transform of the product of the zero-order hold and process transfer function.

Sampling Period (second)	Pulse Transfer Function $G(z) = G_z G_p(z)$
T=0.1	$\dfrac{0.0896(1 + 0.709z^{-1})z^{-4}}{(1 - 0.767z^{-1})(1 - 0.469z^{-1})}$
T=0.15	$\dfrac{0.173(1 + 0.6z^{-1})z^{-3}}{(1 - 0.677z^{-1})(1 - 0.321z^{-1})}$
T=0.3	$\dfrac{0.448(1 + 0.362z^{-1})z^{-2}}{(1 - 0.451z^{-1})(1 - 0.103z^{-1})}$

Table 27.1 Pulse transfer function models of a temperature process

27.4 CONTROL ALGORITHMS

27.4.1 Three Term Controller

The discrete equivalent of the proportional-integral-derivative (PID) controller is given by

$$u_n = k_p e_n + k_i \sum_{j=1}^{n} e_j + k_d(e_n - e_{n-1}) \tag{27.2}$$

where e_n is the sampled data of the system error, u_n the manipulated variable and k_p, $k_i = Tk_p/T_i$ and $k_d = T_d k_p/T$ respectively, the proportional, integral and derivative control gains. The sample time is T.

This equation can be rearranged to give the following pulse transfer function

$$\frac{u}{e}(z) = k_p \frac{(d_0 + d_1 z^{-1} + d_2 z^{-2})}{1 - z^{-1}} \tag{27.3}$$

where

$$d_0 = 1 + k_i/k_p + k_d/k_p$$

$$d_1 = -(1 + 2k_d/k_p)$$

$$d_2 = k_d/k_p$$

Using the procedure suggested by Roberts and Dallard (8) which reduces the number of separate tuning parameters to the one gain term by selecting the sampling period to T=0.1Tu, equation 27.3 reduces to

$$k_p \frac{(2.45 - 3.5z^{-1} + 1.25z^{-2})}{1 - z^{-1}} \tag{27.4}$$

which yields the difference equation

$$u_n = u_{n-1} + k_p(2.45e_n - 3.5e_{n-1} + 1.25e_{n-2})$$

(27.5)

27.4.2 Dahlin Controller

Dahlin's method (9) is a particular case of the general synthesis approach to design where digital controllers are designed directly in the discrete domain (Chapter 13). The process under control, is modelled by a first order or second order transfer function and the desired closed loop transfer function K(z) is arranged to be a first order lag of the form

$$K(z) = \frac{(1 - \exp(-\lambda T))z^{-k-1}}{(1 - \exp(-\lambda T)z^{-1})}$$

(27.6)

The reciprocal of the time constant λ is used as a tuning parameter with larger values giving increasingly tight control.

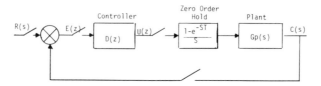

Fig. 27.3 Block diagram of sampled data system for Dahlin design

From Fig. 27.3 the direct synthesis relationship is given by

$$D(z) = \frac{1}{G}(z)\frac{K(z)}{(1 - K(z))}$$

(27.7)

which gives for the Dahlin controller, the pulse transform

$$D(z) = \frac{1}{G}(z)(1 - \beta)\frac{z^{-k-1}}{(1 - \beta z^{-1} - (1 - \beta)z^{-k-1})}$$

(25.8)

where $\beta = \exp(-\lambda T)$

Using the second order model the Dahlin controller equation for a closed loop response with a time constant of 0.15s becomes

$$D(z) = \frac{5.413(1 - 0.767z^{-1})(1 - 0.469z^{-1})}{(1 - 0.513z^{-1} - 0.486z^{-4})(1 + 0.709z^{-1})}$$

(27.9)

Implementing the equation in this form results in unacceptable oscillations of the actuation signal (10). To overcome this the ringing pole at z=-0.709 is removed and the d.c. gain of the controller is adjusted to be the same, which gives

$$3.17\frac{(1 - 0.767z^{-1})(1 - 0.469z^{-1})}{(1 - 0.513z^{-1} - 0.486z^{-4})}$$

(27.10)

$$u_n = 3.17(e_n - 1.236e_{n-1} + 0.36e_{n-2}) + 0.513u_{n-1} + 0.486u_{n-4}$$

(27.11)

27.4.3 Kalman Design

The Dahlin algorithm is based on the specification of the output in response to a setpoint change without any constraints placed on the manipulated variable. An alternative approach originally proposed by Kalman (11) is to design a digital controller with restrictions placed on both the manipulated and controller variables. For example, the specifications for a step change in set point might be for the response to settle at the final value within the actuation signal assuming only a specific number of values before reaching the final value.

This approach allows the designer to take account of load changes and to specify that the error sequence should be zero after a specified number of sample instants, or that the error should reduce in a specified manner. It can be shown that in the case of a second-order system a minimum of two values of manipulated variables are required before the setpoint can be reached. Similarly for a third order system, 3 values etc. Taking account of these restrictions we can write expressions for the controlled and manipulated signal as follows:-

$$C(z) = \sum_{n=0}^{\infty} C_n z^{-n} \tag{27.12}$$

$$U(z) = \sum_{n=0}^{\infty} U_n z^{-n} \tag{27.13}$$

Given the specification that the response should settle to the final value within two sampling periods, it can be shown that (10) for a second order pulse transform model of the form

$$H(z) = \frac{K}{Q}(z) = \frac{\left(\frac{b_1}{b_1+b_2}z^{-1} + \frac{b_2}{b_1+b_2}z^{-2}\right)z^{-k}}{\frac{1}{k(b_1+b_2)}\left(1-\exp\left(-\frac{T}{\tau_1}\right)z^{-1}\right)\left(1-\exp\left(-\frac{T}{\tau_2}\right)z^{-1}\right)} \tag{27.14}$$

that,

$$K(z) = \frac{b_1}{b_1+b_2}z^{-(k+1)} + \frac{b_2}{b_1+b_2}z^{-(k+2)} \tag{27.15}$$

$$Q(z) = \frac{U}{R}(z) = \frac{1}{k(b_1+b_2)}\left(1-\exp\left(-\frac{T}{\tau_1}\right)z^{-1}\right)\left(1-\exp\left(-\frac{T}{\tau_2}\right)z^{-1}\right) \tag{27.16}$$

$$D(z) = \frac{1}{k(b_1+b_2)}\frac{\left(1-\exp\left(-\frac{T}{\tau_1}\right)z^{-1}\right)\left(1-\exp\left(-\frac{T}{\tau_2}\right)z^{-1}\right)}{(1-K(z))} \tag{27.17}$$

A sampling period of 0.3s was selected for the design as a higher sampling frequency resulted in continuous oscillations of its actuation signal. Using equation 27.17

$$D(z) = 1.637\frac{(1-0.553z^{-1}+0.0464z^{-2})}{(1-z^{-1})(1+z^{-1}+0.266z^{-2})} \tag{27.18}$$

$$u_n = 1.637(e_n - 0.553e_{n-1} + 0.0464e_{n-2}) + 0.734u_{n-2} + 0.226u_{n-3} \tag{27.19}$$

And with all the ringing poles removed

$$D(z) = 0.72 \frac{(1 - 0.553z^{-1} + 0.0464z^{-2})}{(1 - z^{-1})} \tag{27.20}$$

$$u_n = 0.72(e_n - 0.553e_{n-1} + 0.0464e_{n-2}) + u_{n-1} \tag{27.21}$$

27.4.4 Model Following Predictor Design

A controller in predictive form requires an estimate of the controlled variable at the (n+k) sampling instant, and based on this the actuation signal is chosen for the output to reach the desired value in as short a time as possible. It has already been seen that the second order plant model is of the form

$$\frac{C}{U}(z) = \frac{(b_1z^{-1} + b_2z^{-2})z^{-k}}{1 + a_1z^{-1} + a_2z^{-2}} = \frac{B}{A}(z)^{-k} \tag{27.22}$$

which can be written in terms of the predicted system output at time (n+k) to give

$$C_{n+k} = -a_1C_{n+k-1} - a_2C_{n+k-2} + b_1u_{n-1} + b_2u_{n-2} \tag{27.23}$$

It is obvious from equation 27.23 that since the terms C(n+k-1), C(n+k-2) are not yet available at time "n", then the equation in its present form is not immediately solveable. To avoid a "k-step" computation loop, it can be shown that

$$C(z)z^k = F(z)C(z) + J(z)B(z)U(z) \tag{27.24}$$

which gives a prediction of C(n+k) based on values of C up to time "n" and previous actuations extending a further (k-1) samples back. The factors F(z) and J(z) are given by expanding 1/A(z) by long division up to z^k.

$$\frac{1}{A}(z) = J(z) + z^{-k}\frac{F}{A}(z) = c_0 + c_1z^{-1} + \dots c_{k-1}z^{-k-1} + z^{-k}\frac{F}{A}(z) \tag{27.25}$$

The design implemented (Fig. 27.4) was based on the model following predictor, where the output response was based on a first order lag with time delay K, which can be written as

$$\frac{C}{W}(z) = \frac{\alpha}{1 - \beta z^{-1}} z^{-k-1} \tag{27.26}$$

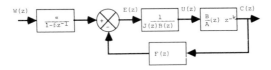

Fig. 27.4 Model following design for plant with time-delay

It proved necessary to increase the sampling period to 0.15s due to the unacceptable actuation signal oscillations that occurred at 0.1s. This resulted in the following equations

$$\frac{C}{U}(z) = \frac{0.173(1+0.6z^{-1})z^{-3}}{(1-0.67z^{-1})(1-0.32z^{-1})}$$ (27.27)

$$J(z) = 1 + 0.992z^{-1} - 0.769z^{-2}$$ (27.28)

$$F(z) = 0.55(1 - 0.3z^{-1})$$ (27.29)

Therefore, the difference equations for the controller are:

$$r_n = 0.632w_n + 0.368r_{n-1}$$

$$e_n = r_n - 0.55c_n + 0.165c_{n-1}$$ (27.30)

$$u_n = 5.77e_n - 1.6u_{n-1} - 1.37u_{n-2} - 0.46u_{n-3}$$

With this implementation an offset occurs in the presence of a disturbance, this can be overcome by using the incremental form of the predictive controller.

A particular form of the general predictor is the one proposed by Smith, which has been covered in Chapter 13. The Smith predictor was implemented with the PID controller specified in equation 27.4.

27.4.5 Pole Placement Controller Design

The pole placement design can be viewed as an extension of the classical root-locus method, where a design is undertaken to position the dominant close-loop poles of the system. In this section, the pole placement design is based on the general linear regulator shown in Fig 27.5, which is of the form

$$J(z)U(z) = -F(z)C(z) + H(z)W(z)$$ (27.31)

where

$$J(z) = 1 + j_1 z^{-1} + j_2 z^{-2} + \ldots + j_p z^{-p}$$

$$F(z) = f_0 + f_1 z^{-1} + f_2 z^{-2} + \ldots + f_f z^{-f}$$

$$H(z) = h_0 + h_1 z^{-1} + h_2 z^{-2} + \ldots + h_i z^{-i}$$

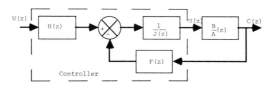

Fig. 27.5 Pole-placement controller

The plant, which will be considered noise free is modelled by

$$A(z)C(z) = z^{-k}B(z)U(z) \tag{27.32}$$

where

$$A(z) = 1 + a_1 z^{-1} + a_2 z^{-2} + \ldots + a_m z^{-m}$$

$$B(z) = b_1 z^{-1} + b_2 z^{-2} + \ldots + b_r z^{-r}$$

By combining equations 27.31 and 27.32 the closed loop transfer function is

$$\frac{C}{W}(z) = BH(z)\frac{z^{-k}}{AJ(z) + BF(z)z^{-k}} = \frac{N(z)z^{-k}}{D(z)} \tag{27.33}$$

and the actuation signal

$$u(z) = \frac{A(z)N(z)}{B(z)D(z)} z^{-k}W(z) \tag{27.34}$$

The closed loop characteristic polynomial

$$AJ(z) + BF(z)z^{-k} = T \tag{27.35}$$

is designed, with an appropriate choice of F,J polynomials to yield the desired closed loop pole set, given by

$$T = 1 + t_1 z^{-1} + t_2 z^{-2} + \ldots + t_c z^{-c} \tag{27.36}$$

so that

$$AJ(z) + BF(z)z^{-k} = T \tag{27.37}$$

It can be shown that (10) in order to obtain a unique solution to this equation the following restrictions are placed on the order of the T, J and F polynomials.

$$c < m + p < m + k + r - 1$$
$$f = m - 1$$
and $$p = k + r - 1$$

The controller parameters can now be obtained by solving a set of linear equations which can be represented in vector-matrix, using the Sylvester type matrix as follows:

$$
\begin{pmatrix}
1 & 0 & . & . & . & 0 & 0 & . & . & . & . & 0 \\
a_1 & 1 & 0 & . & . & 0 & 0 & . & . & . & . & 0 \\
. & a_1 & 1 & 0 & . & 0 & 0 & . & . & . & . & 0 \\
. & . & a_1 & . & . & 0 & b_1 & 0 & . & . & . & . \\
. & . & . & . & . & 0 & . & b_1 & . & . & . & . \\
. & . & . & . & . & 1 & . & . & . & . & . & . \\
a_m & . & . & . & . & a_m & b_r & . & . & . & . & . \\
0 & a_m & . & . & . & . & 0 & b_r & . & . & . & 0 \\
0 & 0 & a_m & . & . & . & . & . & . & . & . & b_1 \\
0 & 0 & 0 & . & . & . & . & . & . & . & . & . \\
0 & 0 & 0 & . & . & . & 0 & . & . & . & . & . \\
0 & 0 & 0 & . & . & a_m & 0 & 0 & . & . & . & b_r
\end{pmatrix}
\begin{pmatrix}
j_1 \\ j_2 \\ . \\ . \\ . \\ j_p \\ f_0 \\ f_1 \\ . \\ . \\ . \\ f_f
\end{pmatrix}
=
\begin{pmatrix}
t_1 & - & a_1 \\
t_2 & - & a_2 \\
t_3 & - & a_3 \\
. \\ . \\ . \\ . \\ . \\ . \\ . \\ . \\ .
\end{pmatrix}
\tag{27.38}
$$

The number of rows is given by k + r + m - 1

In compact matrix form

$$\underline{S}\,\underline{\theta}_c = \underline{R} \tag{27.39}$$

where $\underline{\theta}_c$ is the vector of unknown controller parameters, $\underline{R}$ is a vector of known coefficients of T and A polynomials and $\underline{S}$ is a matrix of known coefficients of A and B polynomials. This equation is then solved provided it is non-singular by inverting $\underline{S}$, so that

$$\underline{\theta}_c = \underline{S}^{-1}\underline{R} \tag{27.40}$$

The solution so far, gives the coefficients of the controller to meet the desired pole requirements, we still have to choose H in order to meet the steady state requirements of the system. The simplest option is to select H to be a scaling factor in order that in steady state c(t) = w(t), so that

$$\frac{C}{W}(z) = BH(z)\frac{z^{-k}}{T(z)} \tag{27.41}$$

and in steady state

$$H(z) = \frac{T}{B}\bigg|_{z=1} \tag{27.42}$$

In general, the pole placement design can give a poor steady state performance in the face of load disturbance. This can be overcome by including an integrator in the design and incorporating it as part of the process.

For the temperature process, cascaded with an integrator, the pulse transform becomes

$$\frac{C}{U}(z) = \frac{(0.173z^{-1} + 0.124z^{-2})z^{-2}}{(1 - 1.992z^{-1} + 1.207z^{-2} - 0.215z^{-3})} \tag{27.43}$$

and given that m=3, r=2 and k=2, then f=2 and p=3, so that the Sylvester matrix is

$$
S = \begin{pmatrix}
1 & 0 & 0 & 0 & 0 & 0 \\
-1.992 & 1 & 0 & 0 & 0 & 0 \\
1.207 & -1.992 & 1 & 0.173 & 0 & 0 \\
-0.215 & 1.207 & -1.992 & 0.104 & 0.173 & 0 \\
0 & -0.215 & 1.207 & 0 & 0.104 & 0.173 \\
0 & 0 & -0.215 & 0 & 0 & 0.104
\end{pmatrix}
$$

The Pole set is specified as

$$
T = 1 - 1.2z^{-1} + 0.48z^{-2}
\tag{27.44}
$$

so that

$$
R = \begin{pmatrix}
t_1 & - & a_1 \\
t_2 & - & a_2 \\
0 & & \\
0 & & \\
0 & &
\end{pmatrix} = \begin{pmatrix}
0.792 \\
-0.727 \\
0 \\
0 \\
0 \\
0
\end{pmatrix}
\tag{27.45}
$$

The inverse of S was then found

$$
\tag{27.46}
$$

$$
S^{-1} = \begin{pmatrix}
1 & 0 & 0 & 0 & 0 & 0 \\
1.992 & 1 & 0 & 0 & 0 & 0 \\
0.8779 & 0.5316 & 0.1157 & -0.1925 & 0.3202 & -0.5326 \\
10.8852 & 8.4416 & 5.116 & 1.1125 & -1.8506 & 3.0784 \\
-9.0900 & -5.9305 & -1.7406 & 2.8955 & 4.7988 & -7.9827 \\
1.8149 & 1.0990 & 0.2392 & -0.3979 & 0.6618 & 8.5144
\end{pmatrix}
$$

Giving

$$
\theta_c = \begin{pmatrix}
-0.792 \\
0.8507 \\
0.3088 \\
2.4840 \\
-2.8878 \\
-6.385
\end{pmatrix}
\tag{27.47}
$$

and

$$
J(z) = 1 + 0.792z^{-1} + 0.8507z^{-2} + 0.3088z^{-3}
\tag{27.48}
$$

$$
F(z) = 2.484 - 2.887z^{-1} + 0.638z^{-2}
\tag{27.49}
$$

Incorporating the integrator in the forward path part of the controller, yields

$$\frac{1}{J(z)(1-z^{-1})} = \frac{1}{1-0.208z^{-1}+0.0587z^{-2}-0.5419z^{-3}-0.3088z^{-4}} \tag{27.50}$$

giving the following difference equations

$$e_n = 0.5987w_n - 2.84c_n + 2.887c_{n-1} - 0.638c_{n-2} \tag{27.51}$$

$$u_n = e_n + 0.208u_{n-1} - 0.0587u_{n-2} + 0.541u_{n-3} + 0.309u_{n-4} \tag{27.52}$$

27.5 IMPLEMENTATION

One of the first critical decisions in the implementation was the choice of number representation. The TMS32010 uses a Q-notation system which is based upon fixed-point two's complement representation of numbers. Out of the 16 binary places possible the MSB is used to represent whether the number is a positive or negative number. This then leaves the programmer to decide where along the remaining 15 binary places he wishes to place the decimal point. Thus if a number has i integer bits, then it also has (15-i) fractional bits and is regarded as a (15-i) bit number. The choice of where to place the decimal point was critical to the performance of the controller, as it proved easy to generate an overflow. Unless this was monitored by checking the overflow register then swings between very large and very small numbers occurred which in turn caused erratic system behaviour. The programmer must either allow enough integer bits in his result to accommodate bit growth due to arithmetic operations or he must be prepared to handle overflows. In this implementation it proved necessary to do both. It was found that by representing all input and output values by Q15 numbers and all coefficients by Q11 numbers then this reduced the possibility of overflow although it still proved necessary to test for this condition and take appropriate action.

The instruction set of the TMS32010 and the manner in which structure can be imposed through the use of macros is one of the TMS32010's greatest assets. In the majority of cases a single instruction will not only allow its own specific function to be completed, but it will also allow the programmer to set up the conditions necessary to execute the next instructions. An example of this would be the LTD instruction which loads the T register with the contents of an address ready for multiplication. The instruction also shifts the data from its present location to the next highest memory location, as well as adding any previous result obtained in the accumulator. Thus the instruction LTD effectively performs the z^{-1} operation and if combined in a loop with the MPY (multiply) instruction then it evaluates the expression.

$$C_n = \sum_{k=0}^{M} b_k e_{n-k} \tag{27.53}$$

which represents in general the structure of the controllers implemented.

Due to the similarity of the structure of the algorithms implemented, the use of macros proved invaluable and aided the efficient development of the software. This required some standardisation of the memory locations used for both the data and parameters, so that the macros developed could be used with all the algorithms.

The development of the software was aided by the use of three software packages available on the VAX 11/785 system.

The packages

(i) The XDS/320 Macro Assemble which translates TMS32010 assembly language into an executive object code.

(ii)   The XDS/320 Linker which apart from producing a final down-loadable object code, also allows a program to be designed and implemented in separate modules which can then be linked to form the complete program.

27.6 APPLICATION TO PROCESS

To demonstrate the performance of the controller implementations, the process was subject to a load disturbance, by opening the vent from a closed position to a vent angle of 30 degrees. To ensure the vent was moved to the same angle in each test, a purpose built clamp was fitted as a stop at the 30 degree position. In order to provide a quantitative measure of the goodness of the system performance, under different control configurations, performance indices based on integral of error squared (IES), and integral of time multiplied by absolute error (ITAE) were evaluated. These were computed over a 15 second time period after the onset of the disturbance. The latter index gave the best selectivity measure to discriminate between the various controller implementations, and so only the result for this index will be presented here.

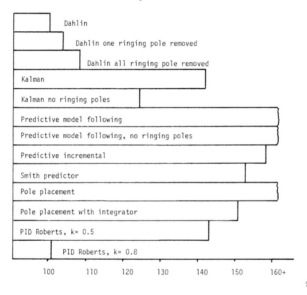

Fig. 27.6 Relative performance of controllers based on ITAE index

It was found that the best results were obtained using the Dahlin Controller. This was true of all the performance indices measured. The indices did not take into account the actuation signal, and so failed to penalise the ringing. The relative merits of the controllers in the face of a load disturbance are best illustrated in Fig. 27.6, where the best Dahlin response is selected as 100% and the other results shown relative to it. The disturbance response for the different controllers are shown in Figures 27.7-27.9. It should be observed that since $(1-z^{-1})$ is a factor of the demoninator polynomial for the Dahlin algorithm, then the controller has integral action which ensures that there is zero offset to a disturbance input.

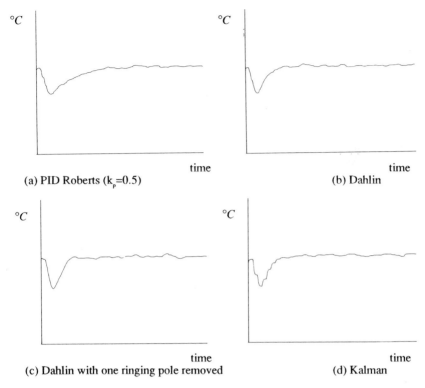

(a) PID Roberts (k_p=0.5)   time

(b) Dahlin   time

(c) Dahlin with one ringing pole removed   time

(d) Kalman   time

Fig. 27.7 Disturbance response with PID, Dahlin and Kalman Controllers
(opening vent)

The slight degradation of responses produced by removing the ringing poles can be seen for both Dahlin and Kalman designs. The Smith predictor performs well as does the PID Roberts when the gain is adjusted correctly. The designs of the predictive model following and pole placement controller which have no integrator in the forward path perform very badly. With the integrator included there is a dramatic improvement, but they still fare worse than the Dahlin implementations.

27.7 CONCLUSIONS

The chapter has shown that the signal processing capability of the TMS320 can be readily harnassed to implement a range of digital controllers. It has demonstrated that the powerful architecture of this DSP lends itself to implement efficiently the kind of algorithms met in discrete controllers and that a great deal of structure is possible with the system. The accuracy obtained depends very much on the number structure used and this required knowledge about the algorithms being implemented. It was not felt that the processor power was in any way stretched. Producing coding in assembly that was robust and error free was time consuming, and any future development will use a high level C compiler.

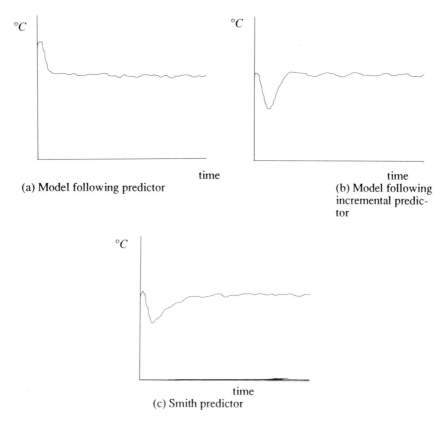

(a) Model following predictor

(b) Model following incremental predictor

(c) Smith predictor

Fig. 27.8 Disturbance response with predictor controllers (opening vent)

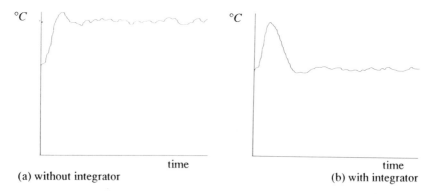

(a) without integrator

(b) with integrator

Fig. 27.9 Disturbance response with pole-placement controller (closing vent)

REFERENCES

1. D Rees and P A Witting:"Use of novel processor architectures for controller realisation":Proc. of EUROCON 84 Conference on Computers in Communication and Control, 1984

2. D Rees:"Controller Implementation using a monolithic signal processor":International Journal of Micro-computer Applications, Vol.4, No.3

3. D Rees:"Controller Implementation using Digital Signal Processors":EE Colloqium on "Novel Architectures and Algorithms for Controllers", 1987

4. W Meshach:"Data Flow IC Makes Short Work Of Tough Processing Chores": Electronic Design, p.161, 1984

5. D T Fitzpatrick et al:"VLSI Implementations of Reduced Instruction Set Computer":VLSI Systems and Computations, Editor H T Kung, Carnegie-Mellon, University, Springer-Verlag, p.327, 1981

6. M S Predade, J M Martins:"Monolithic Signal Processors: a comparative study":Centre de Electronia Aplicada, Proceedings of EUROCON 84, Sept 1984

7. TMS32010 Users Guide: Texas Instruments, 1983

8. Robe ts and Dallard:"Discrete PID Controller with a single tuning parameter":Measurement and Control, Vol.7, pp.T97-T101, Dec 1974

9. E B Dahlin:"Design and Tuning Digital Controllers":Instruments and Control Systems, Vol.41, No.6, 1968

10. K Warwick and D Rees:"Industrial Digital Control Systems":Second Edition, Chapter 4, Peter Peregrinus, 1988

11. Kalman, R E. Discussion following Bergen A R and Ragazzini J R:"Sampled data Processing Techniques for Feedback Control Systems": Trans. AIEE, 263-297, 1954

12. Astrom K J and Wittenmark, B:"Self-tuning controllers based on pole-zero placement", Proc. IEE, Vol.127, Pt D, No.3, pp.120-130, 1980

Chapter 28

Review and outlook for the future

T. S. Durrani

28.1   INTRODUCTION

Signal processing is concerned with the acquisition, abstraction and analysis of information, and is involved with the development of techniques, algorithms and architectures for implementing these function.   As is evident from the earlier chapters, applications of signal processing are vast, and ever increasing.

The seventies were largely concerned with the development of algorithms, and dedicated hardware for signal processing.   With the development of DSP chips there has been a phenomenal growth in signal processing activities, and the focus of attention has turned to real time systems with the ready availability of cheap and reliable devices from Texas Instruments, Fujitsu, NEC, Motorola, amongst others.   These devices, with their support tools offer an easy environment for prototyping, and eventual production.

In this review a brief summary is given of the current trends and new directions in the development of algorithms, architectures and devices for signal processing.

28.2 ALGORITHMS

The thrust of present interest is moving from fixed format, application dependent algorithms, such as the FFT, or digital filters, where the issues of accuracy and stability such as word length effects, or round-off noise were of concern, to more flexible, data dependent algorithms, such as model based spectral analysis methods employing Autoregressive (AR) Moving Average (MA) or ARMA methods, maximum entropy techniques, and adaptive filters, where convergence rates, residual errors, and computing effort are of interest in studying algorithm performance.

Most signal processing algorithms have their origins in linear systems theory and approximation theory, where, in the main, a minimum mean square criterion of performance is used to develop stable, iterative (or block processing) algorithms.   For instance, the DFT of a data sequence is a least mean squares time-domain solution which fits a series of complex sinusoidal harmonics to the data.   Most digital filters are minimum mean sequence error solutions in the frequency domain, to the frequency response of ideal systems. Similarly, adaptive filters arise as iterative solutions to a quadratic error criterion between a desired output and a known output.

More recently attention has turned towards the use of non-linear optimisation criterion, as a number of problems are more suited to this framework, eg finding the maximum likelihood estimates of a set of signals, and new algorithms such as simulated annealing [1], are much in favour.

Simulated annealing maps the cost function associated with an optimisation problem onto the energy of the states of a system, and through the use of an annealing algorithm seeks to find the (global) minimum energy state, ie the configuration that minimises the cost function.   This is particularly important for cost functions with multiple minima.   Here the minimum of a given cost function of many variables is sought.   The variables are subject to inter-dependent constraints, and they interact with each other in complicated ways not unlike the molecules in a physical system.   By appropriately defining an effective 'temperature' for the multivariable system, and imitating the physical annealing process, a diverse collection of combinatorial optimisation problems have been tackled [1].

Conventional minimisation techniques, such as gradient methods (corresponding to rapid cooling or quenching), converge rapidly to the nearest (minimum) solution for a given starting point, moving immediately downhill as far as possible.   For a cost function with multiple minima, this can lead to a local, and not necessarily a global minimum.   The simulated annealing algorithm allows the minimisation process to jump out of a local minimum and approach the global minimum.   This may be explained as follows.

A system in equilibrium at temperature T has its energy probabilistically distributed among all energy states E, with a (Boltzmann) probability distribution given by (k is Boltzmann's constant) :

$$\text{Prob (E)} \approx \exp - E/kT$$

Even at low temperature, there is a chance, however small, of a system being in a high energy state.   Thus there is a corresponding chance for the system to get out of a local energy minimum in favour of a better, more global one.   Initially (for high temperature), the probability is close to unity, and hence the change is accepted.   It is this mechanism which allows the algorithm to pull out of local minima. To assist this pull out, a random element is introduced in the iteration, in that Prob (ΔJ) is compared with random variables uniformly distributed in the interval (0,1). If Prob (ΔJ) is greater than this number, the change in parameter value is retained, if not, the original value is used for the next iteration. The use of Prob (ΔJ) ensures that the system evolves to a Boltzmann distribution corresponding to the lowest energy states, if the analogy is carried forward. The temperature is simply a control parameter, and in the simulated annealing algorithm, the process is first 'melted' by using a high effective temperature and iterations proceed at a given temperature till the system reaches steady state, the temperature is then lowered by slow stages and the procedure repeated, until the system 'freezes', and no further changes in parameter occur. The system has by then reached the global minimum.

28.3 ARCHITECTURE

Architectures for signal processing systems have evolved from conventional structures such as recursive or non-recursive tapped delay line with fixed or adaptive (time varying) weights to lattice filters [2]. Lattice configurations form an important class of architectures for signal processing. They possess regularity of form comprising identical stages (sections), which have orthogonal properties, and involve bounded coefficients. Lattices are thus inherently stable and possess good numerical round-off characteristics. These properties make them particularly attractive for adaptive processing. Lattices arise in the autoregressive modelling of input data, and provide outputs which are the residuals of the models. Figure 1 illustrates an M stage lattice. It consists of two channels corresponding to the

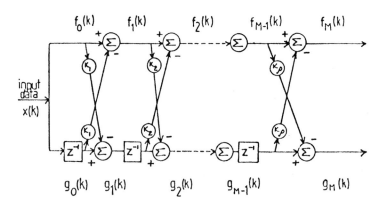

Figure 1 : Lattice Structure

outputs $\{f_m(k)\}$ and $\{g_m(k)\}$ which are referred to as the forward and backward residuals at various stages $\{m\}$. Each stage of the lattice comprises a pair of adders and (for modelling second-order stationary data) a reflection coefficient multiplier $\{K_m\}$ and a delay element. The governing equations for each stage of the lattice are

$$f_m(k) = f_{m-1}(k) - K_m g_{m-1}(k-1)$$
$$g_m(k) = g_{m-1}(k-1) - K_m f_{m-1}(k)$$
$$f_0(k) = g_0(k) = x(k), \quad m = 1, 2, \dots, M,$$

where $\{x)(k)\}$ is the input data sequence, and $\{f_m(k)\}$ and $\{g_m(k)\}$ are the forward and backward residuals at the the lattice stage, for the kth time instant. K_m is the reflection coefficient for the mth stage.

The forward residuals represent the prediction error between the output of the mth stage lattice and the input data, for an mth order autoregressive model of the data; the backward residuals is the prediction of error generated by receiving the order of the AR weights.

Lattices can be multi-channel and multi-dimensional and are finding use in adaptive signal processing, signal modelling, image processing, array processing [3].

In recent years, the demands on speed and performance in signal and image processing have led to new architectures which largely represent parallel processing structures. These include Single Instruction Multiple Data [SIMD] or Multiple Instruction Multiple Data [MIMD] machines. While the former consist of arrays of identical processors with local connectivity, all executing the **same** instruction simultaneously, the later consists of several processors, each with its own control unit, program and data units [4]. In both cases the overall processing task is distributed among all the processors.

Kung [5] introduced a general class of DSP array architectures - VLSI array processors which encompass systolic arrays and wavefront arrays. Systolic arrays which possess the attributes of pipelined processing, local processor connectivity, and close algorithm algorithm/architecture design are finding increasing use in the design of DSP processor arrays composed of discrete processing elements which are particularly suited to execute linear algebraic algorithms. For further information see ref [6]. It is evident that one of the major thrusts in the future will be the mapping of sequential signal processing algorithms onto parallel arrays of processors. The increasing popularity of SIMD, MIMD machines such as the AMT-DAP and the transputer based MEIKO Computing Surface presage new developments in this area. Details of some recent signal processing projects which exploit both fine grain and coarse grain parallelism are available from [7].

Another novel architecture attracting increasing attention is Artificial Neural Networks (ANNs).   These are particularly useful in the areas of speech processing, signal classification, sensor processing and pattern recognition [8,9].

ANNs are a network of interconnections of simple processing units, which can adapt their behaviour (response) according to inputs received during a training phase.   The processing unit or node is a non-linear, element which sums N weighted inputs and passes the result through a non-linearity.   The node is defined in terms of an internal threshold or offset θ_1 and by the type of non-linearity. Typically the output of a simple node is

$$O_i = f\left(\sum_j w_{ij}\, x_j + \theta_i\right)$$

where   f( )   =   functional non linearity

w_{ij} =   weight connecting the jth input to the ith node

θ_i =   threshold (constant)

x_j =   input sequence.

Examples of the non linearity are :

(i) $f(x)$ = $Sgn(x)$

(ii) $f(x)$ = $1/(1 + e^{-x})$

The simple processing elements may be connected together in several layers to yield an architecture called the Multi Layer Perception (MLP), ref [10] which is shown in the figure below :

OUTPUT

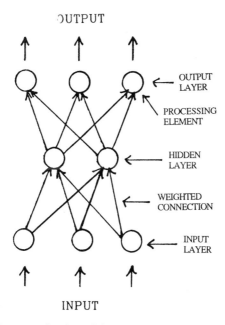

Figure 2 : Typical Multi Layer Perception Network

Every MLP has a input layer, one or more hidden layers, and an output layer.   The processing is performed in the hidden and output layers.   In terms of performance, training signals are applied to the inputs, and propagated through the network to produce output values. These are compared with the desired outputs, and the error used to modify (train) the weights, via a popular algorithm called the Back Propagation Algorithm [12].   Once the network has been trained, it can be used to extract signals from noise offered to the input, or as a pattern classifier where different images (patterns) are applied to the input patterns.   It is now accepted that neural networks will play a significant role in the design and implementation of novel signal processing systems.   The current disadvantage of long learning times, and complex interconnections are being overcome by the use of advanced training algorithms [11], and dedicated VLSI devices [12].

28.4 DEVICES

The explosive growth in dedicated signal processing devices augurs important new developments in signal processing, aided with the ready availability of 1-micron and sub micron technology DSP chips, with clock rates in excess of 30 MHz.   Most DSP devices are based on the RISC architecture, with on-chip memory and fast parallel multiply accumulate units with fast I/O links.   In this context the INMOS transputer with four serial I/O links, and the follow-up INMOS A100 device, and the more recent Motorola DSP 96000 series which includes two fast serial links, and the Texas Instruments TMS 320C30 which also incorporates serial communications links, all contribute to the growing armoury of DSP devices.

The recently announced TMS 320C50 is a complete DSP system on a chip [13] offering a 35ns instruction cycle, with two to four times the performance of fixed point DSPs;   an enhanced instruction set, and source code compatibility with previous generations of TMS 320s. It has a four-wire serial test bus, allowing easy interface to peripherals.

A further innovation which is of direct relevance to real time signal processing is the development of '**mixed signal**' systems on a chip.   This is a relatively new phenomenon which provides a novel means of mixing analogue and digital functions on a chip [14]. Examples are the Brooktree RAMDACs for graphics applications, self calibrating convertors, and data acquisition components from Crystal Semiconductor Corporation.   National Semiconductor plans to release intelligent analog VLSI chips with serial interface, on-board memory and other functions.

These new devices offer challenging opportunities to system designs, both in terms of on chip design of ASIC cells, and their utilisation in practical engineering problems. It is anticipated that over 20,000 'mixed signal' ASICs would be available within the next two years.

The following figure illustrates the configuration of a typical mixed signal IC. Notice the close tie up between the analogue section and the conventional digital microprocessor architecture.

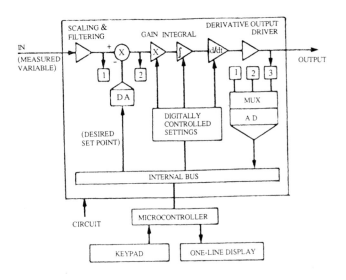

Figure 3 : Typical Mixed Signal ASIC

As a final comment, to ease message passing protocol allocation, data handling and upgrade, the importance of FUTUREBUS and FUTURE BUS + will increase when new signal processing systems are implemented.

28.5 CONCLUSIONS

The above has been a brief journey through some of the new directions in algorithms, architectures and devices for signal processing. This, of necessity, is a personal view point, nevertheless, with significant new developments over the horizon, the subject area of signal processing is set to grow and grow. As a final statement, the integration of fast algorithms, parallel architectures, and high performance multiprocessors, the field of parallel signal processing and its applications is one which will lead to rich rewards.

REFERENCES

[1] KIRKPARTICK S, GELATT Jr C D, and VECCHI M P: 'Optimisation by Simulated Annealing', Science V220, pp671-680, May 1983.

[2] COWAN C F N and GRANT P M [Eds]: 'Adaptive Filters', Prentice Hall, 1985.

[3] LACOUME J L, DURRANI T S and STORA R [Eds]: 'Signal Processing', Proc Nato ASI, Elsevier Science Publishers BV, 1987.

[4] ROBERTS J B G: 'Recent Developments in Parallel Processing', Proc IEEE ICASSP 89, Glasgow UK, pp2461-2467, 23-26 May 1989.

[5] KUNG S Y: 'VLSI Array Processors', Prentice Hall, 1987.

[6] STEWART R W: 'On Parallel and Orthogonal Linear Algebraic Signal Processing', PhD Thesis, University of Strathclyde, 1990.

[7] DTI Centre for Parallel Signal Processing, University of Strathclyde, Glasgow, UK.

[8] Proc IEE 'Artificial Neural Networks' Conference, Savoy Place, London, 16-18 October 1989.

[9] WASSERMAN P D: 'Neural Computing - Theory and Practice', Van Nostrand, 1989.

[10] RUMELHART R D et al: 'Learning Internal Representations by Error Propagation', Chapter 8, Vol 1 in Parallel Distributed Processing, and Rumelhart and McClelland, MIT Press, 1986.

[11] GROSSBERG S: 'Non-Linear Neural Networks - Principles, Mechanisms and Architectures', Neural Networks, Vol 1, pp17-61, 1988.

[12] Micro Devices Data Sheets, no 1210, Fuzzy Set Comparator, July 1989.

[13] Computer Design, Vol 28, No 16, July 1989.

[14] High Performance Systems, Vol 10, No 7, July 1989.

Index